Polyketides: Biosynthesis, Biological Activity, and Genetic Engineering

ACS SYMPOSIUM SERIES **955**

Polyketides

Biosynthesis, Biological Activity, and Genetic Engineering

Agnes M. Rimando, Editor
Agricultural Research Service
U.S. Department of Agriculture

Scott R. Baerson, Editor
Agricultural Research Service
U.S. Department of Agriculture

Sponsored by the
ACS Division of Agricultural and Food Chemistry, Inc.

American Chemical Society, Washington, DC

Library of Congress Cataloging-in-Publication Data

American Chemical Society. Meeting (229th : 2005 : San Diego, Calif.)
Polyketides : biosynthesis, biological activity, and genetic engineering / Agnes M. Rimando, editor, Scott R. Baerson, editor.

p. cm.—(ACS symposium series ; 955)

"Sponsored by the ACS Division of Agricultural and Food Chemistry, Inc."

"Developed from a symposium by the Division of Agricultural and Food Chemistry, Inc., at the 229th National Meeting of the American Chemical Society, San Diego, California, March 13–17, 2005"—Prelim. p.

Includes bibliographical references and index.

ISBN 13: 978–0–8412–3978–4 (alk. paper)

1. Polyketides—Congresses.

I. Rimando, Agnes M., 1957- II. Baerson, Scott R. III. American Chemical Society. Division of Agricultural and Food Chemistry, Inc. IV. Title.

QP752.P65A44 2005
572′.45—dc22 2006043052

The paper used in this publication meets the minimum requirements of American National Standard for Information Sciences—Permanence of Paper for Printed Library Materials, ANSI Z39.48–1984.

Distributed by Oxford University Press

PRINTED IN THE UNITED STATES OF AMERICA

Foreword

The ACS Symposium Series was first published in 1974 to provide a mechanism for publishing symposia quickly in book form. The purpose of the series is to publish timely, comprehensive books developed from ACS sponsored symposia based on current scientific research. Occasionally, books are developed from symposia sponsored by other organizations when the topic is of keen interest to the chemistry audience.

Before agreeing to publish a book, the proposed table of contents is reviewed for appropriate and comprehensive coverage and for interest to the audience. Some papers may be excluded to better focus the book; others may be added to provide comprehensiveness. When appropriate, overview or introductory chapters are added. Drafts of chapters are peer-reviewed prior to final acceptance or rejection, and manuscripts are prepared in camera-ready format.

As a rule, only original research papers and original review papers are included in the volumes. Verbatim reproductions of previously published papers are not accepted.

ACS Books Department

Contents

Indexes

Preface

Polyketides and their derivatives represent one of the most important classes of natural products, comprising approximately 20% of the world's top-selling pharmaceuticals, with combined revenues of over $18 billion per year. Antibiotics such as tetracycline, erythromycin, and nystatin are just a few of the important examples of polyketides that have benefited our society. In addition, polyketides serve important roles in the life cycles of many producing organisms, such as protection from invasion by pathogens, deterring predators, as well as providing many of the essential pigments found in plant species necessary for their survival. As the structural complexity of polyketides often renders their production via synthetic chemistry impractical, a further understanding of the genes and enzymes involved in their biosynthesis will directly impact future efforts toward the generation of novel compounds with beneficial properties, and provide an efficient means for their large-scale production. For these reasons as well as others, research on polyketides has been intensive and fast-paced. The present time is thus a particularly opportune one for assembling a volume covering recent results and perspectives from leading researchers around the world.

This volume was developed from a symposium that took place at the 229th ACS National Meeting on March 13–17, 2005 in San Diego, California. The original symposium presentations addressed a wide range of polyketide-producing plant, bacterial, and fungal systems and we have endeavored to maintain this scope in this volume. The included chapters provide discussions on the underlying genetics and biochemistry of important polyketide biosynthetic pathways found in nature, mechanistic insights obtained from studying the crystal structures of cloned polyketide synthases, new approaches for the identification of genes encoding novel polyketide synthase enzymes, and biotechnological advances toward the development and large-scale production of novel polyketides. We hope that this collection of chapters prepared by an international group of experts will be a valuable resource for investigators working on the biology and chemistry of polyketides as well as those drawn by curiosity wishing to improve their understanding of these fascinating and diverse compounds.

We greatly appreciate the contribution of the authors, and we acknowledge the valuable critiques of the reviewers.

Agnes M. Rimando
Scott R. Baerson
Natural Products Utilization Research Unit
Agricultural Research Service
U.S. Department of Agriculture
P.O. Box 8048
University, MS 38677

Polyketides: Biosynthesis, Biological Activity, and Genetic Engineering

Introduction

Chapter 1

A Plethora of Polyketides: Structures, Biological Activities, and Enzymes

Scott R. Baerson and Agnes M. Rimando

Natural Products Utilization Research Unit, Agricultural Research Service, U.S. Department of Agriculture, P.O. Box 8048, University, MS 38677

Polyketides represent a family of highly structurally diverse compounds, all produced via iterative decarboxylative condensations of starter and extender units, analogous to the biosynthesis of fatty acids. Polyketides have been shown to play important roles in the life cycles of producing organisms, as well as serving as chemical defense agents. Because a large number of polyketide-derived compounds are biologically active, they have also provided the basis for many important pharmaceuticals of enormous commercial and therapeutic value. Furthermore, the genetic and mechanistic diversity of polyketide synthase enzyme complexes involved in their biosynthesis in different organisms almost rivals the complexity of the molecules themselves. In this chapter, a brief overview is provided on these major subject areas, to serve as an entry point for readers exploring the subsequent chapters in this proceedings volume.

Introduction

In their simplest form, polyketides are natural compounds containing alternating carbonyl and methylene groups ('β-polyketones'). The biosynthesis of polyketides begins with the condensation of a starter unit (typically, acetyl-CoA or propionyl-CoA) with an extender unit (commonly malonyl-CoA or methylmalonyl-CoA, followed by decarboxylation of the extender unit (*1*, *2*) (**Fig. 1**). Repetitive decarboxylative condensations result in lengthening of the polyketide carbon chain, and additional modifications such as ketoreduction, dehydratation, and enoylreduction may also occur (discussed below).

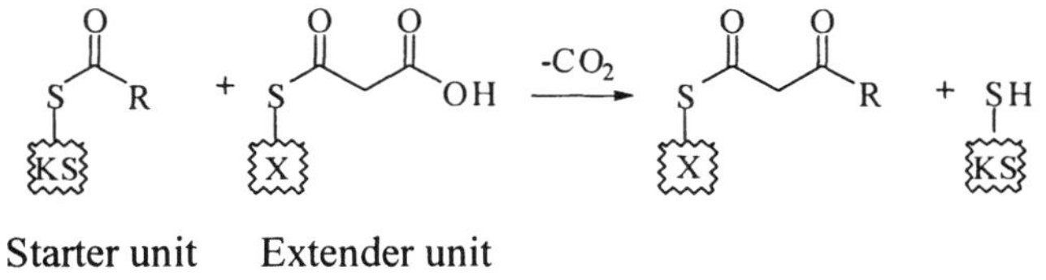

Starter unit   Extender unit

Figure 1. General condensation reaction in polyketide biosynthesis. The starter units are attached to thiol groups of the ketosynthase (KS), and extender units to thiol groups of either acyl carrier protein or acetyl coenzyme A (X).

Although the majority of polyketides are apparently produced by microbes (both bacteria and fungi), polyketides and their derivatives are ubiquitous, and are also produced by a host of other organisms including plants (e.g., flavonoids), insects (e.g., hydroxyacetophenones), mollusks (e.g., haminols), sponges (e.g., mycothiazole), algae (e.g., bromoallene acetogenins), lichens (e.g., usnic acid), and crinoids (e.g., polyhydroxyanthraquinones). Overall, polyketides represent the largest class of natural products and the most diverse in structure and function. Different classes of compounds have been grouped on the basis of common structural features, however due to their immense diversity, a unified classification scheme has yet to emerge. One major distinction that has been drawn is between those compounds derived from unreduced polyketone chains that are largely aromatic, and those in which the carbonyl functionalities are mostly reduced (*3*).

As a result of their intensive investigation over many years, polyketides and their derivatives have taken center stage in the quest for new antibiotics and therapeutic agents. Approximately 1% of the 5,000 to 10,000 known polyketides possess drug activity (*4*), and polyketides comprise 20% of the top-selling pharmaceuticals with combined worldwide revenues of over USD 18 billion per year (*3*). Some important examples include antibiotics such as tetracycline, erythromycin, nystatin, avermectin, and spiramycin, the anticancer agent doxorubicin, the hypocholesterolemic agent lovastatin, and the immunosuppressant rapamycin (**Fig. 2**). In addition, polyketide-derivatives such as the insecticide spinosyn A have proven useful to agricultural pest management control programs. It should also be noted, while a large percentage of polyketides do have medicinal or commercial value, a significant number, particularly the mycotoxins (*e.g.*, fumonisins and aflatoxins), pose severe hazards to animal and human health (*5*, *6*, *7*).

Biological Roles of Polyketide-Derived Compounds

The exact role for many polyketides in the life cycle of producing organisms are not known, however a large number appear to primarily serve as a means of chemical defense, conferring a competitive advantage to the producer (*8*). Interestingly, several cases of polyketides serving as anti-feedants have been documented. For example, polyketide sulfates of the crinoids *Comatula pectinata* and *Comanthera perplexa* provide a feeding deterrent to fish and prevent predation of these organisms (*9*). Likewise, toxic acetogenins produced by the American paw paw tree, *Asimina triloba,* provide a general deterrence to herbivores, and their accumulation in the larvae and adults of the Zebra swallowtail butterfly (*Eurytides marcellus*) which feed on *A. triloba*, protects them in turn from predation by birds (*10*).

One interesting class of polyketide-derived compounds, the stilbenes, are phenolic phytoalexins shown to play an important role in defense against fungal pathogens. The stilbene resveratrol has been the focus of genetic engineering experiments in a variety of plant species, including papaya, alfalfa, and wheat, where its engineered production has led to increased resistance against specific fungal pathogens (*11*, *12*, *13*). Plants can also produce polyketide-derived chemicals that suppress the growth of other plant species, a phenomenon known as allelopathy. One such compound, sorgoleone, is related to a family of plant-specific phenolic lipids and constitutes the major phytotoxic constituent exuded from the roots of *Sorghum bicolor* (*14*). Sorgoleone likely accounts for much of the allelopathy of root exudates produced by grain sorghum, which is considered an agronomically important characteristic for reducing the potential for weed infestations in cropping systems (*15*).

In addition to serving as chemical defense agents, polyketides also play essential roles in the growth and development of different organisms. For example, volatile bicyclic acetals from caddisflies serve as important intra- and

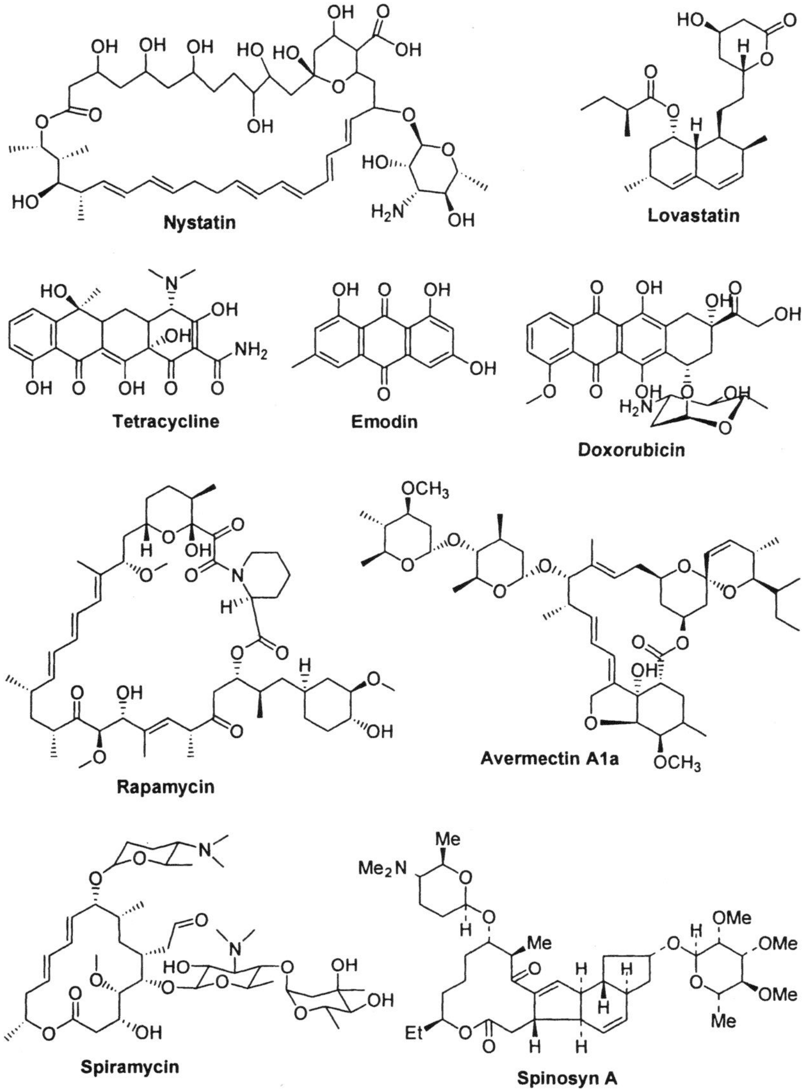

Nystatin

Lovastatin

Tetracycline

Emodin

Doxorubicin

Rapamycin

Avermectin A1a

Spiramycin

Spinosyn A

Figure 2. Representative polyketides with medicinal and pesticidal properties.

interspecific chemical communication signals (*16*). Dictyostelium spp. utilize polyketides as differentiation-inducing signaling factors during development, and recently a new prespore cell differentiation inducing factor (PSI-2), as well as two stalk cell differentiation factors (DIF-6 and DIF-7) have been identified (*17*). In plants, anthocyanins are polyketide-derived pigments responsible for many of the colors found in flowers and fruits, which serve as attractants for pollinators, UV protectants, as well as fulfilling several other important biological roles (*18*). Polyketides also contribute to fragrances associated with specific plants species, such as the polyketide p-hydroxyphenylbut-2-one, which provides the characteristic aroma of ripe raspberry fruits and serves as an additional attractant for pollinators (*19*).

The Biosynthesis of Polyketides

The variation in the organization and function of genes encoding components of polyketide synthase (PKS) enzyme complexes are nearly as diverse as the compounds they produce. While basic classification schemes have emerged in recent years (*e.g.*, *20*), specific cases are frequently discovered which underscore the difficulty in assigning categorizations to this family of enzymes (*21*). Nevertheless, at a minimum all PKS enzymes possess a ß-keto synthase activity which catalyzes the formation of a polyketide intermediate via repeated decarboxylative condensation reactions, analogous to the chain elongation reactions performed by fatty acid synthase (FAS) enzymes. In contrast with fatty acid synthesis however, where the condensation reaction is followed by consecutive keto reduction, dehydration and enoyl reduction, polyketide synthases can lack some or all of these reduction or dehydration activities (**Fig. 3**). Thus, PKS intermediates may possess unreduced keto groups within the nascent chain, hydroxyl groups formed by ketoreductase activity, double bonds due to dehydratase activity, or fully reduced alkyl functions via enoylreductase activity. In this respect, the simplest polyketide synthases and fatty acid synthases represent opposite ends of an enzymatic spectrum, producing products possessing keto groups on alternating carbon atoms (unreduced polyketides) at one end, and saturated fatty acids at the other (*21*).

Additional complexity occurs at the level of the ‘starter’ and ‘extender’ units used for constructing the polyketide scaffold, where the typically 2-4 carbon building blocks such as acetyl-, malonyl-, and propionyl-CoA are selectively used by different PKSs, thus increasing the repertoire of potential products formed. For example, results obtained from work with plant enzymes have shown that larger and more complex ‘starter’ units such as phenylpropanoid- as well as fatty acyl-CoAs can also serve as efficient substrates (*22*, *23*, *24*).

Further variation in product complexity occurs due to differences in the number of condensation steps performed, enzyme stereospecificity, as well as

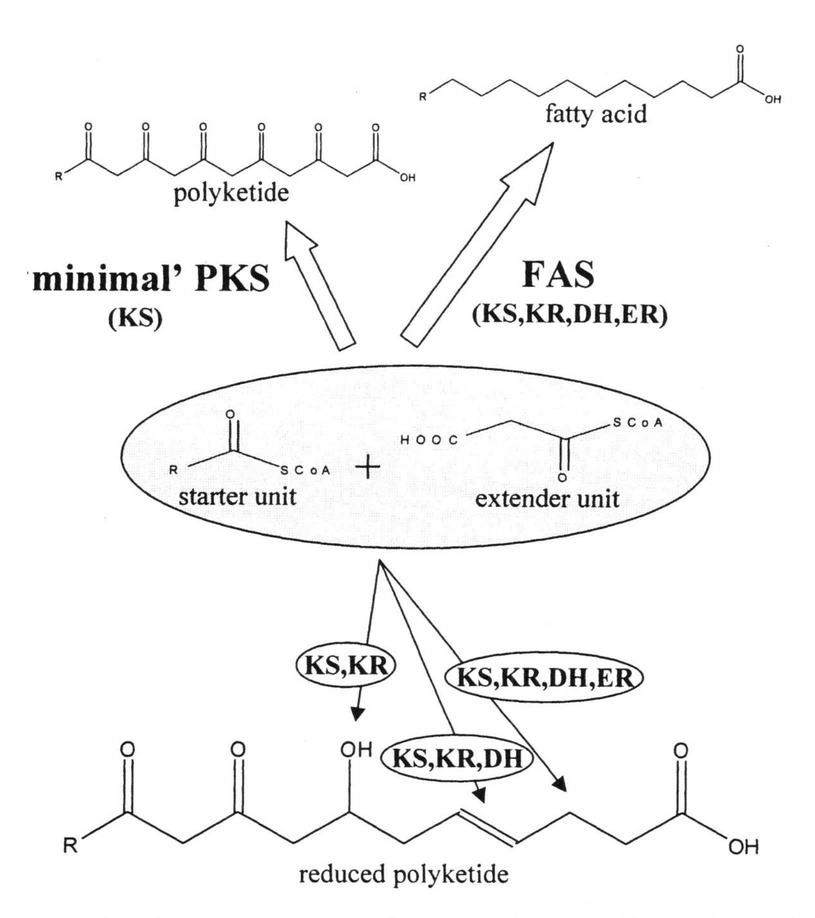

Figure 3. Relationship between polyketide and fatty acid biosynthesis. The simplest ('minimal') PKSs possess ketosynthase activity and produce linear polyketide products. In contrast, FASs also catalyze successive ketoreduction-dehydration-enoyl reduction reactions following each condensation. Diverse PKSs may perform none, part, or all of this reductive sequence. KS, ketosynthase; KR, ketoreductase; DH, dehydratase; ER, enoyl reductase.

different chain termination mechanisms which include alternative intramolecular cyclization reactions (*20*, *25*). As is apparent from this list of variables, PKSs represent a complex and widely divergent family of enzymes, thus one can readily envision their involvement in the biosynthesis of the vast array of polyketide-derived structures found in nature.

Three major themes broadly describe the organization of polyketide synthases from diverse organisms, based in part on the well-established paradigms for FAS complexes (**Fig. 4**). Type I and type II PKSs are structurally analogous to the type I FASs found in fungi, animals, and certain monocotyledonous plants, and the type II FASs associated with bacteria and all plant species, respectively (*21*, *26*). Moreover, it is highly likely that the type I and II PKSs are evolutionarily related to their type I and II FAS counterparts. For example, recent phylogenetic studies focusing on the more highly conserved keto synthase domain sequences of bacterial PKSs have demonstrated the common evolutionary history for these enzymes, and further suggest that evolution of type I PKSs has occurred largely through multiple gene duplication, gene loss, and horizontal transfer events from a common ancestor (*e.g.*, *27*). This relationship is perhaps self-evident, given that the boundaries for active site domains within type I PKSs were initially determined based on primary sequence comparisons with their animal FAS counterparts (reviewed in *3*).

Type III polyketide synthases represent an evolutionarily distinct sub-type, and are less structurally and catalytically complex than the type I and II enzymes. In general, the type III enzymes have no significant similarity at the primary sequence level to either FAS or other PKS sub-types, and likely arose via an entirely distinct evolutionary path (*24*). Initially, type III PKSs were designated as 'plant-specific', however in more recent years type III synthases have also been characterized from bacteria (*e.g.*, *28*). Additionally, the presence of related sequences identified within the sequenced genomes of numerous bacterial species strongly suggests that the 'blueprint' for type III enzymes pre-dates their adaptation for use by land plants (*24*, *25*).

In the following paragraphs a brief overview of some of the basic features of each PKS type is provided, however those seeking more detailed information concerning these enzymes are encouraged to consult the many excellent review articles available (*e.g.*, *20*, *24*, *25*, *29-31*).

Type I Polyketide Synthases

As mentioned, the genetic organization of type I PKSs parallels that of type I FAS enzyme complexes, in that these multimeric complexes are comprised of large, multifunctional subunits possessing all of the active sites required for

Type I (erythromycin A):

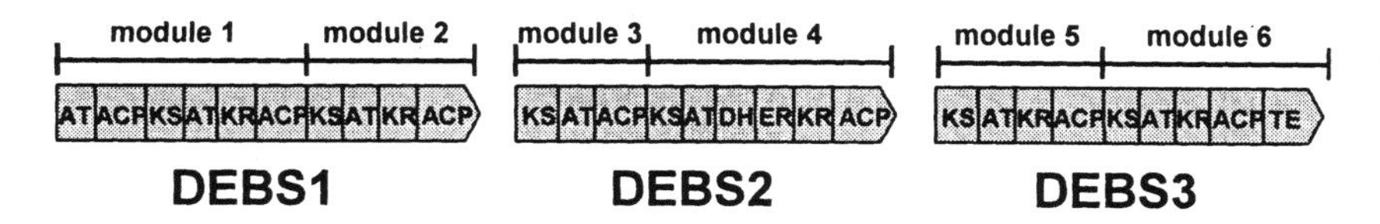

Type II (actinorhodin):

Type III (tetrahydroxychalcone)

Figure 4. Organization of representative type I, II, and III polyketide synthases. Upper: modular arrangement of DEBS1,2,3 subunits; Center: orientation and arrangement of open reading frames in actinorhodin gene cluster; Lower: chalcone synthase subunit. AT, acyltransferase; ACP, acyl carrier protein; KS, ketosynthase; KR, ketoreductase; DH, dehydratase; ER, enoyl reductase; TE, thioesterase; TA, tailoring enzyme; R/T, regulatory/transport related; AR, aromatase; CY, cyclase.

polyketide biosynthesis within discrete catalytic domains (**Fig. 4**). Type I PKSs can act in either a processive (similar to type I FASs), or noniterative manner. One particularly intriguing class of noniterative type I PKSs are referred to as 'modular' PKSs (*20*), which posses multiple catalytic domains (active sites) organized into 'modules', and each enzyme subunit may be comprised of several modules.

Modular polyketide synthases synthesize nascent polyketide scaffolds in an assembly line-like process, with each module participating in a single round of chain elongation and modification. At a minimum, all modules posses a ß-keto synthase catalytic domain, an acyl transferase domain, and an acyl carrier protein domain. In general, specialized modules exist at the N-terminus of specific subunits to facilitate starter molecule loading ('loading modules') possessing additional acyltransferase and acyl carrier protein domains, as well as thioesterase domains associated with the last module to facilitate product offloading (**Fig. 4**). The modular type I PKS complexes can also attain an astonishing level of complexity - for example one constituent polypeptide (MLSA1) associated with the type I PKS from *Mycobacterium ulcerans* involved in the biosynthesis of the macrolide toxin mycolactone contains 9 modules, more than 50 discrete catalytic domains, and has a mass exceeding 2 megadaltons (*32*). Perhaps the most extensively characterized type I PKS is the modular 6-deoxyerythronolide B synthase from *Saccharopolyspora erythraea*, responsible for the biosynthesis of the aglycone core of the clinically important macrolide antibiotic erythromyacin A (*33*). The heteromultimeric 6-deoxyerythronolide B synthase (DEBS) complex is comprised of three approximately 350 kDa subunits (designated DEBS1, 2, and 3), each possessing two modules, some of which contain additional ketoreductase, enoylreductase, or dehydratase catalytic domains (**Fig. 4**). Once released from the complex by thioesterase, the product 6-deoxyerythronolide B is subsequently modified by tailoring enzymes to yield the final macrolide antibiotic structure.

Type II Polyketide Synthases

Type II polyketide synthases, also referred to as bacterial aromatic polyketide synthases (*23*) are involved in the biosynthesis of a number of clinically important bacterial aromatic polyketides products exhibiting antitumor or antibiotic activity, such as doxorubicin and oxytetracycline. As mentioned, type II synthases are evolutionarily and structurally related to type II FASs, which occurr as heteromultimeric complexes. In contrast to type I synthases however, where multiple catalytic sites occur within a given subunit, the polypeptides associated with type II synthase complexes are typically monofunctional and dissociable (*20*, *30*).

Type II PKS complexes are comprised at a minimum of four types of subunits encoded by discrete open reading frames: acyl carrier protein, ketosynthase α, ketosynthase β (also referred to as 'chain length factor'), and a malonyl-CoA acyltransferase responsible for loading acyl-CoA extender units on to the acyl carrier protein subunit (*34*; **Fig. 4**). Additional subunits containing ketoreductase, cyclase, or aromatase activity may also occur in more complex type II synthases. Typically, the four core subunits (acyl carrier protein, ketosynthase α, ketosynthase β, and malonyl-CoA acyltransferase) participate in the iterative series of condensation reactions until a specified polyketide chain length is achieved, then folding and cyclization reactions yielding the final aromatic polyketide products are catalyzed by the associated aromatase and cyclase activities.

Interestingly, although hundreds of different bacterially-derived aromatic polyketide products have been identified to date, all correspond to a limited number of basic structural themes, as post-PKS tailoring activities involved in modifications such as methylation, glycosylation, oxidation, and reduction are apparently responsible for much of the structural diversity observed for this class of polyketides (*35*). The fact that most of the type II PKS subunits, post-PKS tailoring enzymes, host resistance factors, and regulatory proteins associated with a given polyketide biosynthetic pathway are typically encoded within tightly-linked biosynthetic gene clusters has greatly facilitated their cloning and subsequent functional characterization (*31*). One of the more extensively studied biosynthetic pathways involving a type II PKS is that of the aromatic benzoisochromanequinone antibiotic, actinorhodin, produced by *Streptomyces coelicolor* (**Fig. 4**). The gene cluster for actinorhodin spans an approximately 22 kb chromosomal region, and contains 22 open reading frames encoding the various pathway-associated functions, organized within a series of mono- and polycistronic operons (*36-38*).

Type III Polyketide Synthases

Type III polyketide synthases are responsible for the biosynthesis of a vast number of plant-derived natural products, including flavonoids derived from the important branch metabolite 4',2',4',6'-tetrahydroxychalcone, the product of the enzyme chalcone synthase (*39*). Because chalcone synthase was the first type III enzyme discovered, and a second flavonoid pathway type III enzyme, stilbene synthase, was discovered shortly thereafter, type III PKSs are also collectively referred to as the 'chalcone synthase/stilbene synthase superfamily' of enzymes (*24, 25*).

The type III plant and bacterial synthases feature the least complex architecture among the three PKS types, occurring as comparatively small homodimers possessing subunits between 40-45 kDa in size. As in the case for type II enzymes, type III PKSs catalyze iterative decarboxylative condensation reactions typically using malonyl-CoA extender units, however in contrast to type II synthases, the subsequent cyclization and aromatization of the nascent polyketide chains occurs within the same enzyme active site (*25*). Also unique to this family of PKSs, free CoA thioesters are used directly as substrates (both starter an extender units) without the involvement of acyl carrier proteins.

Type III synthases, as a whole, employ a wider spectrum of physiological starter molecules than their type I and II counterparts, including a variety of aromatic and aliphatic CoA esters such as coumaryl-CoA, methyl-anthraniloyl-CoA, as well as the recently identified medium- and long-chain fatty acyl-CoA ester starters used by certain bacterial and plant type III enzymes involved in the biosynthesis of phenolic lipids (*22*, *24*, Cook et al., *unpublished results*). The most extensively studied type III enzyme, chalcone synthase (**Fig. 4**), uses 4-coumaryl-CoA as the starter unit and catalyzes three successive condensation reactions with malonyl-CoA as the extender. Cyclization and aromatization of the linear tetraketide intermediate is performed within the same active site, yielding the final product 4',2',4',6'-tetrahydroxychalcone.

Conclusions

The importance of polyketides in nature and medicine has fueled an intensive effort to gain further understanding of their biological properties, and of the genetic and biochemical mechanisms underlying their biosynthesis. Current screening efforts ongoing in both the private and public research sectors to exploit natural biodiversity will undoubtedly result in the discovery of many new commercially valuable polyketide-derived compounds. Moreover, recent progress in understanding the genetic organization and catalytic mechanisms employed by the corresponding biosynthetic enzymes have created new tools with which to generate novel polyketide structures that would be difficult, if not impossible to synthesize directly. In addition, recent scientific breakthroughs in high-throughput sequence analysis, bioinformatics, and metabolomics will further accelerate the discovery of new enzymes and compounds in this field, which will likely remain at the forefront of natural products research in the years to come.

References

1. Hrannueli, D.; Perić, N.; Borovička, B.; Bogdan, S.; Cullum, J.; Waterman, P. G.; Hunter, I. A. *Food Technol. Biotechnol.* **2001**, *39*, 203.
2. *IUPAC Compendium of Chemical Terminology*; 2nd edn., McNaught, A. D.; Wilkinson, A., Eds.; Blackwell Scientific Publications, **1997**; 464 pp.
3. Weissman, K. J.; Leadlay, F. *Nat. Rev. Microbiol.* **2005**, *3*, 925.
4. Koskinen, A. M. P.; Karisalma, K. *Chem. Soc. Rev.* **2005**, *34*, 677.
5. Cousin, M. A.; Riley, R. T.; Pestka, J. J. In *Foodborne Pathogens*; Fratamico, P. M.; Bhunia, A. K.; Smith, J. L. Eds.; Caister Academic Press, Wymondham, UK, 2005; p. 163.

6. Oswald, I. P.; Marin, D. E.; Bouhet, S.; Pinton, P.; Taranu, I.; Accensi, F. *Food Addit. Contam.* **2005**, *22*, 354.
7. Hussein, H. S.; Brasel, J. M. *Toxicology.* **2001**, *167*, 101.
8. Pfeifer, B. A.; Khosla, C. *Microbiol. Mol. Biol. Rev.* **2001**, *65*, 106.
9. Rideout, J. S.; Smith, N. B.; Sutherland, M. D.; *Experientia*, **1979**, *35*, 1273.
10. Martin, J. M.; Madigosky, S. R.; Stephen, R.; Gu, Z.; Zhou, D.; Wu, J.; McLaughlin, J. L.; *J. Nat. Prod.* **1999**, *62*, 2.
11. Serazetdinova, L.; Oldach, K. H.; Loerz, H. *J. Plant Physiol.* **2005**, *162*, 985.
12. Zhu, Y. J.; Tang, C. S.; Moore, P.; Acta Horticulturae **2005**, *692* (Proceedings of the IInd International Symposium on Biotechnology of Tropical and Subtropical Species), 107.
13. Hipskind, J. D.; Paiva, N. L. *Mol. Plant Microbe Interact.* **2000**, *13*, 551.
14. Einhellig, F. A.; Souza, I. F, *J. Chem. Ecol.* **1992**, *18*, 1.
15. Duke, S.O. *Trends Biotechnol.* **2003**, *21*, 192-195.
16. Bergmann, J.; Löfstedt, C.; Ivanov, V. D.; Francke, W. *Tetrahedron Lett.* **2004**, *45*, 3669.
17. Serafimidis, I.; Kay, R. R. *Dev. Biol.* **2005**, *282*, 432.
18. Winkel-Shirley, B. *Plant Physiol.* **2001**, *126*, 485-493.
19. Kumar, .; Ellis, B. E. *Phytochemistry* **2003**, *62*, 513.
20. Khosla, C.; Gokhale, R.S.; Jacobsen, J.R.; Cane, D.E. *Annu. Rev. Biochem.* **1999**, *68*, 219-253.
21. Hopwood, D.A. *Chem. Rev.* **1997**, 97, 2465–2497.
22. Funa, N.; Ohnishi, Y.; Fujii, I.; Shibuya, M.; Ebizuka, Y.; Horinouchi S. *Nature* **1999**, *400*, 897-899.
23. Abe, I.; Watanabe, T.; Noguchi, H. *Phytochemistry* **2004**, *65*, 2447-2453.
24. Schröder, *J. Nat. Struct. Biol.* **1999**, *6*, 714-716.
25. Austin, M.B.; Noel, J.P. *Nat. Prod. Rep.* **2003**, *20*, 79-110.
26. McCarthy, A.D.; Hardie, D.G. *Trends Biochem. Sci.* **1984**, 9, 60-63.
27. Jenke-Kodoma, H.; Sandmann, A.;Mueller, R.; Dittman, E.; *Mol. Biol. Evol.* **2005**, *22*, 2027.
28. Funa, N.; Ozawa, H.; Hirata, A.; Horinouchi, S. *Proc. Natl. Acad. Sci.* **2006**, *103*, 6356-6361.
29. Shen, B. *Curr. Opin. Chem. Biol.* **2003**, *7*, 285-295.
30. Staunton, J. & Weissman, K. J. *Nat. Prod. Rep.* **2001**, *18*, 380–416.
31. Hopwood, D.A.; Khosla, C. *Ciba. Found. Symp.* **1992**, *171*, 88-106.
32. Stinear, T.P.; Mve-Obiang, A.; Small, P.L.; Frigui, W.; Pryor, M.J.; Brosch, R.; Jenkin, G.A.; Johnson, P.D.; Davies, J.K.; Lee, R.E.; Adusumilli, S.; Garnier, T.; Haydock, S.F.; Leadlay, P.F.; Cole, S.T. *Proc. Natl. Acad. Sci.* **2004**, *101*, 1345–1349.
33. Cortes, J.; Haydock, S,F,; Roberts, G.A.; Bevitt, D.J.; Leadlay, P.F. *Nature* **1990**, *348*, 176-178.

34. Wendt-Pienkowski, E.; Huang, Y.; Zhang, J.; Li, B.; Jiang, H.; Kwon, H.; Hutchinson, C.R.; Shen, B. *J. Am. Chem. Soc.*, **2005**, *127*, 16442 –16452.
35. Rix, U.; Fischer, C.; Remsing, L.L.; Rohr, J. *Nat. Prod. Rep.* **2002**, *5*, 542-580.
36. Fernández-Moreno, M.A.; Caballero, J.L.; Hopwood, D.A.; Malpartida, F. *Cell*, **1991**, *66*, 769–780.
37. Fernández-Moreno, M. A., Martinez, E., Boto, L., Hopwood, D. A. & Malpartida, F. *J. Biol. Chem.* **1992**, *267*, 19278–19290.
38. Fernández-Moreno, M.A.; Martinez, E.; Caballero, J.L.; Ichinose, K., Hopwood, D.A.; Malpartida, F. *J. Biol. Chem.* **1994**, *269*, 24854–24863.
39. Reimold, U.; Kroger, M.; Kreuzaler, F.; Hahlbrock, K. *EMBO J.* **1983**, *2*, 1801-1805.

Polyketide Biosynthesis in Plants and Microorganisms

Chapter 2

Hygromycin A Biosynthesis

Nadaraj Palaniappan and Kevin A. Reynolds

Department of Chemistry, Portland State University, P.O. Box 751, Portland, OR 97207–0751

Hygromycin A, an antibiotic produced by *Streptomyces hygroscopicus*, is an inhibitor of the bacterial ribosomal peptidyl transferase. The antibiotic binds to ribosome in a distinct but overlapping manner with other antibiotics, and offers a different template for generation of new agents effective against multidrug-resistant pathogens. Reported herein are the results from a series of stable isotope incorporation studies which have demonstrated the biosynthetic origins of the three distinct structural moieties which compromise hygromycin A. The 31.5 kb hygromycin A biosynthetic gene cluster has been identified, cloned and sequenced. It contains 29 genes whose predicted products can be ascribed roles in hygromycin A biosynthesis, regulation, and resistance. A series of gene deletion studies have been carried out, and provided both intermediates in the biosynthetic pathway and a new analog, 5''-dihydrohygromycin A. The convergent biosynthetic pathway established for hygromycin A offers significant versatility for applying the techniques of combinatorial biosynthesis and mutasynthesis to produce new antibiotics which target the ribosomal peptidyl transferase activity.

Introduction

Hygromycin A [**1**] (Figure 1) was first isolated in 1953 from the fermentation broth of several strains of *Streptomyces hygroscopicus* (*1*). In addition to several streptomycetes strains, the biosynthesis of hygromycin A was also noted in *Corynebacterium equi* (*2*). Initial reports indicated that hygromycin A possessed a relatively broad spectrum of activity against gram-positive and gram-negative bacteria (*1, 3*). Almost three decades later, Guerrero and Modolell demonstrated that the mode of action was inhibition of the ribosomal peptidyl transferase activity. Early studies also demonstrated that hygromycin A blocked the binding of either chloramphenicol or lincomycin to the ribosomes (*4*), and bound more tightly than chloramphenical. More recent footprinting experiments have shown that macrolides only block binding of hygromycin A to the ribosome if they contain a mycarose unit (*5*). Crystallographic evidence indicates that in such macrolides, the C5-disaccharide group extends from the polypeptide exit channel into peptidyl transferase center (*6*). Hygromycin A thus offers a distinct carbon skeleton and binding mode from other antibiotics that target the bacterial ribosome. As such, it represents a promising starting point for generating new antibiotics to treat infections with drug resistant pathogens.

Recently, hygromycin A has gained renewed interest due to its hemagglutination inactivation activity (*7*) and high antitreponemal activity (*8*) which has led to the possible application of hygromycin A-related compounds for the treatment of swine dysentery, a severe mucohemorrhagic disease thought to be caused by *Serpulina* (*Treponema*) *hyodysenteria* (*8, 9*). Hygromycin A has also been reported to possess an immunosuppressant activity in the mixed poor lymphocyte reaction, but does not work via suppression of interleukin 2 production (*10*). Most recently, methoxyhygromycin A [**2**] (Figure 1), an analog of hygromycin A produced in the same fermentation broth (*7*), has been shown to have herbicidal activity and has led to the suggestion that it could be developed as a biological agent for weed control (*11*).

Biosynthesis of Hygromycin A

Chemical Structure

Hygromycin A [**1**] is classified chemically as a member of the aminocyclitols antibiotic family represented by streptomycin and kanamycin. Its structure determined by degradation and spectral analyses, (*12, 13*) revealed that it consists of three unusual moieties, namely 5-dehydro-α-L-fucofuranose, (*E*)-3-(3,4-dihydroxyphenyl)-2-methylacrylic acid, and an aminocyclitol (2L-2-amino-2-deoxy-4,5-*O*-methylene-*neo*-inositol). The 5-dehydro-α-L-fucofuranose moiety is attached by a glycosidic linkage of the 4-hydroxy position of (*E*)-3-

(3,4-dihydroxyphenyl)-2-methylacrylic acid, whereas the aminocyclitol moiety is linked by an amide bond to the acid.

In mid 1980s, the potential effectiveness of hygromycin A in controlling swine dysentery, a severe mucohemorrhagic disease common in the swine industry, led to a renewed interest in this class of compounds (7, 8, 14), and semi-synthetic and synthetic programs based on hygromycin A and its biological properties (*14-16*). This work led to the production of over 100 analogs and the determination of their biological activities, both in terms of minimum inhibitory concentrations (MICs) for *S. hyosodysenteriae* and their ability to inhibit protein synthesis in an *E. coli* cell-extract. The resulting structure-activity relationship (SAR) revealed that the unique aminocyclitol moiety is an important component for antibacterial activity, while the 5-dehydro-α-L-fucofuranose moiety is not essential and can be replaced with a hydrophobic allyl group. Reduction in antibacterial activity was also observed following replacement of the methyl group of (*E*)-3-(3,4-dihydroxyphenyl)-2-methylacrylic acid with a propyl, allyl, or hydrogen moiety (*14, 17*). For the most part, these structural analogs were generated by a semi-synthetic method using hygromycin A as a starting point. The total synthesis of hygromycin A and C2"epi-hygromycin A has also been reported (*15, 18*). Several multi-step syntheses of 2L-2-amino-2-deoxy-4,5-*O*-methylene-*neo*-inositol have also been reported (*19*). The most recent synthesis was enantioselective, and was accomplished in 14 steps with an overall yield of 12 % (*20*).

In our work, we have been interested in the biosynthetic origins of hygromycin A, and in developing a complementary approach to synthetic efforts by engineering the production of hygromycin A analogs in *Streptomyces hygroscopicus*.

Biosynthesis of the 5-dehydro-α-L-fucofuranose Moiety

The stereochemical configuration at the C2''position of hygromycin A indicates that mannose, rather than glucose is the more immediate precursor of the 5-dehydro-*α*-L-fucofuranose-derived moiety. NMR analyses, based on a feeding study using [1-^{13}C] mannose, have shown a specific 3-fold enrichment at C1'' (103.7 ppm) of hygromycin A (Figure 2) (*21*). Furthermore, no labeling was noted for the aminocyclitol moiety of hygromycin A from mannose. The aminocyclitol moiety is, however, efficiently labeled from [1,2-$^{13}C_2$]-D-glucose. These results indicate that, under these experimental conditions, the carbon flux is from glucose towards mannose. Consistent with this hypothesis, 1% intact incorporation of C-1" and C-2" from [1,2-$^{13}C_2$] -D- glucose clearly indicated that the labeling of 5-dehydro-*α*-L-fucofuranose moiety is consistent with a pathway in which glucose is converted via primary metabolism to mannose-1-phosphate

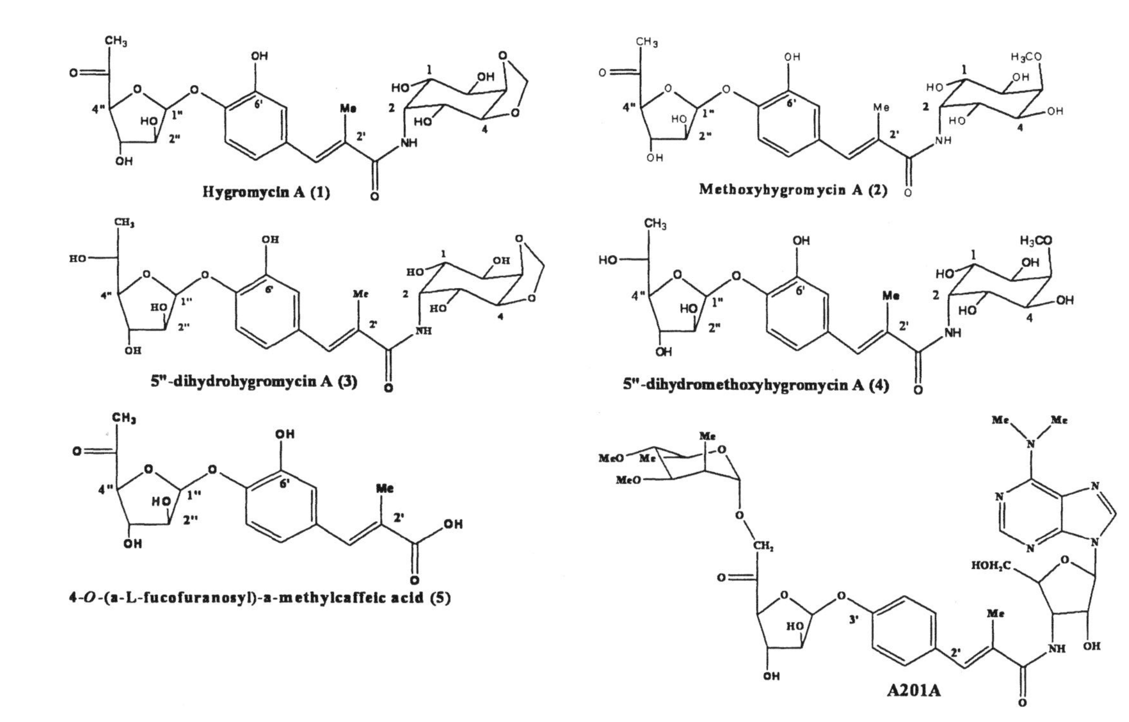

Figure 1. Structures of hygromycin A, hygromycin A analogues, and antibiotic A201A.

(*21*). In subsequent steps this intermediate is likely converted to an activated nucleoside-diphosphate, (NDP)-mannose, followed by dehydration to generate an NDP-4-keto-6-deoxy-D-mannose, a reaction catalyzed by GDP-D-mannose-4,6-dehydratase (MDH). This intermediate would then be converted to GDP-L-fucose by an NDP-L-fucose synthetase, which catalyzes epimerization at the C-3 and C-5 positions of the hexose ring and an NADPH-dependent reduction at the C-4 position (*22, 23*). In the proposed pathway (Figure 2), conversion of the NDP-L-fucose (a pyranose) to the furanose, and oxidation of the 5-keto group would provide the activated 5-dehydro-α-L-fucofuranose for attachment to the C4 hydroxyl of (*E*)-3-(3,4-dihydroxyphenyl)-2-methylacrylic acid-derived moiety of hygromycin A.

The mutase responsible for the ring contraction of NDP-L-fucose likely follows the mechanism established for the primary metabolic enzyme, UDP-galactopyranomutase (*24, 25*). To the best of our knowledge, no rearrangements of this type have been reported previously for antibiotic biosynthetic pathways. Similarly, the antibiotic A201A (Figure 1), with significant structural similarities to hygromycin A, contains a moiety that is likely derived from 5-keto-D-arabino-hexofuranose. No biosynthetic experiments have been reported for the antibiotic A201A, but it is likely that this moiety is also derived from NDP-mannose in a pathway involving a similar ring contraction and oxidation at C5. The details of the multi-step conversion of the probable precursor mannose to the 5-keto-D-arabino-hexofuranose moiety for this antibiotic still remain unclear. Furthermore, obvious gene candidates potentially encoding proteins capable of catalyzing this ring contraction could not be identified within either the partial sequence data available for the gene cluster of antibiotic A201A, or within the hygromycin A biosynthetic gene cluster.

Biosynthesis of the Aminocyclitol Moiety

It has previously been proposed that the unusual aminocyclitol of hygromycin A is formed via a pathway which involves *myo*-inositol as an intermediate (*26*). A related pathway has been characterized for the *scyllo*-inosamine-derived moiety of streptomycin (*27*). In the latter case, oxidation of the C2 hydroxyl of *myo*-inositol (derived from C5 of glucose), and subsequent transamination (catalyzed by L-glutamine: *scyllo*-inosose amino transferase) (*27*) yields *scyllo*-inosamine. A similar process occurring at the C5 position of *myo*-inositol (derived from C2 of glucose) would be predicted to generate *neo*-inosamine-2 (Figure 2). Feeding studies using [1,2-$^{13}C_2$]-D-glucose have previously demonstrated clear and unequivocal labeling at the C1 (71.6 ppm) and C2 positions (50.7 ppm) of hygromycin A (*21*). Labeling of the aminocyclitol portion of hygromcyin A is consistent with a pathway in which glucose is converted to glucose-6-phosphate, and then to 1L-*myo*-inositol-phosphate by 1L-myo-inositol-1-phosphate synthase (*28*). Dephosphorylation of this moiety by 1L-*myo*-inositol-1-phosphatase would generate *myo*-inositol, which could be converted to *neo*-inosamine-2 via an oxidation and aminotransferase step at the C-5 position, differing from the well-established *scyllo*-inosamine pathways for streptomycin antibiotics (*27, 29*). Consistent with this proposal, there was no detectable labeling at the C5 or C6 position of the aminocyclitol moiety that would be contributed by the divergent pathways leading from *myo*-inositol to *neo*-inosamine-2 and *scyllo*-inosamine (*21*). A low level of intact labeling (J 44.8 Hz) (0.2%) at C3 (72.6 ppm) and C4 (78.2 ppm) of hygromycin A was however noted in the biosynthetic experiment, suggesting that an alternative process or pathway must contribute, albeit to a lesser degree, to the formation of *neo*-inosamine-2 from glucose.

In the same biosynthetic experiment, the origin of the methylene group that bridges the C4 and C5 hydroxyl groups in hygromycin A was investigated by a feeding study with [*methyl*-^{13}C]-L-methionine (*21*). The ^{13}C NMR analysis of the resulting hygromycin A revealed a single 25-fold enrichment for the methylene bridge of aminocyclitol, consistent with the involvement of *S*-adenosylmethionine in the generation of this moiety. It remains to be determined at which stage during biosynthesis the methylene group is introduced onto the aminocyclitol ring.

Biosynthesis of the (*E*)-3-(3,4-dihydroxyphenyl)-2-methylacrylic Acid Moiety

Hygromycin A and A201A (Figure 1) share some common structural features, particularly within the central portion of these molecules. In the structurally-related puromycin, the α-methyl substituent of this central unit is replaced with an amine, and it has been proposed that this moiety is derived from tyrosine (30). Despite these structural similarities, it has become clear that a more complex process must provide the moieties for hygromycin A and A201A. Feeding studies using [3-^{13}C]tyrosine, [3-^{13}C]phenylalanine, and [carboxy-^{13}C]benzoic acid clearly demonstrated that these compounds were not precursors for the (E)-3-(3,4-dihydroxyphenyl)-2-methylacrylic acid moiety of hygromycin A. However, a similar feeding experiment with [carboxy-^{13}C]-4-hydroxybenzoic acid showed a specific incorporation of 6% at the C3' position of hygromycin A, suggesting its involvement in the biosynthesis of this moiety (21). In addition, a feeding study with [2,3-$^{13}C_2$]propionate led to 1% intact incorporation of ^{13}C label at both the C2' and the C2' methyl positions of hygromycin A. Conversely, no incorporation at the C2' methyl position of hygromycin A was observed in the methionine incorporation experiment, ruling out an S-adenosylmethionine-

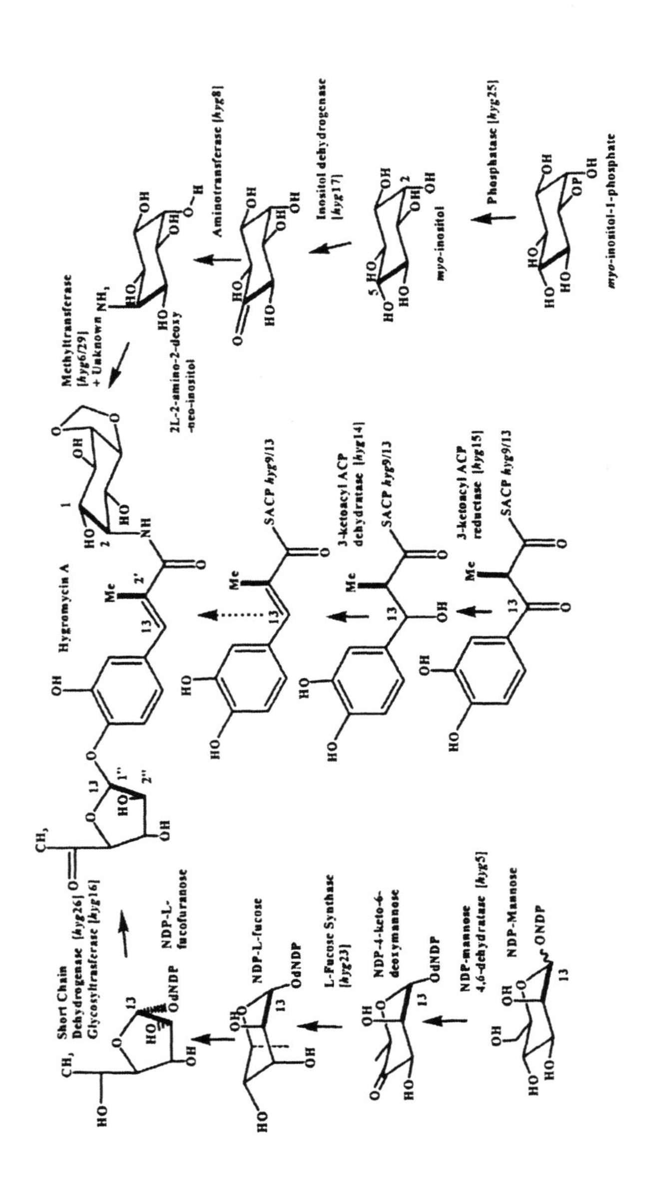

Hygromycin A
Methyltransferase [hyg6/29] + Unknown
2L-2-amino-2-deoxy-neo-inositol
Aminotransferase [hyg8]
Inositol dehydrogenase [hyg17]
myo-inositol
Phosphatase [hyg25]
myo-inositol-1-phosphate
SACP hyg9/13
3-ketoacyl ACP dehydratase [hyg14]
3-ketoacyl ACP reductase [hyg15]
Short Chain Dehydrogenase [hyg26]
Glycosyltransferase [hyg16]
NDP-L-fucofuranose
NDP-L-fucose
L-Fucose Synthase [hyg23]
NDP-4-keto-6-deoxymannose
NDP-mannose 4,6-dehydratase [hyg5]
NDP-Mannose

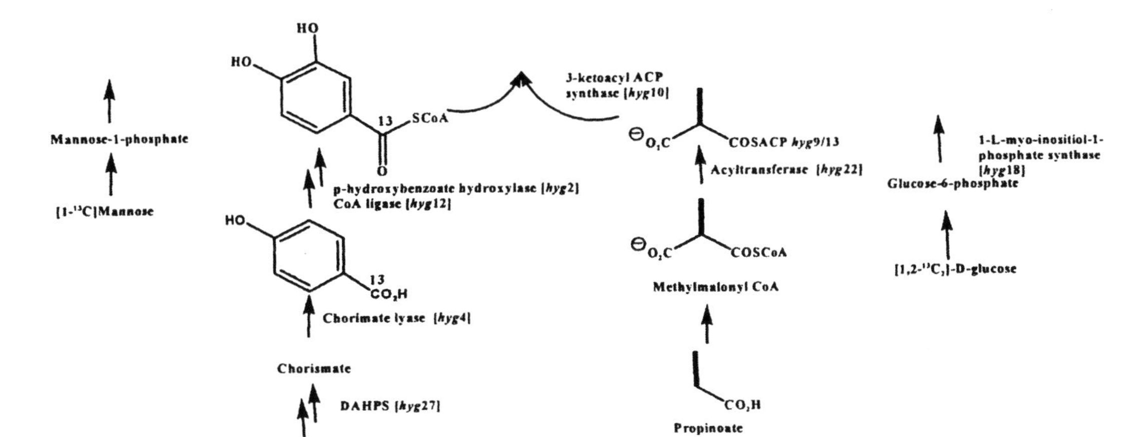

Figure 2. Proposed pathways for hygromycin A biosynthesis in S. hygroscopicus NRRL 2388 and putative assignments for the products of the corresponding gene cluster.

dependent methylation process. These observations suggested a polyketide synthase-type decarboxylative condensation between a 4-hydroxybenzoyl-CoA or 3,4-dihydroxylbenzoyl-CoA starter unit, which is elongated by a methylmalonyl CoA-derived extender unit to generate a 3-ketoacyl thioester product. Further processing by an 3-ketoacyl thioester reductase and a 3-hydroxyacyl thioester dehydratase would then yield the (*E*)-3-(3,4-dihydroxyphenyl)-2-methylacrylic acid moiety of hygromycin A (Figure 2) *(21)*. Consistent with this proposed biosynthetic pathway, our analysis of the hygromycin A biosynthetic gene cluster (see below) and of the published A201A antibiotic gene cluster revealed gene products which can clearly be assigned to individual roles in this biosynthetic process. Included in the hygromycin biosynthetic gene cluster is *hyg2*, which encodes a protein with homology to 4-hydroxybenzoate hydroxylase and which is tentatively assigned a role in the formation of 3,4-dihydroxybenzoic acid (Figure 2). It is not clear if the hydroxylation occurs directly with 4-hydroxybenzoic acid or at a later step in the biosynthesis of the (*E*)-3-(3,4-dihydroxyphenyl)-2-methylacrylic acid moiety. This hydroxylation is not required in antibiotic A201A biosynthesis, and our analysis revealed that there is no *hyg2* homolog encoded by the corresponding biosynthetic gene cluster.

Overall, this set of biosynthetic studies clearly demonstrated mannose, glucose, 4-hydroxybenzoic acid, propionate, and methionine as precursors for the assembly of the three unique structural moieties of hygromycin A. These moieties are likely predominantly assembled as separate entities, and then linked using an amide synthetase and glycosyl transferase. Such a convergent biosynthetic route offers tremendous potential for structural modification via the techniques of mutasynthesis and combinatorial biosynthesis. In a step towards this direction, we have cloned and sequenced the biosynthetic gene cluster which is responsible for the production of hygromycin A in *S. hygroscopicus NRRL 2388*.

Hygromycin A Biosynthetic Gene Cluster

A number of different approaches for identifying the hygromycin A biosynthetic gene cluster were considered. At the time this project was initiated, a partial sequence of the antibiotic A201A biosynthetic gene cluster had been reported *(31)*. However, this analysis showed only the *ard1* and *ard2* resistance determinants, and the *ataP3*, *ataP5*, *ataP4* and *ataP7* open reading frames, which were likely involved in the formation of the N^6,N^6-dimethyl-3'-amino-3'-deoxyadenosine moiety, which is not present in hygromycin A (a partial sequence of the ataPKS1 gene was also reported). Thus we took a different approach using a PCR-based method to identify a gene encoding a putative NDP-mannose 4,6-dehydratase (MDH), which we predicted would be required for the committed step of the pathway generating the 5-dehydro-α-L-fucofuranose moiety. Genes encoding proteins with homology to MDH have been identified in the nystatin, candicidin, and other antibiotic biosynthetic gene clusters (*31-33*). A blast search for similar genes in the genome sequences of *S. coelicolor* A3(2)(*34*) and *S. avermitilis* MA-4680 (*35*) failed to identify sequences with significant homology to MDH, suggesting that this approach would be selective.

Accordingly, a pair of degenerate primers based on highly conserved motifs, identified by creating an alignment of predicted MDHs, was used to amplify a portion of *hyg5* from genomic DNA of the hygromycin A producer *S. hygroscipicus* NRRL 2388. Sequencing confirmed that the PCR product encoded a protein with homology to putative MDH enzymes. The partial *hyg5* fragment was then used to screen a cosmid library of *S. hygrocospicus* NRRL 2388 to isolate the hygromycin A biosynthetic gene cluster. By the above approach, a 31.5 kb genomic DNA region covering the hygromycin A biosynthetic gene cluster (*hyg*) was identified. Analysis of the gene cluster revealed 29 ORFs, putatively involved in hygromycin A resistance, as well as the regulation and biosynthesis of the three key moieties of hygromycin A (Figure 2 and 3B). After this work had been completed, more sequence data from the antibiotic A201A biosynthetic gene cluster became available (accession number: X84374). Comparison of the two clusters revealed that there are 15 putative ORFs within the *hyg* gene cluster with homologs in the antibiotic A201A gene cluster (Table 1 and Figure 3). As shown in Figure 2, putative homologs for the *hyg* gene products predicted to be involved in the biosynthesis of the central (*E*)-3-(3,4-dihydroxyphenyl)-2-methylacrylic acid moiety, are found in the A201A gene cluster (Table 1). As mentioned, the exception is the absence of a *hyg2* homolog, consistent with the structural differences between the central cores of hygromycin A and A201A. There are also homologs of *hyg19*, *hyg21*, and *hyg29* in the A201A gene cluster, which are predicted to be involved in antibiotic resistance. The *hyg21* gene product is predicted to be a phosphotransferase, and has 56% identity at the amino acid level with the *ard2* gene product from the A201A cluster. It has been shown that Ard2 protein catalyzes an ATP-dependent phosphorylation of the C2 hydroxyl group in the furanose moiety of A201A, thereby inactivating the antibiotic (*36*). Hyg21 may provide resistance to hygromycin A by a similar mechanism.

Of particular interest to us is the 3-ketoacyl ACP synthetase homolog which we have assigned a role in catalyzing the condensation of 3,4-hydroxybenzoyl CoA with methylmalonyl ACP, the critical elongation step in the biosynthesis of the central hygromycin core. Hyg10 (and the AtaPKS3 homolog from the antibiotic A201A cluster) encodes a β-ketoacyl synthase (KS) with low homology to discrete KS proteins in type II PKSs, but not KS domains in modular type I PKSs. A phylogenetic analysis with known β-ketoacyl synthases

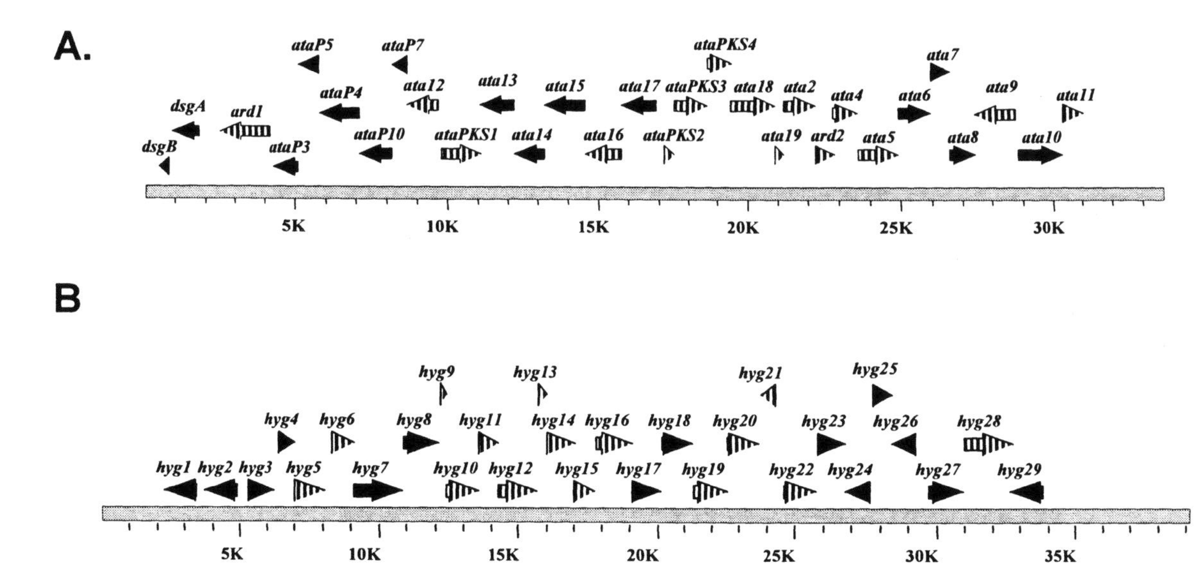

Figure 3. Biosynthetic gene clusters of (A) the aminonucleoside antibiotic A201A of S. capreolus NRRL 3817 (accession number: X84374), and (B) hygromycin A of S. hygroscopicus NRRL 2388. The horizontally shaded arrows represent the homologous genes present both of the antibiotic gene clusters.

revealed that Hyg10 and AtaPKS3 form a separate cluster from type II PKSs, bacterial and plant type III PKSs, and the KAS 1 nodulation protein (Figure 4). This observation suggested that the β-ketoacyl synthases of the *hyg* and A201A gene clusters might represent a unique family of KS proteins. A multiple sequence alignment of the predicted Hyg10 and AtaPKS3 with type I PKS KS proteins, led to the same conclusion. This analysis revealed the presence of the two highly conserved active site histidines (*37*), which are presumably required for catalyzing the methylmalonyl ACP decarboxylation. However, in both Hyg10 and AtaPKS3 a serine residue is observed in place of a highly conserved nucleophilic cysteine residue (required for formation of the acyl thioester intermediate in the catalytic cycle of these enzymes). While serine acts as a nucleophile in generating enzyme bound acyl ester intermediates for acyltransferases, we are unaware of such a role for KS proteins or domains. The *hyg11* gene product exhibited 48% identity to *ataPKS4* in the antibiotic A201A gene cluster, and very low amino acid identity to several putative type II KS proteins found in the NCBI non-redundant peptide sequence database (http://ncbi.nlm.nih.gov). The highly conserved catalytic triad of KS proteins and domains was not observed in either Hyg11 or A201A, thus no clear role could be assigned based on this sequence analysis.

Production of 5″-dihydrohygromycin Analogues

A genetic experiment was used to confirm that the *hyg* biosynthetic gene cluster is responsible for hygromycin A biosynthesis. Allelic replacement of the

Table 1. Comparison of homologous genes from the hygromycin A and A201A antibiotic gene clusters

Hyg gene	*A201A gene*	*Proposed function*	*Homologous protein*	*Identity/ similarity (%)*	*Accession Number*
hyg5	*ata12*	Mannose dehydratase	Ata12	45/58	CAD27644
hyg6	*ata11*	Methyltransferase	Ata11	31/42	CAD62205
hyg9	*ataPKS2*	ACP	AtaPKS2	34/55	CAD62191
hyg10	*ataPKS3*	3-ketoacyl ACP synthase	AtaPKS3	54/64	CAD62192
hyg11	*ataPKS4*	Unknown	Ata9PKS4	48/59	CAD62193
hyg12	*ata18*	CoA-ligase	Ata18	48/61	CAD62194
hyg13	*ata19*	ACP	Ata19	43/64	CAD62195
hyg14	*ata2*	3-ketoacyl ACP dehydratase	Ata2	43/59	CAD62196
hyg15	*ata4*	3-ketoacyl ACP reductase	Ata4	54/67	CAD62198
hyg16	*ata4*	Glycosyltranferase	Ata5	63/74	CAD62199
hyg19	*ata9*	Transmembrane protein	Ata9	50/63	CAD62203
hyg20	*ata16*	Transglucosylase	Ata16	60/71	CAD62189
hyg21	*ard2*	Phosphotranseferase	Ard2	56/66	CAD62197
hyg22	*ataPKS1*	Acyltransferase	AtaPKS	45/52	CAD27643
hyg29	*ard1*	ABC transporter	Ard1	78/86	CAA59109

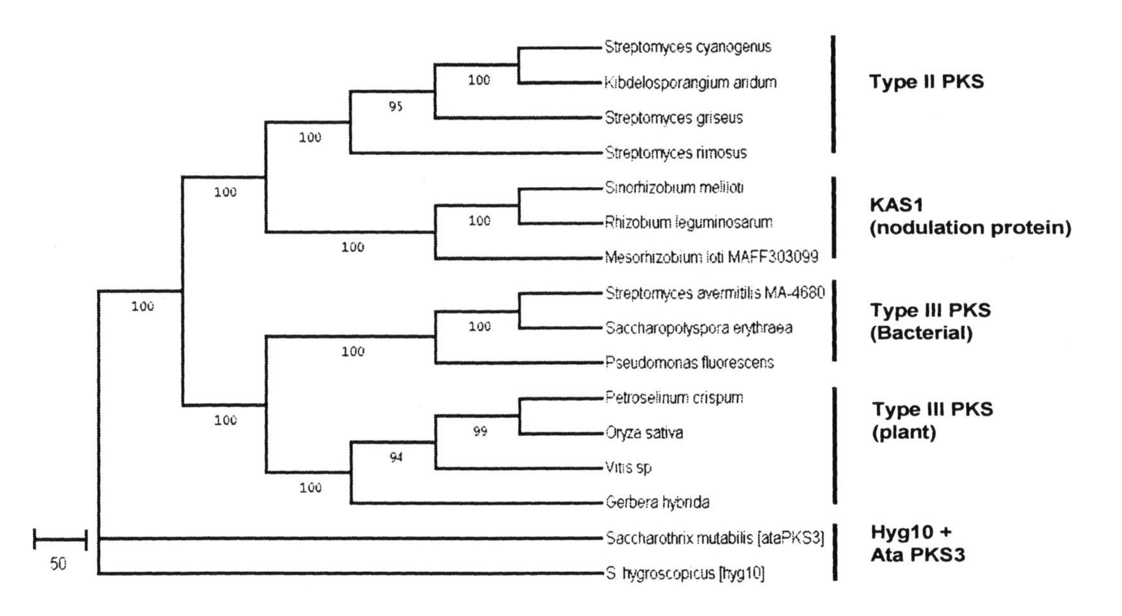

Figure 4. Phylogenetic analysis of various KS proteins from microorganisms and plants. Bootstrap values (shown as percentages) based on a neighborhood-joining algorithm are indicated at the tree nodes.

hyg26 gene in *S. hygroscopicus* by the *aac(3)IV* resistance marker (conferring apramycin resistance) and *oriT* led to the SCH30 mutant. Hyg26, which has homology to a family of short-chain dehydrogenases, is proposed to catalyze the final step in the biosynthetic pathway of the 5-dehydro-α-L-fucofuranose moiety by oxidation of NDP-L-fucofuranose (Figure 2). The SCH30 mutant makes 5"-dihydrohygromycin products, consistent with the role assigned to Hyg26. As shown in Figure 5A, the HPLC analyses of *S. hygroscopicus* NRRL 2388 clearly reveals the presence of hygromycin A (**1**) and methoxyhygromycin A (**2**) (the 5" epimeric forms of these two compounds appear as smaller shoulder peaks at a slightly earlier retention time). Production levels under the fermentation conditions used were typically 350 mg/L of methoxyhygromycin A, and 880 mg/L of hygromycin A. In contrast to the wild type strain, the SCH30 mutant generated no detectable levels of hygromycin A or methoxyhygromycin A, but three new peaks of hygromycin A analogs were observed (**3**, **4** and **5** in Figure 5B). The three new fermentation products were purified and characterized by LC-MS and NMR (^{13}C and ^{1}H) analyses, and shown to be 5"-dihydrohygromycin A (**3**), 5"-dihydromethoxyhygromycin A (**4**), and (*E*)-3-(3-hydroxy-4-*O*-α-fucofuranosylphenyl)-2-methylacrylic acid (**5**).

These compounds represent shunt metabolites of biosynthetic pathway intermediates and have not previously been identified, providing the first insight into the ordering of the sequence of steps which comprise the convergent hygromycin A biosynthetic pathway. In particular, the detection of compound **5** lacking the aminocyclitol moiety demonstrates that the glycosidic linkage in hygromycin A can be formed in the absence of the amide linkage. Additional experiments and analyses have suggested that formation of the glycosidic bond of hygromycin may actually be a prerequisite for formation of the amide linkage.

Concluding Comments

The biosynthetic origins of hygromycin A and the corresponding biosynthetic gene cluster have been revealed. Analysis of this gene cluster, and of the products generated by the SCH30 mutant, support both the convergent pathway model and many of the individual steps proposed for hygromycin A biosynthesis. There are several unique features of both the structure of hygromycin A and its biosynthetic pathway which warrant further investigation. We have also established a genetic system for manipulating the hygromycin biosynthetic gene cluster in *S. hygroscopicus*, and have shown that this can be used to provide significant yields of hygromycin A analogs. These, and potentially other hygromycin A analogs, could thus be produced using a cost-effective fermentation method, and serve as useful starting points for producing molecules with either clinical or agricultural applications.

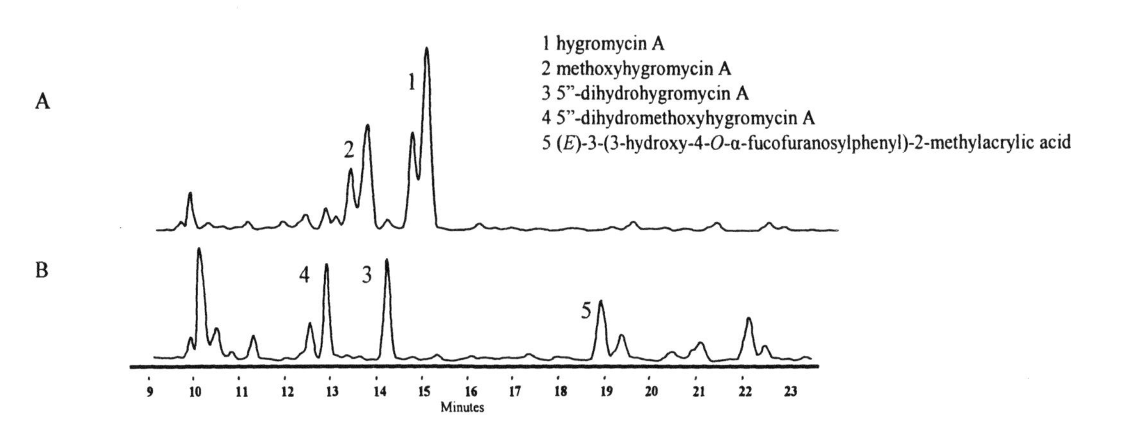

Figure 5. HPLC analyses of the fermentation broth of the wild type (A), and mutant SCH 30 strain (B).

References

1. Pittenger, R. C.; Wolfe, R. N.; Hoehn, P. N.; Daily, W. A.; McGuire, J. M. *Antibiot. Chemother.* **1953**, *3*, 1268.
2. Wakisaka, Y., Koizumi, K., Nishimoto, Y., Kobayashi, M., Tsuji, N. *J. Antibio.t (Tokyo)* **1980**, *33*, 695.
3. Mann, R. L.; Bromer, W. W. *J. Am. Chem. Soc.* **1958**, *80*, 2714.
4. Guerrero, M. D.; Modolell, J. *Eur. J. Biochem.* **1980**, *107*, 409.
5. Poulsen, S. M.; Kofoed, C.; Vester, B. *J. Mol. Biol.* **2000**, *30*, 471.
6. Hansen, J. L.; Ippolito, J. A.; Ban, N.; Nissen, P.; Moore, P. B.; Steitz, T. A. *Mol. Cell* **2002**, *10*, 117.
7. Yoshida, M.; Takahashi, E.; Uozumi, T.; Beppu, T. *Agric. Biol. Chem.* **1986**, *50*, 143.
8. Omura, S.; Nakagawa, A.; Fujimoto, T.; Saito, K.; Otoguro, K.; Walsh, J. C. *J. Antibiot. (Tokyo)* **1987**, *40*, 1619.
9. Nakagawa, A.; Fujimoto, T.; Omura, S.; Walsh, J. C.; Stotish, R. L.; George, B. *J. Antibiot. (Tokyo)* **1987**, *40*, 1627.
10. Uyeda, M.; Mizukami, M.; Yokomizo, K.; Suzuki, K. *Biosci. Biotechnol. Biochem.* **2001**, *65*, 1252.
11. Lee, H. B.; Kim, C.-J.; Kim, J.-S.; Hong, K.-S.; Cho, K. Y. *Lett. Appl. Microbiol.* **2003**, *36*, 387-391.
12. Mann, R. L.; Wolf, D. O. *J. Am. Chem. Soc.* **1957**, *79*, 120.
13. Kakinuma, K.; Sakagami, Y. *Agric. Biol. Chem.* **1978**, 42, 279.
14. Hayashi, S. F.; Norcia, L. J.; Seibel, S. B.; Silvia, A. M. *J. Antibiot. (Tokyo)* **1997**, *50*, 514.
15. Chida, N.; Ohtsuka, M.; Nakazawa, K.; Ogawa, S. *J. Org. Chem.* **1991**, *56*, 2976.
16. Jaynes, B. H.; Elliott, N. C.; Schicho, D. L. *J Antibiot (Tokyo)* **1992**, *45*, 1705.
17. Hecker, S. J.; Minich, M. L.; Werner, K. W. *Bioorg. Med. Chem. Lett.* **1992**, *2*, 533.
18. Trost, B. M.; Dirat, O.; Dudash, J., Jr., Hembre, E.J.; *Angew. Chem.* **2001**, *40*, 3658.
19. Trost, B. M.; Dudash, J.; Hembre, E. J. *Chem. Eur. J.* **2001**, *7*, 1619.
20. Donohoe, T. J.; Johnson, P. D.; Pye, R. J.; Keenan, M., *Org. Lett.* **2005**, *7*, 1275.
21. Habib, E.-S. E.; Scarsdale, J. N.; Reynolds, K. A. *Antimicrob. Agents Chemother.* 2003, *47*, 2065.
22. Jarvinen, N.; Maki, M.; Rabina, J.; Roos, C.; Mattila, P.; Renkonen, R. *Eur. J. Biochem.* **2001**, *268*, 6458.
23. Somers, W. S.; Stahl, M. L.; Sullivan, F. X. *Structure* **1998**, 6, 1601.

24. Zhang, Q.; Liu, H. *J. Am. Chem. Soc.* **2001**, *123*, 6756-6766.
25. Sanders, D. A.; Staines, A. G.; McMahon, S. A.; McNeil, M. R.; Whitfield, C.; Naismith, J. H. *Nat. Struct. Biol.* **2001**, *8*, 858.
26. Piepersberg, W., In *Biotechnology of Antibiotics*, Strohl, W. E., Ed. Marcel Dekker: New York, **1997**; p 81.
27. Ahlert, J.; Distler, J.; Mansouri, K.; Piepersberg, W. *Arch. Microbiol.* **1997**, *168*, 102.
28. Majumder, A. L.; Johnson, M. D.; Henry, S. A. *Biochim Biophys Acta* **1997**, *1348*, 245.
29. Walker, J. B. *J. Bacteriol.* **1995**, *177*, 818.
30. Tercero, J. A.; Espinosa, J. C.; Lacalle, R. A.; Jimenez, A. *J. Biol. Chem.* **1996**, 271, 1579.
31. Saugar, I.; Sanz, E.; Rubio, M., A; Espinose, J. C.; Jimenez, A. *Eur. J. Biochem.* **2002**, *269*, 5527.
32. Brautaset, T.; Sekurova, O. N.; Sletta, H.; Ellingsen, T. E.; Strom, A. R.; Valla, S.; Zotchev, S. B. *Chem. & Biol.* **2000**, *7*, 395.
33. Campelo, A. B.; Gil, J. A. *Microbiology* **2002**, *148*, 5.
34. Bentley, S. D.; Chater, K. F.; Cerdeno-Tarraga, A. M.; Challis, G. L.; Thomson, N. R.; James, K. D.; Harris, D. E.; Quail, M. A.; Kieser, H.; Harper, D.; Bateman, A.; Brown, S.; Chandra, G.; Chen, C. W.; Collins, M.; Cronin, A.; Fraser, A.; Goble, A.; Hidalgo, J.; Hornsby, T.; Howarth, S.; Huang, C. H.; Kieser, T.; Larke, L.; Murphy, L.; Oliver, K.; O'Neil, S.; Rabbinowitsch, E.; Rajandream, M. A.; Rutherford, K.; Rutter, S.; Seeger, K.; Saunders, D.; Sharp, S.; Squares, R.; Squares, S.; Taylor, K.; Warren, T.; Wietzorrek, A.; Woodward, J.; Barrell, B. G.; Parkhill, J.; Hopwood, D. A. *Nature* **2002**, *417*, 141.
35. Omura, S.; Ikeda, H.; Ishikawa, J.; Hanamoto, A.; Takahashi, C.; Shinose, M.; Takahashi, Y.; Horikawa, H.; Nakazawa, H.; Osonoe, T.; Kikuchi, H.; Shiba, T.; Sakaki, Y.; Hattori, M. *Proc. Natl. Acad. Sci. U S A.* **2001**, *98*, 12215.
36. Barrasa, M. I.; Tercero, J. A. ; Jimenez, A. *Eur. J. Biochem.* **1997**, *245*, 54.
37. Huang, W.; Jia, J.; Edwards, P.; Dehesh, K.; Schneider, G.; Lindqvist, Y., *EMBO J.* **1998,** *17*, 1183.

Chapter 3

The Biosynthesis of Polyketides, Tetramic Acids, and Pyridones in Fungi

Russell J. Cox

School of Chemistry, University of Bristol, Cantock's Close, Bristol BS8 1TS, United Kingdom (r.j.cox@bris.ac.uk)

The biosynthesis of compounds containing polyketides and amino acids by fused PKS-NRPS systems in filamentous fungi is discussed.

Fungi are often overlooked as sources of biologically active natural products. However, filamentous fungi are prodigious producers of secondary metabolites and especially compounds of high structural complexity. In particular, fungi produce numerous polyketide derived compounds. A number of fungal polyketides are used directly in medicine and agriculture, such as griseofulvin **1** produced by *Penicillium griseofulvum* (a potent systemic antifungal drug) (1). Other compounds have been the basis for the development of more active or selective compounds. Examples include lovastatin **2a** and compactin **2b** produced by *Aspergillus* species which are potent cholesterol lowering compounds in human medicine (2) and strobilurin **3** (3), produced by the basidiomycete *Strobilurus tenacellus* which is the basis for the development of a number of potent agricultural fungicides. Other compounds, such as fusarin C **4** (4) and aflatoxin B1 **5** (5) are potent environmental toxins with both acute and chronic effects. Yet others have potent biological activities which are not yet exploited, such as the squalestatins **6** (6) (inhibitors of squalene synthase) and the cytochalasins such as cytochalasin D **7** (7) from *Zygosporium masonii* a potent inhibitor of actin filament formation (Figure 1).

There have been significant advances in the understanding and exploitation of the biosynthesis of similar compounds in bacteria, particularly the actinomycetes where combinatorial biosynthesis can be used to manipulate

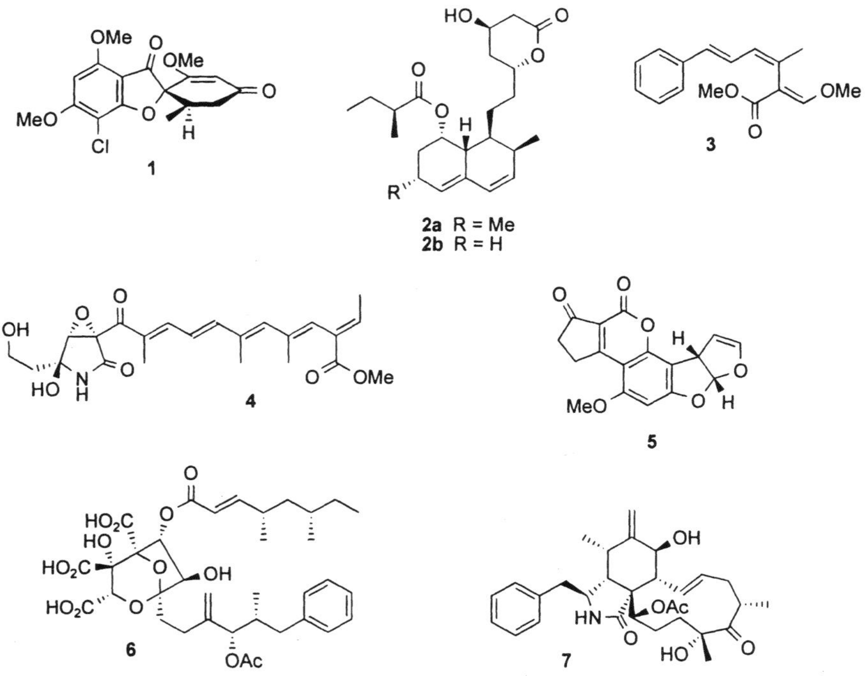

Figure 1. Polyketides and polyketide-amino acids from fungal species.

biosynthetic genes to produce new compounds (8). In contrast, progress in understanding and exploiting fungal biosynthetic systems has lagged many years behind. This has been for two main reasons. Firstly molecular genetic methods have not been well developed for the very wide range of fungal species involved in bioactive metabolite biosynthesis. Secondly, in the case of fungal polyketides, it is not clear how the synthases are programmed. For example, in the actinomycetes, polyketide synthases are often large multimodular proteins which operate as well-understood production-lines. In these cases it is possible to predict product structure, at least of the core polyketide, from gene sequence. This is because, in general, each module catalyses a single cycle of chain extension and contains only the modification enzymes used in that cycle, and the modules act sequentially. In the case of fungal polyketides, however, the synthases consist of single modules which act *iteratively* so it is neither possible to predict how many extension cycles will be catalysed, or which modification enzymes will act in each cycle. Thus understanding the programming of fungal iterative polyketide synthases is a significant challenge. We are interested in the mechanism of programming of iterative polyketide synthases and we wished to investigate them further. However, there appeared to be no general method for obtaining fungal PKS genes, and we thus had to develop this methodology.

Strategy

In the early 1990s the Bristol group began a long-term project to clone and characterise fungal polyketide synthases. At the outset of the project it was considered that fungal polyketides fell into three clear structural types: unreduced compounds such as orsellinic acid **8** (9), tetrahydroxynaphthalene **9** and YWA1 **10** (10); partially reduced compounds such as 6-methyl salicylic acid **11** (11); and highly reduced compounds such as lovastatin **2a**, squalestatin S1 **6**, T-toxin **12** (12) and fumonisin **13**. We wanted to develop a general method to clone polyketide synthase genes from any given filamentous fungus producing any specified compound. We planned to use a two-stage process. In the first stage, we would use degenerate oligonucleotides as primers in PCR reactions with genomic DNA from the fungi in question. The primers would be designed to amplify fragments of specific fungal polyketide synthases. In the second stage, the products of the PCR reactions would be labelled and used to probe genomic DNA or cDNA libraries in an attempt to obtain full PKS genes.

At the outset of our work, the only fungal PKS genes available in public databases encoded tetrahydroxynaphthalene synthase (THNS) from *Colletotrichum lagenarium* (13), wA synthase from *A. nidulans* (14) (both of the unreduced type) and 6-methyl salicylic acid synthase (6-MSAS) from *P. patulum* (15) and *A. terreus* (16) (of the partially reduced type). Two sets of degenerate oligonucleotide primers were designed which were biased towards each of these sets. The pair LC1/LC2c were designed to be selective for the unreduced PKS, while LC3/LC5c were designed to be selective for partially reduced PKS (Figure 3) (17). Later, as further sequences became available we also designed oligonucleotide primers selective for highly reduced PKS (KS3/KS4c) (18).

These degenerate oligonucleotides were used in PCR reactions with gDNA from a variety of filamentous fungi. In all cases, after optimisation, the reactions yielded DNA fragments of the expected sizes (*ca* 700-750 bp). The fragments were cloned and sequenced and then compared to known sequences. Thus PKS fragments obtained with the LC1/2c primers most closely matched known PKS involved in unreduced polyketide biosynthesis, fragments from the LC3/5c experiments most closely matched other MSAS sequences and fragments from the reactions run with degenerate primers based on HR PKS sequences (KS3/KS5c) most closely matched LNKS and TTS. This analysis confirmed our original hypothesis that there was a relationship between DNA sequence and chemical structure (18). The primers were also selective: they did not amplify sequences which cross-hybridised with DNA probes from other classes.

In further work, we focussed on the *C*-methyltransferase (*C*-MeT) domain. Fungi add methyl groups to the growing polyketide chain from *S*-adenosyl methionine (SAM), catalysed by a *C*-MeT domain. This contrasts to bacterial PKS which usually use a methylated extender unit (*i.e.* methylmalonate). Once again, degenerate oligonucleotides were designed to be selective for *C*-MeT domains, rather than the more common *O*- and *N*-methyl transferases common in primary

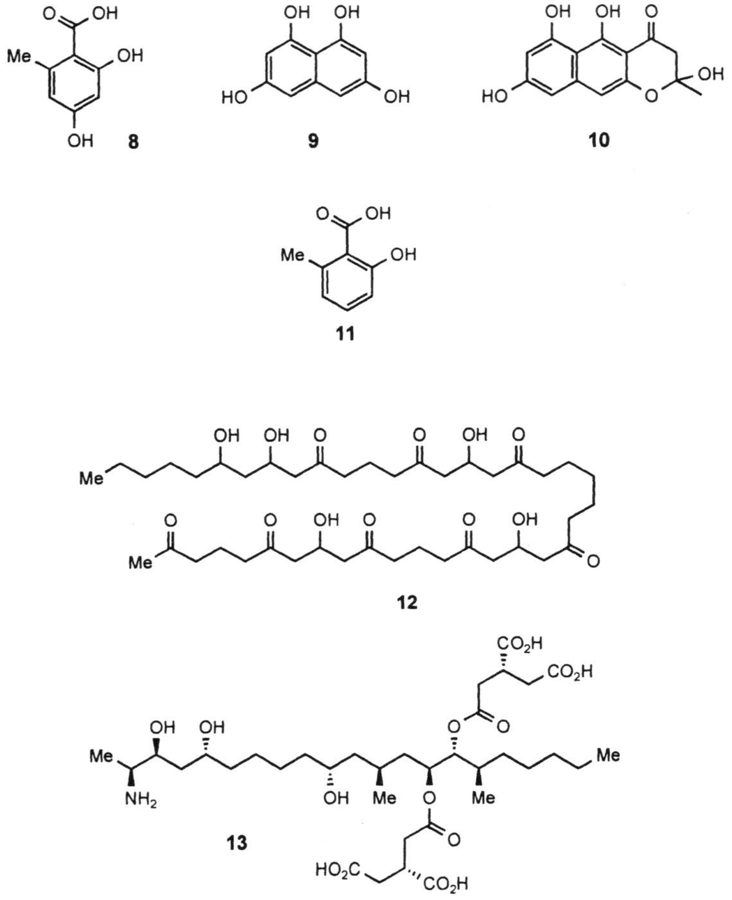

Figure 2. The three structural types of fungal polyketide.

metabolism (Figure 3). In PCR reactions with genomic DNA from a variety of filamentous fungi we obtained *ca* 320 bp products. Sequencing indicated that these all belonged to a family of *C*-MeT domains which are distinct from the *O*- and *N*-methyl transferases.

The Squalestatin PKS

The squalestatins (also known as zaragozic acids) are a family of potent squalene synthase inhibitors with potential use as anticholesterol compounds.

LC1 (5' KS NR primer)

```
Primer sequence     GAT CCI AGI TTT TTT AAT ATG
Degeneracy = 32       C     C         C   C   C
```

LC2c (3' KS NR primer)

```
Primer sequence      GT ICC IGT ICC GTG CAT TTC
Degeneracy = 4                      A       C
```

LC3 (5' KS PR primer)

```
Primer sequence     GCI GAA CAA ATG GAT CCI CA
Degeneracy = 8            G   G       C
```

LC5c (3' KS PR primer)

```
Primer sequence      GT IGA IGT IGC GTG IGC TTC
Degeneracy = 4                      A       C
```

***C*-MeT1 (SAM binding site primer)**

```
Primer sequence     GAA ATI GGI GGI GGI ACI GG
Degeneracy = 4        G           C
```

***C*-MeT2c (LDKS type active site primer)**

```
Primer sequence      AT IAG TTT ICC ICC IGG TTT
Degeneracy = 8            A C               C
```

***C*-MeT3c (LNKS type active site primer)**

```
Primer sequence      AC CAT TTG ICC ICC IGG TTT
Degeneracy = 4              C               C
```

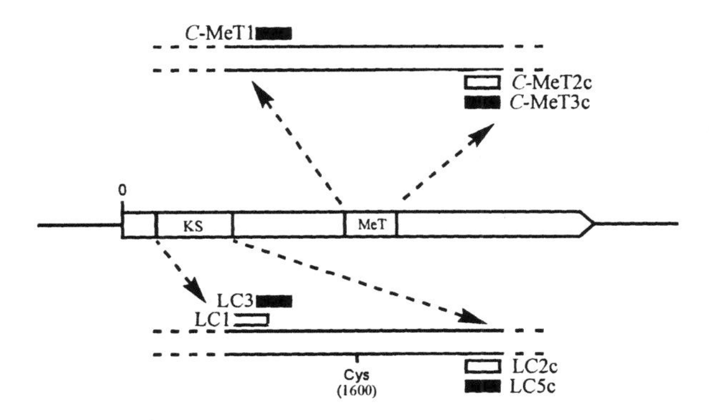

Approximate predicted PCR product size:
LC1/2c - 700 bp; LC3/5c - 680 bp;
C-MeT1/2c - 320 bp; *C*-MeT1/3c - 320 bp.

Figure 3. Degenerate PCR primers selective for fungal PKS sequences.

They are produced by Phoma and other species of filamentous fungi and clearly belong to the highly reduced class of fungal polyketides. Biosynthetic feeding studies show that the main carbon backbone of squalestatin S1 **6** consists of head-to-tail acetate units, with pendant carbons derived from the methyl group of SAM (19). The main chain is a hexaketide derived from a benzoate starter unit, while the C-6 ester side chain is also a polyketide derived from four acetates and two SAM derived methyl groups (Figure 4).

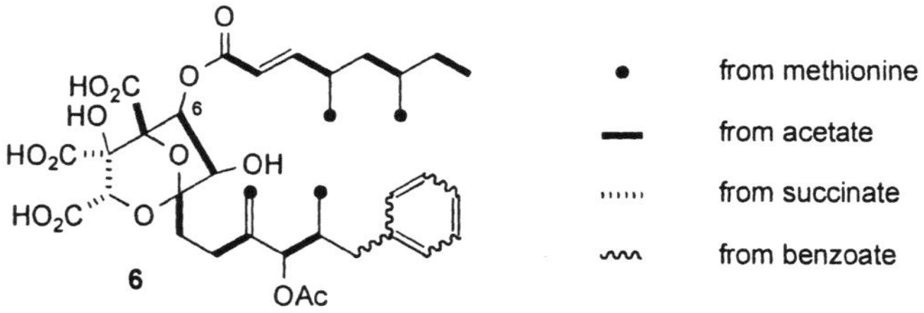

Figure 4. origin of carbon atoms in Squalestatin S1 ***6****.*

We set out to clone the squalestatin S1 **6** PKS genes from two organisms: *Phoma sp.* C2932 and an unidentified strain MF5453. Because both of the polyketide chains of **6** are methylated we reasoned that *C*-MeT probes would give selectivity and discriminate against fungal FAS and non-methylated PKS genes. Thus PCR reactions using *Phoma* gDNA and degenerate *C*-MeT primers yielded *ca* 320bp products which were confirmed as fragments of PKS genes by sequence analysis. These PCR products were then used as radio-labelled probes to screen both gDNA and cDNA libraries of *Phoma sp.* C2932 and MF5453. We ensured that the cDNA libraries were constructed using cells harvested during maximum S1 production. This strategy was adopted to further increase selectivity for the S1 biosynthetic genes as we reasoned that the cDNA library should be enriched for S1 PKS transcripts relative to other PKS genes (20).

We isolated several overlapping clones from these libraries which hybridised strongly to the labelled *C*-MeT probes. Sequence analysis of hybridised plaques revealed the presence of a 4.8 Kb fragment of a Type I PKS, containing the polyadenylated 3' end of the sequence. This clone contained (5'→3') a truncated fragment of a dehydratase domain (DH) and complete *C*-MeT, enoyl reductase (ER), keto-reductase (KR) and acyl carrier protein (ACP) domains. The 5' portion of the PKS was then obtained using a process involving the rapid amplification of cDNA ends (RACE) to produce an overlapping clone, again of 4.8 Kb containing β-ketoacyl synthase (KS), acyl transferase (AT), dehydratase (DH) and *C*-MeT domains. A contiguous, full-length clone was then constructed by digest and religation of the two cDNA fragments.

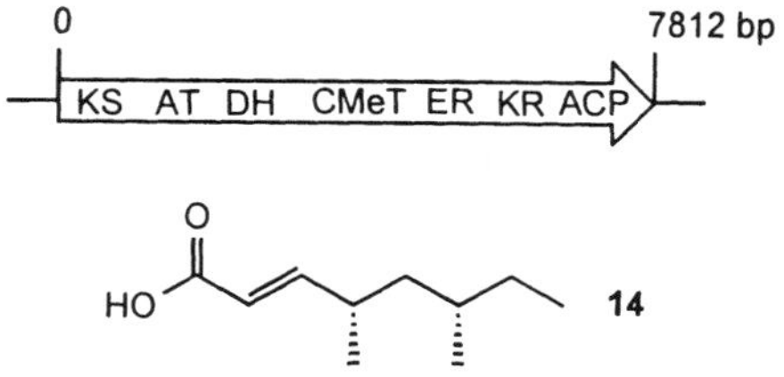

Figure 5. Domain analysis of Phoma PKS1 (SQTKS).

The reconstructed 7812 bp PKS (Phoma PKS1) contained KS, AT, DH, *C*-MeT, ER, KR and ACP domain encoding regions (Figure 5). The translated gene, encoding a polypeptide of 2604 amino acids, shows high end-to-end amino acid homology with other known fungal PKS involved in the biosynthesis of highly reduced polyketides such as the lovastatin diketide synthase (LDKS, 59% identity), and the compactin diketide synthase (60% identity).

Close inspection of individual domains suggested they were catalytically competent. For example in the ER domain, the conserved NADPH binding motif (LxHx(G/A)xGGVG) was present (LIHAASGGVG). This contrasts with the situation in other fungal PKS genes such as the lovastatin nonaketide synthase (LNKS) where the ER domain is probably inactive due to deleterious mutations in the NADPH binding region (2).

We then exploited the *Aspergillus oryzae* fungal expression system which has been used by Ebizuka and Fujii for the successful expression of fungal unreduced PKS (21). Thus Phoma-PKS1 was inserted into the expression vector pTAex3 to form a 15747 bp expression construct in *Escherichia coli*. *A. oryzae* strain M-2-3 (*arg*B^{-}) protoplasts were transformed with the expression vector and selected on arginine deficient plates. True transformants were selected by repeated sub-cloning on arginine deficient media, and further confirmed by colony PCR using Phoma-PKS1 specific primers and by Southern blotting.

The pTAex3 expression system utilises the *amy*B promoter which is repressed by glucose and induced by starch. *A. oryzae* transformants were grown in the presence of starch. Organic extracts of the medium from the WT and transformant strains were then analysed by RP-HPLC. Of five transformants shown to have integrated the Phoma PKS1, one strain showed the presence of a new compound in the HPLC trace. This compound was isolated and purified by repeated RP-HPLC. Full structural characterisation revealed the new compound to be the doubly methylated unsaturated acid **14** (Figure 5). Comparison of optical rotation data with synthetic material proved it to be the 4*S*,6*S* enantiomer, and thus chain B of squalestatin. We thus named this PKS SQTKS (squalestatin tetraketide synthase).

The Fusarin Synthase

Fusarin C **4** is a mycotoxin produced by a number of *Fusarium* species. These organisms are often significant pests of cereals. Isotopic labelling studies indicated that it is biosynthesised from a polyketide chain fused to a C_4N unit most probably derived from an amino acid. Once again the polyketide moiety is methylated with carbon atoms from SAM (22).

We used the same strategy as described for SQTKS to rapidly clone *ca* 26Kb of gDNA from *F. moniliforme* containing the fusarin synthase and several other ORFs. The fusarin synthase (FUSS) is unusual because it combines an iterative PKS with one module of a non-ribosomal peptide synthase (NRPS). However, the PKS portion of the synthase shows high similarity to the small number of other known fungal PKS genes, for example LNKS.

Our analysis, thus far, shows that the FUSS PKS consists of the catalytic domains: β-ketoacylsynthase (KS), acyltransferase (AT), dehydratase (DH), *C*-methyltransferase (*C*-MeT), enoylreductase (ER), ketoreductase (KR) and acyl carrier protein (ACP). The PKS is fused to condensation (C), adenylation (A) and thiolation (T) domains which are terminated by a reduction (R) domain (Figure 6).

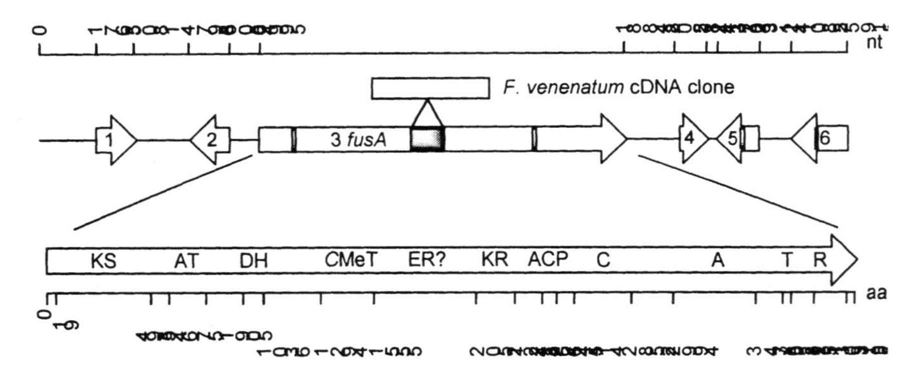

Figure 6. Part of the fusarin C gene cluster.

The PKS most likely makes a heptaketide (*i.e.* 6 cycles) with four appended methyl groups from SAM. The PKS must use the KS, AT, KR, DH and ACP domains in every cycle, but methylation only occurs after cycles 1, 2, 3, and 5 (Scheme 1). Enoyl reduction occurs in none of the cycles - revealed by the polyunsaturated structure of fusarin C and consistent with the fact that the NADPH binding region of the ER appears to include a number of deleterious mutations. The polyketide intermediate **15** probably remains attached to the PKS ACP at the end of its synthesis (Scheme 1).

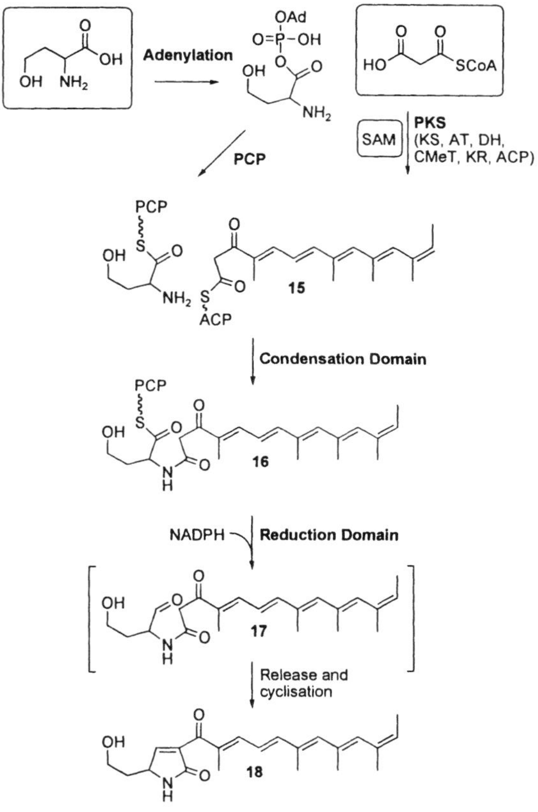

Scheme 1. Proposed mode of action of FUSS.

Meanwhile, in parallel, homoserine is recognised and selected by the adenylation domain of the NRPS and adenylated using ATP. The acyl adenylate then probably reacts with the free thiol of the PCP. The condensation domain could then catalyse a peptide forming reaction between the amine of the enzyme-bound homoserine moiety and the thiol ester carbon of the ACP-bound polyketide, leading to the peptide intermediate **16**. At this stage the intermediate would still be attached to the synthase *via* a thiolester linkage to the PCP domain. Final reduction of this thiolester would then allow release from the synthase, but the released aldehyde **17** is also likely to undergo cyclisation to form the 5-membered ring **18**. Whether this cyclisation is spontaneous or catalysed is not known. Compound **18** must be subjected to further oxidative reactions in order to reach fusarin C. Genes encoding oxidative enzymes are present in the fusarin cluster, but their exact roles remain to be elucidated.

The Tenellin Synthase

The insect pathogenic filamentous fungus *Beauveria bassiana* makes the 2-pyridone tenellin **19** (Figure 7) in high yield. Feeding studies established that tenellin is derived from tyrosine (or phenylalanine) and a pentaketide (23).

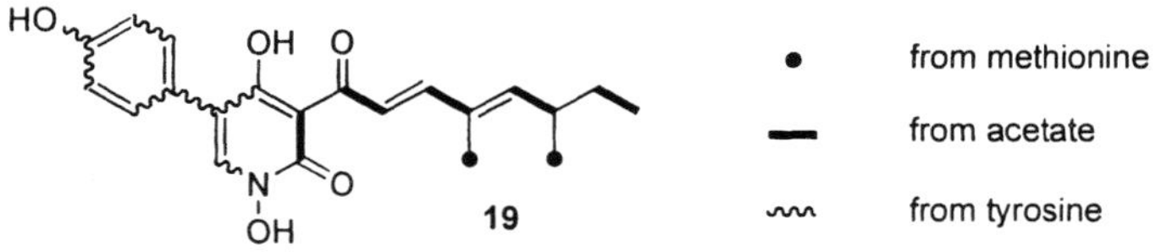

Figure 7. Tenellin and the biosynthetic origin of C and N atoms.

Again, the *C*-MeT PCR primers were used to obtain a gene probe from gDNA, which was then used to isolated a fused iterative Type I PKS / NRPS from *B. bassiana* which appears to be involved in tenellin biosynthesis (preliminary evidence from knockout experiments). The PKS/NRPS (hereafter called TENS) is highly homologous to FUSS, but appears to have a functional ER domain. In this synthase the PKS extends 4 times. It methylates and fully reduces after the first condensation, methylates but does not enoyl-reduce after the second condensation, does not methylate or enoyl reduce after the third condensation and does no post-condensation processing at all after the final condensation. Here, evidence from feeding experiments shows that the selected amino acid is tyrosine or phenylalanine (24).

By analogy to the fusarin synthase, it would be expected that this synthase would produce an intermediate peptide such as **20** (Scheme 2) as a precursor of a 5-membered ring rather than the 6-membered pyridone of tenellin itself. This is supported by isotopic feeding experiments which suggest that tyrosine or phenylalanine are not rearranged before incorporation into tenellin (25). Reductive release would form **21a**. Cytochrome P_{450} genes in the vicinity of the TENS gene may be responsible for oxidation of the ring to give **21b**. In terms of subsequent chemistry this makes sense as there must be further *oxidative* ring expansion steps to form the 6-membered pyridone ring. This hypothesis remains to be proven by further experimentation.

Other products of fungal PKS NRPS - Relationships and Hypotheses.

Intriguingly LNKS, one of the best investigated fungal polyketide synthases has long been known to possess NRPS domains (2). Similarly to FUSS and TENS, it contains a seemingly intact condensation domain after the PKS ACP, and the *N*-terminal region of the adenylation (A) domain. However the catalytic

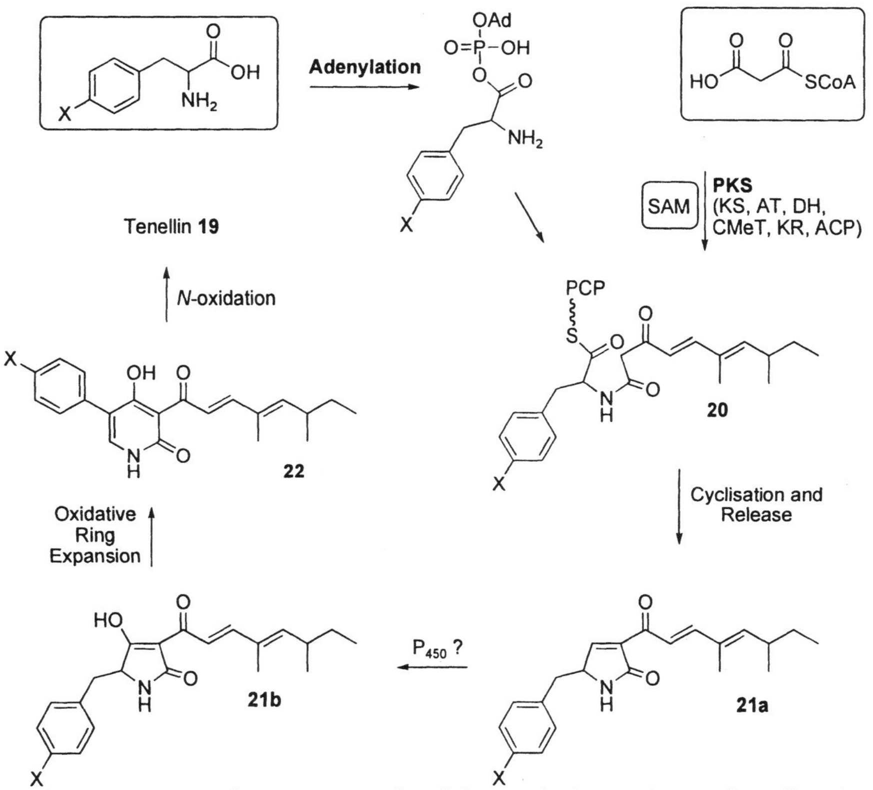

*Scheme 2. Proposed reactions catalysed during the biosynthesis of tenellin **19**.*

parts of the A domain are missing, as are the thiolation (T) and reduction (R) domains. The existence of compounds combining amino acids and polyketides closely related to the structure of lovastatin suggests that perhaps once LNKS was a fully-fledged PKS-NRPS. The tetramic acid equisetin **30** from *Fusarium heterosporum* (26) has recently been shown to arise *via* PKS-NRPS biosynthesis. It is not difficult to imagine how other compounds such as zopfiellamide A **32** (from *Zopfiella latipes*) could arise by the action of very similar synthases (Figure 8). It has been suggested that LNKS may have evolved from a synthase very similar to the equisetin or zopfiellamide synthases. Interestingly Diels Alder cyclisations are likely post-PKS modifications in these cases. Diels Alder cyclisation between the PKS and the newly formed heterocycle could also explain the origin of deoxaphomin **28** (a precursor of the cytochalasins) (27) which could arise from the Diels Alder reaction of the polyketide diene with the heterocyclic ring itself.

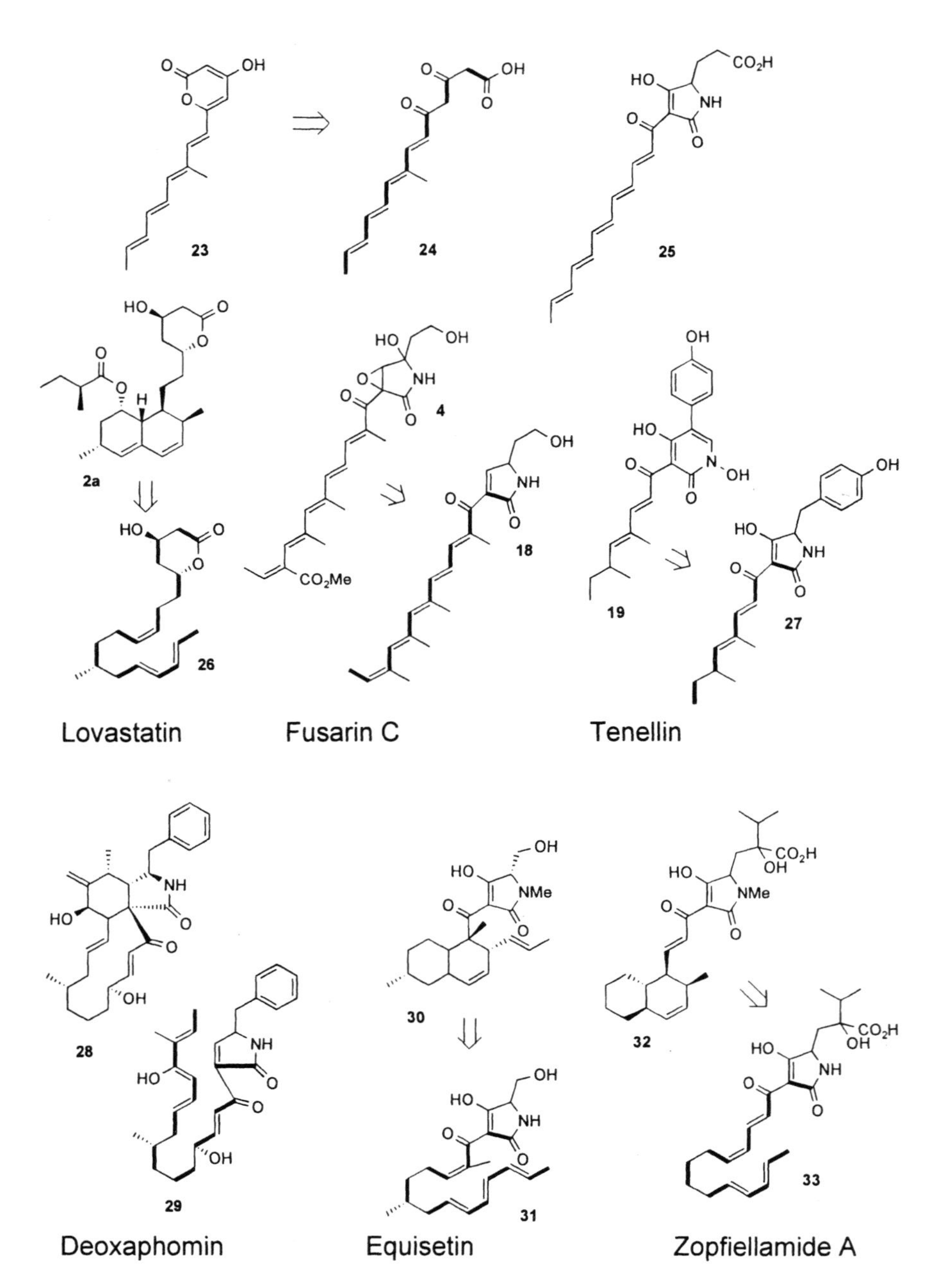

Figure 8. Hypothesised and observed PKS-NRPS precursors.

When LNKS was expressed in the absence of its usual partner genes, it was shown to produce the polyene-pyrone **23** - a heptaketide (2) indicating that the ER domain of LNKS is inactive. When a separate ER from the lov cluster (*lovC*) was expressed in tandem with LNKS, correct reduction and chain extension was observed. The fusarin C polyketide fragment **18** is remarkably similar to the LNKS polyunsaturated heptaketide **24** (Figure 8). This is in agreement with the fact that there appears to be no *lovC* homologue in the FUSS cluster and the ER of FUSS contains deleterious mutations at the NADPH binding site. However, the equisetin **30** synthase, which produces a decalin similar to that of lovastatin, does not appear to have a closely associated *lovC* gene in the biosynthetic cluster, and presumably its ER is fully active (28). The presence or absence of *lovC* homologues in biosynthetic gene clusters may not therefore be indicative of whether a fungal PKS-NRPS makes the simpler polyunsaturated compounds such as fuligorubin A (29) **25** (from *Fuligo septica*) and fusarin C **4** from *Fusarium moniliforme* or the cyclic types including equisetin **30**, deoxaphomin **28** (and related cytochalasins and chaetoglobosins) and zopfiellamide **32**. It is not yet clear whether the tenellin synthase possess a *lovC* homologue, but preliminary sequence analysis suggests that the TENS ER should be active, similarly to the equisetin ER domain.

A further interesting question surrounds the activity of the NRPS R domains. During Fusarin C **4** biosynthesis, reduction of the thiolester **16** (Scheme 1) probably leads to an aldehyde which can then cyclise to form the 1,5-dihydro-pyrrole-2-one skeleton **18** which is at the correct oxidation level for subsequent epoxidation. A similar reduction would be predicted in the case of deoxaphomin **28** which would lead to the potential precursor **29** (Figure 8). However, in the case of equisetin where the R domain is apparently present on the synthase, reductive cyclisation would lead to formation of a compound which would require subsequent oxidation to form the observed tetramic acid - a seemingly redundant process (28). In this case it is possible either that the R domain of the equisetin synthase is inactive, or that it is active and another enzyme reoxidises to give the observed tetramic acid of **36**. In the case of tenellin biosynthesis, the ring of **21a** formed by the PKS-NRPS needs to undergo further oxidation to **21b** before ring expansion, if the R domain is again active (Scheme 2). Thus in the cases of both tenellin and equisetin the action of the R domain seems redundant. Thus more experimental work is required to delineate the activity of R domains in each case.

Future Prospects

Genomics approaches are now discovering polyketide and non-ribosomal peptide synthases much more rapidly that the gene probing approaches discussed above. However genome sequencing methodology is currently limited in scope due to cost and at present there is still a requirement for directed gene probing methods in particular species associated with particular compound production. The major challenge, which is brought into sharper relief by the numbers of synthases discovered through genomics, is that it is not possible to predict the structure of the compound produced by any given fungal biosynthetic gene cluster. While it is possible to predict whether a polyketide will belong to the unreduced, partially reduced or highly reduced families, it is not possible to predict chain length, methylation and reduction patterns or cyclisations. In the case of the post PKS genes, general classes can be predicted, such as enoyl reductases or cytochrome P_{450} oxidases, but their exact chemo-selectivity cannot be predicted. For example, it is clear that the product of the ace1 PKS-NRPS protein in *Magnaporthe grisea* is involved in avirulence signalling between rice plants and the fungal host, but the chemical structure cannot be predicted (30). Thus there is not yet a strong link between the possession of fungal genetic sequence and knowledge of chemical structures. Expression and directed knockout strategies, as described above for SQTKS and the fusarin synthase will therefore remain of key importance.

Acknowledgements

The work carried out in Bristol could not have been completed without the participation of the following coworkers and students: Lewis Bingle, Tom Nicholson, Frank Glod, Ying Zhang, Deirdre Hurley, Kate Harley, Dr Zhongshu Song and Dr Kirstin Eley. Particular thanks are due to my academic coworkers Dr Colin Lazarus, Dr Andy Bailey and Prof. Tom Simpson. Funding has been generously provided by The University of Bristol, EPSRC, BBSRC, and GlaxoWellcome.

References

1. A. E. Oxford. H. Raistrick and P. Simonart, *Biochem. J.*, **1939**, *33*, 240.
2. J. Kennedy, K. Auclair, S. G. Kendrew, C. Park, J. C. Vederas and C. R. Hutchinson, *Science*, **1999**, *284*, 1368.
3. G. Schram., W. Steglich, T. Anke and F. Oberwinkler, *Chemische Berichte-Recueil*, **1978**, *111*, 2779-2784.
4. W. C. A. Gelderblom, W. F. O. Marasas, P. S. Steyn, P. G. Thiel, K. J. Vandermerwe, P. H. Vanrooyen, R. Vleggaar and P. L. Wessels, *J. Chem. Soc. Chem. Commun.*, **1984**, 122-124.
5. R. E. Minto and C. A. Townsend, *Chem. Rev.*, **1997**, *97*, 2537.
6. A. Baxter, B. J. Fitzgerald, J. L. Hutson, A. D. Mccarthy, J. M. Motteram, B. C. Ross, M. Sapra, M. A. Snowden, N. S. Watson, R. J. Williams and C. Wright, *J. Biol. Chem.*, **1992**, *267*, 11705.

7. J. C. Vederas and C. Tamm, *Helv. Chim. Acta* **1976**, *59*, 558.
8. J. Staunton and K. J. Weissman, *Nat. Prod. Rep.* **2001**, *18*, 380.
9. L. Reio, *J. Chromatography*, 1958, **1**, 338; K. Mosbach, *Z. Naturforsch B.*, **1959**, *14*, 69.
10. A. Watanabe, I. Fujii, U. Sankawa, M. E. Mayorga, W. E. Timberlake and Y. Ebizuka, *Tetrahedron Lett.*, **1999**, *40*, 91.
11. P. Dimroth, H. Walter, and F. Lynen, Eur. J. Biochem., **1970**, *13*, 98.
12. K. J. Tegtmeier, J. M. Daly and O. C. Yoder, *Phytopathology*, **1982**, *72*, 1492.
13. Y. Takano, Y. Kubo, K. Shimizu, K. Mise, T. Okuno and I. Furusawa, *Mol. Gen. Genet.*, **1995**, *249*, 162.
14. M. E. Mayorga and W. E. Timberlake, *Mol. Gen. Genet.*, **1992**, *235*, 205.
15. J. Beck, S. Ripka, A. Seigner, E. Schiltz and E. Schweizer, *Eur. J. Biochem.*, **1990**, *192*, 487.
16. I. Fujii, Y. Ono, H. Tada, K. Gomi, Y. Ebizuka and U. Sankawa, *Mol. Gen. Genet.*, **1996**, *253*, 1.
17. L. E. H. Bingle, T. J. Simpson And C. M. Lazarus, *Fungal Genet. Biol.*, **1999**, *26*, 209.
18. T. P. Nicholson, B. A. M. Rudd, M. Dawson, C. M. Lazarus, T. J. Simpson and R. J. Cox, *Chem. Biol.*, **2001**, *8*, 157.
19. K. M. Byrne, B. H. Arison, M. Nallinomstead and L. Kaplan, *J. Org. Chem.*, **1993**, *58*, 1019; C. A. Jones, P. J. Sidebottim, R. J. P. Cannell, D. Noble and B. A. M. Rudd, *J. Antibiot.*, **1992**, *45*, 1492.
20. R. J. Cox, F. Glod, D. Hurley, C. M. Lazarus, T. P. Nicholson, B. A. M. Rudd, T. J. Simpson, B. Wilkinson and Y. Zhang, *Chem. Commun.*, **2004**, 2260.
21. H. F. Tsai, I. Fujii, A. Watanabe, A. H. Wheeler, Y. C. Chang, Y. Yasuoka, Y. Ebizuka and K. J. Kwon-Chung, *J. Biol. Chem.*, **2001**, *276*, 29292.
22. P. S. Steyn and R. Vleggaar, *J. Chem Soc., Chem. Commun.*, **1985**, 1189.
23. J. L. C. Wright, L. C. Vining, A. G. Mcinnes, D. G. Smith and J. A. Walter, *Can. J. Biochem.*, **1977**, *55*, 678.
24. M. C. Moore, R. J. Cox, G. R. Duffin and D. O'Hagan, *Tetrahedron*, **1998**, *54*, 9195.
25. R. J. Cox and D. O'Hagan, *J. Chem. Soc., Perkin Trans. 1.*, **1991**, 2537.
26. J. W. Sims, J. P. Fillmore, D. D. Warner and E. W. Schmidt, *Chem. Commun.*, **2005**, 186.
27. J-L Robert and C. Tamm, *Helv. Chim. Acta*, **1975**, *58*, 2501.
28. Prof. Eric Schmidt, University of Utah, *personal communication.*
29. W. Steglich, *Pure & App. Chem.*, **1989**, *61*, 281.
30. H. U. Bohnert, I. Fudal, W. Dioh, D. Tharreau, J. L. Notteghem and M. H. Lebrun, *Plant Cell*, **2004**, *16*, 2499.

Chapter 4

Searching for Polyketides in Insect Pathogenic Fungi

Donna M. Gibson[1], Stuart B. Krasnoff[2], and Alice C. L. Churchill[2]

[1]Plant Protection Research Unit, U.S. Plant, Soil, and Nutrition Laboratory, Agricultural Research Service, U.S. Department of Agriculture, Tower Road, Ithaca, NY 14853
[2]Department of Plant Pathology, Cornell University, Tower Road, Ithaca, NY 14853

Among the major microbial taxa, fungi are surpassed only by actinomycetes as a source of therapeutic agents, especially antibiotics. It has been argued that there is a higher likelihood of discovering producers of novel antibiotics among parasitic microorganisms because of the unique adaptive roles adduced to secondary metabolites during the infection, both in overcoming host defenses and in competing with other micro-organisms for host resources. The extensive polyketide synthase sequence data available for fungi now makes molecular screening for novel bioactive polyketides a viable alternative to traditional screening approaches. The development, application, and utility of one such screening approach to predict chemistries in insect pathogenic fungi are described herein.

Traditional drug discovery efforts from microorganisms have typically relied on mass screening of extracts from large, environmentally disparate populations to detect promising biological activity and identify novel chemistries. Microbes are a proven source of biologically active compounds (*1-6*), where actinomycetes are the leading producers of useful natural products, followed by fungi and eubacteria. The fungi, in particular, represent a large genetic resource estimated at 1.5 million species worldwide, of which only 65,000 species (or 5%) have been described (*2, 7*). Fungi could prove to be a richer source of novel compounds than bacteria. Of the top 20 selling pharmaceuticals in 1995, 6 were fungal products (*4*), including the antibiotic penicillins and cephalosporins, and the cholesterol-lowering lovastatins. While known fungi are a proven source of natural products, the majority represent an immense unexploited and underutilized resource for novel metabolites of biomedical and agricultural importance.

The environmental factors that induce production of secondary metabolites are poorly known, as are the roles that these natural products perform for each producing organism. Fungi in ecological niches associated with other organisms, such as in animals, soil, dung, or plants, are interactive. These interactions, especially for pathogens, endophytes, and saprophyes, are likely to be mediated by small molecules. Ecological theory predicts that metabolic diversity will be promoted by two key factors: 1) high levels of environmental stress, and 2) intense and frequent interactions with other organisms (*8*). So, it is expected that organisms that meet one or both of these criteria would be likely to yield biologically active secondary metabolites. Such might be the case for entomopathogenic (insect pathogenic) fungi.

Entomopathogenic Fungi And Their Relationship To Hosts

Pathogens of insects and other invertebrates have been identified in more than 100 fungal genera, with more than 700 species recognized that attack insect hosts (*9, 10*). These fungi are also taxonomically diverse, since the genera are found scattered throughout the various classes of fungi, but mostly among the Hyphomycetes and Zygomycetes. The genera include both perfect (sexual) and imperfect (asexual, conidial, or vegetative) states, and each state can potentially produce distinct sets of secondary metabolites.

Fungal pathogens of invertebrates may be a particularly rich source of novel metabolites since aspects of their pathobiology indicate the presence of pathogen-released chemicals affecting invertebrate behavior and movement (*11*). Insect pathogenic fungi are characterized by their mode of action in invading the host organism: causing early death of the host, impairing host activities without causing early death, or attacking only old or weakened host insects. Upon penetration and colonization of the insect, it is predicted that the fungus evades, overwhelms, or subverts the host's immune system by producing toxins, by changing morphology, or by unknown, but presumably biochemical, mechanisms (*11*). The exoskeleton of infected insects is often preserved while the fungus undergoes extensive reproductive development within the insect body. The resting stages of insect pathogenic fungi may persist for long periods in the soil, suggesting that protective antibiotics are part of the suite of secondary metabolites produced during proliferation within the insect.

Exploiting the Resources of the ARS Collection of Entomopathogenic Fungal Cultures (ARSEF)

The ARSEF culture collection is considered to be one of the world's most complete sources of insect fungal cultures for use in biocontrol, bioprospecting, and taxonomic research (http://arsef.fpsnl.cornell.edu/). With over 7,000 accessions from 480 taxa collected from over 1,000 hosts and world-wide locations, the culture collection represents a unique opportunity to evaluate the hypothesis that pathogens are novel sources of biologically active compounds.

Secondary Metabolites of Entomopathogenic Fungi

The existing literature on toxins of entomopathogenic fungi, although substantial, deals with relatively few species (*6, 12*). Virtually nothing is known of the secondary chemistry of many fungal pathogens of invertebrates. Of those toxins that have been described, the three principal classes of chemistries are polyketides (PK), nonribosomally synthesized peptides (NRSP) , and alkaloids.

Polyketides in entomopathogenic fungi

The occurrence of polyketides and polyketide-like compounds from entomopathogenic fungi has been reported (reviewed in (*6*). For instance, *Metarhizium flavoviride* produces viridoxins, novel diterpene derivatives of polysubstituted gamma-pyrones, which exhibit insecticidal activity against Colorado potato beetle (*13*). The viridoxins resemble polyketide-type compounds, although their mode of synthesis is not known, and are closely related chemically to phytotoxins produced by plant pathogenic fungi in the genus *Colletotrichum*. One of these phytotoxins, colletochin, which differs from the viridoxins by the absence of an α-hydroxy acid moiety, is much less toxic to insects than the viridoxins (*13*). This preliminary observation of a structure/activity relationship suggests that *M. flavoviride* produces a toxin with targeted activity against insects. This is an encouraging example of the potential for finding compounds with high target-selectivity among the secondary metabolites of entomopathogenic fungi.

Nonribosomally derived peptides in entomopathogenic fungi

The genus *Tolypocladium*, which includes several species pathogenic to invertebrates, produces at least four types of biologically active NRSPs: the potent mitochondrial ATPase inhibitory efrapeptins (*14*), the immunosuppressive cyclosporins (*15, 16*), the antibacterial amino isobutyric acid (Aib)-containing cicadapeptins (*17*), and the amino acid-derived diketopiperazines (*18*). The asexual stage of *Tolypocladium* has been linked definitively to the sexual stage of the insect pathogenic fungus *Cordyceps* (*19*). *Paecilomyces*, which includes numerous entomopathogenic species, produces the NRSPs called leucinostatins, another class of mitochondrial ATPase inhibitory peptides (*16*), as well as beauvericin (*6, 12*). *Metarhizium* spp. and *Aschersonia* spp. produce destruxins, a class of cyclic depsipeptides (*20, 21*). These NRSPs activate calcium channels in insect muscle, causing paralysis when injected into caterpillars and adult flies (*22*). Observations that destruxins cause untimely degranulation of meocytes in crayfish, which are comparable to leucocytes in other animals, suggest that they may act to reduce immunological competence in arthropod hosts (*23-26*).

Alkaloids in entomopathogenic fungi

A number of fungal pathogens of invertebrates (e.g., *Cordyceps*, *Torrubiella*, *Hypocrella*, and *Beauveria*) belong to the same family as the plant pathogenic *Claviceps* species. *Claviceps* spp. produce the ergot peptide alkaloids or ergopeptines (*27, 28*), which include ergotamine, ergobasine, and the precursor for semisynthetic production of methylergobasine. Methylergobasine is used medically to stimulate labor and uterine contractions (*29*). These taxonomic relationships suggest that many insect pathogenic fungi have the potential to synthesize novel peptide alkaloids.

Convergence Of Chemistry, Biology And Genetics In The Search For Novel Polyketide Chemistries

In contrast to traditional screening methods, the tools of molecular biology offer a directed approach to screening unknown germplasm libraries for the potential to produce novel chemistries.

Polyketides and Polyketide Synthases of Fungi

Polyketides (PK) are a large group of secondary metabolites with chemically diverse structures and a broad range of biological activities (*30*). This chemical family includes the well-known antibiotics, tetracycline, erythromycin, and avermectin, as well as immunosuppressants such as rapamycin and FK506, and phytotoxins such as T-toxin and AAL-toxin (*30*). Mithramycin, chromomycin, and olivomycin are clinically important anti-tumor agents, which share a common aromatic aglycone that is derived from a single polyketide backbone (*31*).

All PK synthesis is orchestrated by polyketide synthases (PKSs), multimeric enzymes that function much like fatty acid synthases. Chemical diversity is due to the action of PKSs in controlling the number and type of carboxylic units added and the extent of reduction and stereochemistry of the α-keto group at each condensation, and to subsequent structural modifications that add additional chemical diversity to the products (*32*). PKSs are classified into various types based on the molecular architecture of the genes (*32, 33*), but highly conserved domains are shared among all PKS genes, most notably within the ketosynthase (KS) domain (*32*) that may be useful for targeting PKSs in diverse organisms (*34-36*). A number of fungal PKS genes have been cloned, starting with the *MSAS* gene from *Penicillium patulum* that synthesizes the antibiotic 6-methylsalicylic acid (*37*). Several PKS genes that synthesize pathogenicity-related toxins have been cloned from plant pathogenic fungi (*38, 39*), as well as those genes involved in the synthesis of aflatoxins and fumonisins (*40-43*). Hendrickson et al. (*44*) cloned the lovastatin biosynthesis cluster, containing two PKS genes, from *Aspergillus terreus.*

A Molecular Screening Strategy For Novel Polyketides

The genetic capacity to synthesize at least some families of natural products can be assessed directly. The rationale behind this approach is the following: 1) biosynthesis of polyketides requires at least one PKS, and 2) PKS genes contain conserved core sequences that can be used to clone gene fragments by amplification via the polymerase chain reaction (PCR). Since it is relatively easy to detect and clone fragments of PKS genes from fungal genomic DNA, it is feasible to screen fungal strains for the presence of such genes. The existence of these common consensus sequence motifs can be used to construct primers to PCR amplify PKS-like domains from multiple, genetically diverse organisms (*34-36, 45-48*).

Screening for PKS-encoding genes

A group of 157 isolates from the ARSEF collection was selected for this study (*47*). This group represents seven classes of fungi from 36 geographically diverse locations. Individuals in this group were isolated from 14 different host organisms, with the vast majority being arthropods. Fungi were grown in Saboraud dextrose liquid medium supplemented with 0.5% yeast extract and harvested 1- 5 weeks after initiation, once sufficient growth had occurred. Mycelia were harvested, and then processed for isolation of genomic DNA. Using this approach, high quality DNA suitable for PCR screening was obtained from 149 isolates.

Five degenerate primer pairs were designed *(47),* using comparative information from conserved regions of ketosynthase (KS) domains of fungi and bacteria. Using low stringency amplification conditions for PCR, isolates were then scored based on the amplification of a single band of expected size (~0.3kb), as seen in Table I. Three of the five primer pairs amplified products of the expected size from at least 50 individual isolates. While four isolates were positive with all primer pairs for a 0.3 kb band, there were 27 isolates that produced a putative KS fragment with only a sole pair of primers.

Table I. Summary of PCR screening

	Primer Pair				
No. of isolates amplified[a]	KS1/2	FKS2/3	FKS2/4	FKS1/3	FKS1/4
Total	53	55	51	29	16
Exclusive[b]		9	7	8	21

[a] Number of isolates from which a PCR product of expected size was amplified (300 bp)

[b] Number of isolates amplified solely by a single primer pair

Applied Microbiology and Biotechnology. Reproduced from reference 47 with kind permission of Springer Science and Business Media.

Molecular analysis of PCR products

PCR products from 92 amplicon-positive isolates were cloned and sequence data were obtained from at least one clone per isolate; approximately 2/3 of the isolates (66 total) contained PKS sequences based on inferred amino acid homology. The remaining sequences from 26 isolates were not representative of KS domains and did not correspond with any single class of genes by blastx analysis.

To evaluate whether these isolates contained multiple PKS domains, we sampled two isolates using a set of primers designed to amplify fragments of genes specific for different polyketide chemistries (*34*). Using this approach, four different PKS fragments were identified from *Verticillium coccosporum* (ARSEF 2064) using the KS and FKS primer pairs. When evaluated with the LC set of primers (designed for reduced, partially reduced, or unreduced PKS gene products)(*34*), three additional unique sequences were identified. *Tolypocladium inflatum* (ARSEF 3790) contained four KS fragments when the KS and FKS primer pairs were used, while the LC primers allowed the identification of another PKS fragment. These data indicated that many, if not all, of these fungi are likely to contain multiple PKS genes; the choice of specific primer pairs may be a useful tool to detect genes encoding different chemistries. Recent fungal genome sequencing efforts, described more fully below, have confirmed these observations (*49*). Although the work presented here is, by no means, an exhaustive search, it is clear that these fungi will be expected to have the capacity to produce a multitude of diverse and interesting polyketides.

Sequence comparisons of PCR fragments from entomopathogenic fungi

Cladistic analysis was used to evaluate the relatedness of 76 KS fragments from insect pathogenic fungi with KS domains from both fungi and bacteria deposited in GenBank, as seen in Figure 1. Three major clades resulted, in which there are six subclades composed only of sequences derived from insect-pathogenic fungi, and two subclades consisting of both reference sequences and those from insect-pathogens. Within the clades comprised of both reference sequences and those from insect pathogenic fungi, some subclades contain only sequences from insect pathogens. It should be noted that this analysis bears no relationship to the evolutionary history and accepted phylogeny of these organisms. There are instances in which related species are found together within a subclade, as well as instances where a single species contains multiple PKS-like sequences found within several subclades.

This distinct grouping of PKS gene fragments from insect pathogenic fungi might be indicative of different chemistries present in these organisms, separate from those groupings that contain sequences of known function. For examples, one clade contained gene fragments from actinomycetes, eubacteria and fungi, of which some are known to be involved in macrolide and aromatic PK biosynthesis. This analysis is based on a short sequence region and, thus, the data had a greater "noise level" that could potentially overemphasize variation, magnifying small differences within the region. The KS domain is, however, the most highly conserved within the PKS gene, and the differences may indicate functionality.

There is little information on the chemistry profiles of many of the isolates in this study, which include fungi existing as parasites or saprophytes in association with insects, nematodes, and other fungi. This study represents our first attempt in evaluating whether unknown classes of PKS genes are functionally linked to the specialized needs of entomopathogenic fungi. In order to predict whether the cladistic analysis will be useful as a tool to evaluate chemical profiles, we have begun to examine isolates for the presence of polyketides as described in the following section.

Polyketide Chemistries of Selected Entomopathogenic Fungi

Inferred homology of PCR-amplified PKS fragments to genes of known function might be a useful approach to eliminate some genes from future consideration in the search for novel chemistries. However, relatively few PKS genes with known function have been described and minor differences in nucleotide or amino acid sequence would have significant effects on PKS enzymatic activity and compound structure. Homology relationships might be better used to guide priorities rather than eliminate genes from consideration. Conversely, subclades consisting of genes of unknown function might be novel sources of new chemistries. To address these considerations, we used the cladistic analysis in Figure 1 as a guide to test this approach to bioprospecting by examining subclades for known genes and gene products and the relative relationship between isolates within a subclade.

Phomalactone-producing Hirsutella thompsonii isolates

One small subclade contained several isolates of *H. thompsonii* (ARSEF 2464, 3323, and on a linked subclade, ARSEF 257), a pathogen of citrus rust mite, as well as other insects and nematodes (*50*). Of the four isolates of *H. thompsonii* in the study, ARSEF 256 and 2464 were PCR-positive with all five primer pairs, while ARSEF 257 and 3323 were positive with three of the five primer pairs tested.

At least three of the isolates (256, 257, and 3323) are known producers of phomalactone, a tetraketide depicted in Figure 2, having insecticidal activity against apple maggot fly, *Rhagoletis pomonella* (IC_{50} = 468 μg/ml)(*51*). In addition, phomalactone inhibited conidial germination (IC_{50}= 450 μg/ml) of several entomopathogenic fungi (*Beauveria bassiana, Metarhizium anisopliae, Tolypocladium geodes*, and *T. cylindrosporum*), although no self-inhibition occurred.

Phomalactone has been reported from a number of other fungi, including *Nigrospora* sp. (*52*) and *Phoma minispora* (*53, 54*); activity against fungi, bacteria, and a protozoan was reported (*53*). Phomalactone is quite similar chemically to asperlin produced by several species of *Aspergillus - A. nidulans, A. elegans, A. carneus*, and *A. caespitosus* (*53, 55, 56*); *A. caespitosus* also produced acetylphomalactone and an asperlin epimer (*55*). Phomalactone, as well as asperlin and related compounds with phytotoxic activity, has been reported from *Macrophomina phaseolina* (*57*), *Nigrospora sacchari* (*58*), *N. sphaerica* (*59*), and *Verticillum chlamydosporium* (*60*). Phomalactone isolated from *Nigrospora sphaerica* showed activity against *Phytophthora infestans* and *P. capsici* with a MIC of 2.5 mg/L, and reduced late blight symptoms in potato at 100 and 500 mg/L (*59*). Phomalactone esters, isolated from a *Phomopsis* sp., inhibited production of lipopolysaccharide-induced cytokines, tumer necrosis factor α and interleukin 1-β, with IC_{50} values of 80 nM and 190 nM, respectively, in cell cultures (*61*). Nematicidal activity against *Meloidogyne incognita* from *V. chlamydosporium* extracts was due to the presence of phomalactone (*60*).

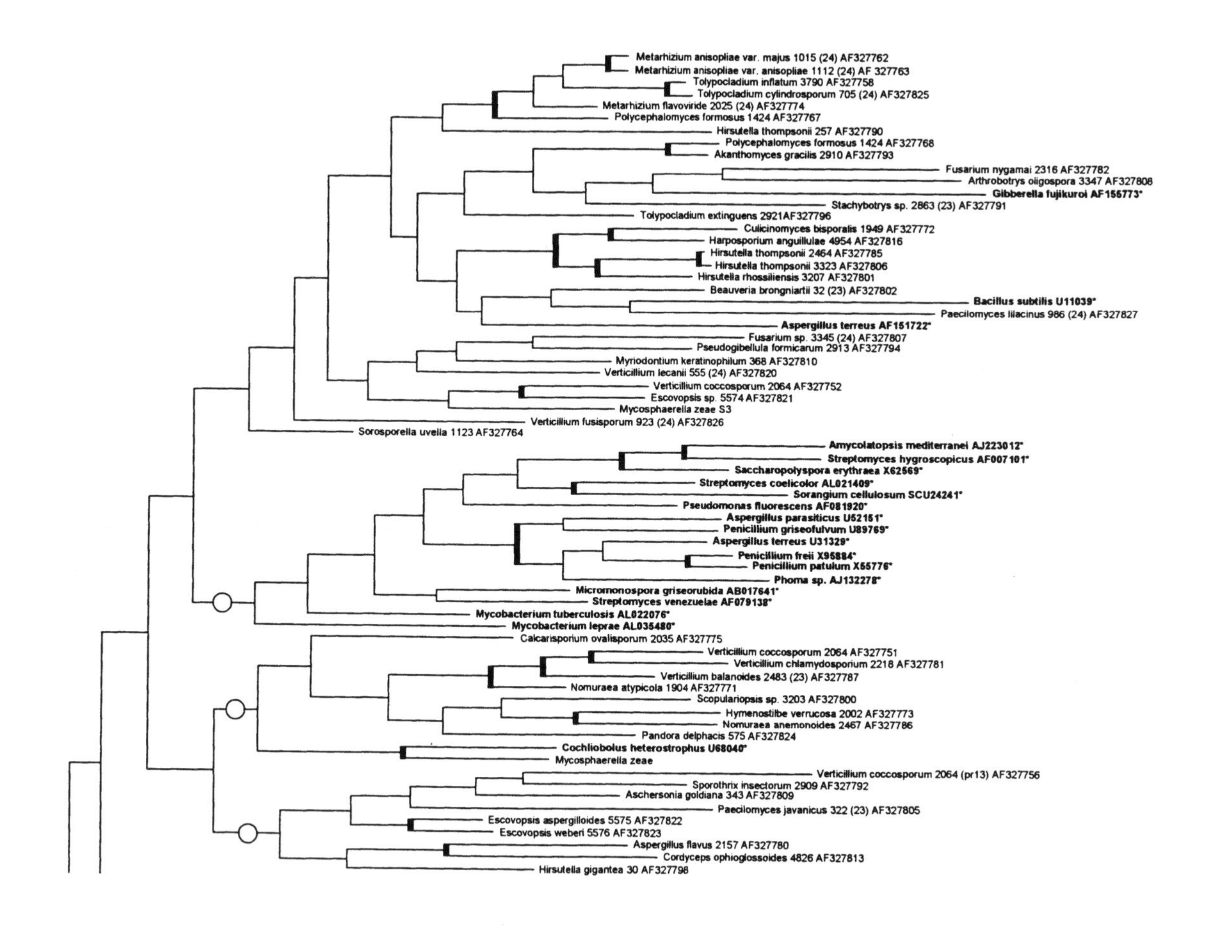
Metarhizium anisopliae var. majus 1015 (24) AF327762
Metarhizium anisopliae var. anisopliae 1112 (24) AF 327763
Tolypocladium inflatum 3790 AF327758
Tolypocladium cylindrosporum 705 (24) AF327825
Metarhizium flavoviride 2025 (24) AF327774
Polycephalomyces formosus 1424 AF327767
Hirsutella thompsonii 257 AF327790
Polycephalomyces formosus 1424 AF327768
Akanthomyces gracilis 2910 AF327793
Fusarium nygamai 2316 AF327782
Arthrobotrys oligospora 3347 AF327808
Gibberella fujikuroi AF155773*
Stachybotrys sp. 2863 (23) AF327791
Tolypocladium extinguens 2921AF327796
Culicinomyces bisporalis 1949 AF327772
Harposporium anguillulae 4954 AF327816
Hirsutella thompsonii 2464 AF327785
Hirsutella thompsonii 3323 AF327806
Hirsutella rhossiliensis 3207 AF327801
Beauveria brongniartii 32 (23) AF327802
Bacillus subtilis U11039*
Paecilomyces lilacinus 986 (24) AF327827
Aspergillus terreus AF151722*
Fusarium sp. 3345 (24) AF327807
Pseudogibellula formicarum 2913 AF327794
Myriodontium keratinophilum 368 AF327810
Verticillium lecanii 555 (24) AF327820
Verticillium coccosporum 2064 AF327752
Escovopsis sp. 5574 AF327821
Mycosphaerella zeae S3
Verticillium fusisporum 923 (24) AF327826
Sorosporella uvella 1123 AF327764
Amycolatopsis mediterranei AJ223012*
Streptomyces hygroscopicus AF007101*
Saccharopolyspora erythraea X62569*
Streptomyces coelicolor AL021409*
Sorangium cellulosum SCU24241*
Pseudomonas fluorescens AF081920*
Aspergillus parasiticus U52151*
Penicillium griseofulvum U89769*
Aspergillus terreus U31329*
Penicillium freii X95884*
Penicillium patulum X55776*
Phoma sp. AJ132278*
Micromonospora griseorubida AB017641*
Streptomyces venezuelae AF079138*
Mycobacterium tuberculosis AL022076*
Mycobacterium leprae AL035480*
Calcarisporium ovalisporum 2035 AF327775
Verticillium coccosporum 2064 AF327751
Verticillium chlamydosporium 2218 AF327781
Verticillium balanoides 2483 (23) AF327787
Nomuraea atypicola 1904 AF327771
Scopulariopsis sp. 3203 AF327800
Hymenostilbe verrucosa 2002 AF327773
Nomuraea anemonoides 2467 AF327786
Pandora delphacis 575 AF327824
Cochliobolus heterostrophus U68040*
Mycosphaerella zeae
Verticillium coccosporum 2064 (pr13) AF327756
Sporothrix insectorum 2909 AF327792
Aschersonia goldiana 343 AF327809
Paecilomyces javanicus 322 (23) AF327805
Escovopsis aspergilloides 5575 AF327822
Escovopsis weberi 5576 AF327823
Aspergillus flavus 2157 AF327780
Cordyceps ophioglossoides 4826 AF327813
Hirsutella gigantea 30 AF327798

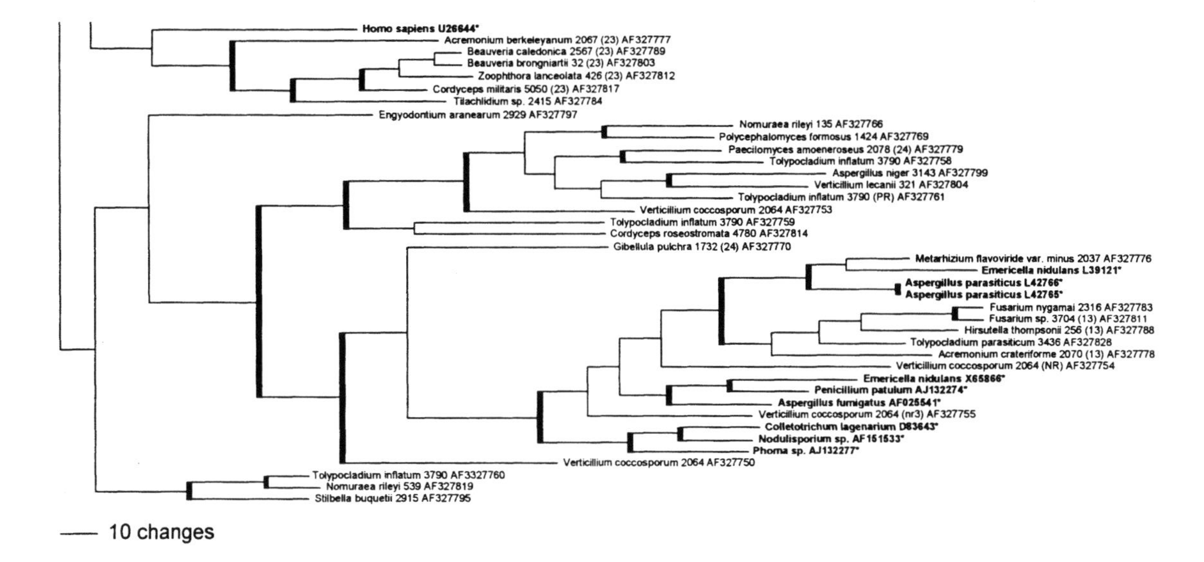

Figure 1. Sequence relatedness among putative KS domains of PKS gene sequences from insect pathogenic fungi and selected KS domains from GenBank. Names of source taxa are followed by their ARSEF accessions number and a GenBank sequence Accession No. Primer designation is given for those sequences not amplified by primers KS1/KS2. Applied Microbiology and Biotechnology: Reproduced from reference 47 with kind permission of Springer Science and Business Media.

Emodins from Escovopsis sp.

Several isolates of *Escovopsis* sp. (ARSEF 5575, *E. aspergilloides* and 5576, *E. weberii*) grouped in a small subclade. *Escovopsis* sp. are associated with fungus-ant colony complexes, where it acts as a mycoparasite to the cultivated fungal cultivars employed as nutrient sources by the ant colony (*62, 63*). When we evaluated *Escovopsis* sp. extracts for biological activity, potent antibacterial and antifungal activity was detected, and following bioassay-guided fractionation, the activity resided in a well-characterized group of compounds, the emodins, as seen in Figure 2. Emodins are a family of polycyclic aromatic ketides, reported from a wide variety of fungi and plants, known to exhibit not only antimicrobial activity, but also antiviral, anti-inflammatory, and anticancer activity (*30*).

Helvolic acid from Metarhizium sp.

Helvolic acid, depicted in Figure 2, was detected in *Metarhizium anisopliae*, ARSEF isolate 1015 (*64*); another member of the subclade, *M. anisopliae* (ARSEF 1112), has significant PKS sequence relatedness. Helvolic acid, and the closely related compound, fusidic acid, are both members of a small class of antibiotics called fusidanes. They are considered to be steroidal in nature, and are derived via squalene cyclization reactions. These compounds inhibit prokaryote and eukaryote protein synthesis via inhibition of the ADP-ribosylation of protein elongation factor 2 (*65, 66*). Although the fusidanes are considered to be steroid terpenoids, it is possible that they are synthesized at least in part via a polyketide pathway. For example, the polyether polyketide monensin contains genes in the PKS cluster that are highly homologous to genes encoding the enzymes involved in double bond migrations in steroid biosynthesis (*67*).

Cephalosporium sp. are known producers of helvolic acid as well as fusidic acid (*3*). Helvolic acid has also been reported from the opportunistic human pathogen *Aspergillus fumigatus* (*3*) and the hypocrealean fungus *Sarocladium oryzae* (*68*). The extent to which these steroidal fusidanes are present in other entomopathogenic fungi is unknown at present.

2-Pyridones from Akanthomyces gracilis and other entomopathogenic fungi

Akanthomycin (8-methyl pyridoxatin) as shown in Figure 2, and a related known compound, cordypyridone C (*69*), were found in extracts of the entomopathogenic fungus *Akanthomyces gracilis* (ARSEF 2910) (*70*). The compounds showed antibacterial activity against *Staphlococcus aureus*, causing growth inhibition at an applied dose of 250 ng. Akanthomycin and

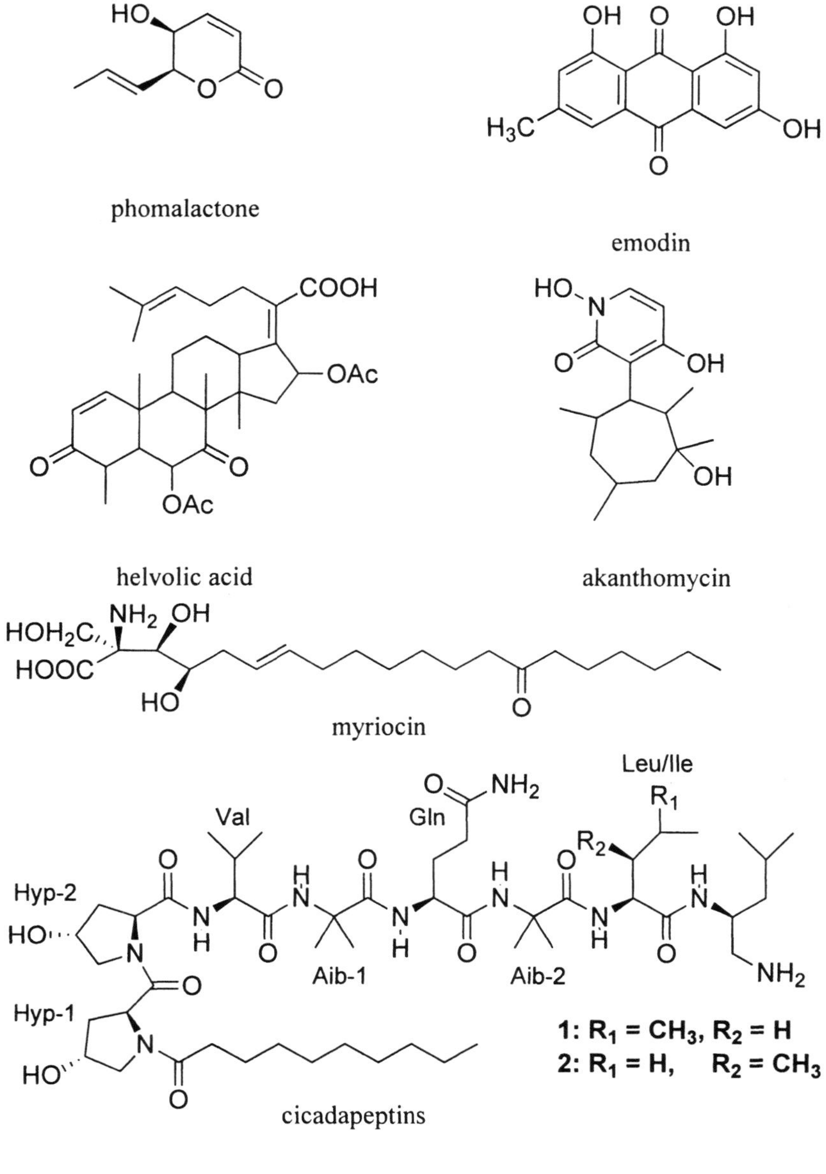

HO
O
O

phomalactone

OH O OH
H_3C
OH
O

emodin

COOH
OAc
O
O
OAc

helvolic acid

HO
N
O
OH
OH

akanthomycin

NH_2 OH
HOH_2C
HOOC
HO
O

myriocin

Leu/Ile
O
NH_2
R_1
Val
Gln
R_2
O
H
O
O
H
Hyp-2
N
N
N
N
N
HO
N
H
O
H
O
O
N
H
O
Aib-1
Aib-2
NH_2
O
Hyp-1
N
HO
O

1: $R_1 = CH_3$, $R_2 = H$
2: $R_1 = H$, $R_2 = CH_3$

cicadapeptins

Figure 2. Natural products described from insect pathogenic fungi.

cordypyridone C are structurally related to a large group of 2-pyridones that have been isolated from other entomopathogenic fungi, including tenellin and bassianin from *Beauveria bassiana* (*71, 72*) and militarinone C from *Paecilomyces militaris* (*73*).

Song et al. (*74*) suggested that the 2-pyridones might be derived through ring expansion of the tetramic acid portion of the fusarins as supported by recent work with militarinone C (*75*). They identified the genes encoding fusarin C biosynthesis as an iterative Type I polyketide synthase linked to a single nonribosomal peptide synthetase module from *Fusarium moniliforme* and *F. venenatum*; this work is described more fully in chapter xx of this volume. Fusarins are potent mycotoxins that have been reported from a large number of fungi, including the plant pathogens *Fusarium moniliforme* and *F. graminearum*. Acyl-tetramic acids, of which fusarin C is a member, include representatives that can be found in slime molds (*76*) and marine and terrestrial fungi (*77-79*). If the hypothesis of Song et al. is true, it would be expected that the production of akanthomycin and cordypyridone C would be derived via a PKS-NRPS complex as described for fusarin C biosynthesis.

The few examples described above are supportive of entomopathogenic fungi as a source for new natural products, as well as the expectation that polyketides will be well represented among the chemistries to be found. Although the direct linkage between the gene fragment identified via the PKS molecular screen and the final chemical product has not been established in studies so far, it is highly likely that polyketide synthases are involved in the biosynthesis of these secondary metabolites. As molecular tools are better established for these genetically intractable fungi, directed gene knockout strategies can be used to establish the chemical identities of natural products synthesized by these diverse PKS pathways.

Can Molecular Screening Strategies Be Useful In The Search For Natural Products?

A molecular screening approach based on conserved KS domains can detect PKS genes across a diverse group of organisms from which the chemical potential for PK synthesis is largely unknown. These genome-based approaches include analyses of cyanobactera and dinoflagellates (*80*), Streptomyces (*35*), lichenized fungi (*46, 81, 82*) and other fungi (*34, 36, 45, 47, 49*). In our study with entomopathogenic fungi, we demonstrated a putative KS domain in 66 isolates out of an initial 157 isolates (*47*). In a more intensive PCR analysis with two isolates and a group of primers specific for variable reduced chemistries (*34*), we identified additional KS domains, underscoring the fact that the data to date on PKS genes in entomopathogenic fungi are underreported.

Limitations of the PCR Strategy

Although the PCR-based screening approach can identify highly conserved domains, these genome studies are limited by primer design. For instance, we obtained negative results with the four Oomycetes evaluated in our study, so either KS domains were not detected or may be not be present in this group of organisms. By slightly altering primer sequences, the other 40% of isolates from which PKS gene fragments were not cloned are likely to be identified as PKS+ strains. Although the PCR-based screening allows the rapid identification of PKS gene fragments within a large population of isolates, it is not an exhaustive analysis of the distribution and number of PKS genes represented within the group or each fungus. Primer design is a major limitation, since the PCR experiments can produce artifacts, can underrepresent or not detect the genes present, as well as detect pseudogenes or defective genes within the organisms being screened. There is a high similarity between fatty acid synthases and PKS, leading to some false positives, but polyunsaturated fatty acids, like eicosapentaeenoic acid, appear to be synthesized via PKS in cold marine organisms (*83*). Although a recent study employed PCR-detection and restriction analysis to dereplicate known chemistries, they also conclude that the metabolic potential of the Actinomycetes used is underestimated by this approach (*48*).

Our recent analyses of the chemical profiles of *Cordyceps heteropoda*, an insect pathogen of cicadas, is possibly illustrative of the limitations of the PCR-screening approach (*17*). *C. heteropoda* (ARSEF 1880) was used in the Lee et al. study (*47*), yet PKS gene fragments were not isolated from this fungus. Fermentation extracts contained both a family of novel nonribosomal peptides with *n*-decanoic acid as an N-terminal blocking group, as well as the known antifungal and immunosuppressive compound, myriocin (= thermozymocidin) (*84*), as depicted in Figure 2. The biosynthetic pathways of myriocin and cicadapeptins have not been described, so it is unclear as to whether a PKS-like gene is involved in the production of either compound. Myriocin consists of a long alkyl side chain to which is attached amino, alcohol, and carboxyl groups, mimicking sphingosine; the primary target site is serine palmitoyltransferase, the enzyme catalyzing the first step in biosynthesis of sphingolipids, compounds associated with the proliferation of T cells (*85, 86*). Animal transplantation studies are being conducted with a synthetic analog, FTY6720, of the natural product (*87-89*).

Inherent Versus Expressed Chemical Diversity

Progress on fungal genomes has been relatively slow in comparison to bacterial genomes since the release of the *Saccharomyces cerevisiae* genome in 1996. Other fungal genomes currently available in public databases include

Schizosaccharomyces pombe, Aspergillus nidulans, Neurospora crassa, the human pathogens *Candida albicans* and *Cryptococcus neoformans*, *Coprinus cinereus*, and the plant pathogens *Magnaporthe grisea, Fusarium graminearum*, and *Ustilago maydis*. From information available, it appears that fungal genomes have a high gene density coupled with low proportions of repetitive sequences (*90*). To date, no insect pathogenic fungus has been submitted for complete sequencing.

In the past five years, however, the availability of full genome sequences has also pointed out the presence of multiple PKS and NRPS genes in fungi for which no known function or chemical product is known (for fungal genomes, see www.broad.mit.edu/annotation/fgi/). It is readily apparent that the potential chemical diversity of an organism is much broader than originally realized and that many products have yet to be identified. In a phylogenomic analysis of KS domains from 5 taxonomically diverse pathogenic and saprobic ascomycetes, eight groups of KS domains resulted, of which the majority of KS domains encoded enzymes with no known function (*49*). Although molecular tools are able to identify KS domains in organisms, the sequencing information only gives sparse clues as to what the ultimate products of these genes may be. Using a reverse chemical genetic approach, a peptide natural product was predicted from a trimodular NRPS sequence in *Streptomyces coelicolor*, and finally isolated as a tetrapeptide product (*91*). Further studies will be needed in order to assign individual functionality to the specific KS domains present in fungi.

The fact that the fungal genomes examined to date harbor an unexploited biochemical diversity presents an incredible opportunity for natural product chemists and molecular biologists alike. Microbes need to adjust and adapt to changing environmental conditions, and this flexibility to respond to change should be reflected by the metabolic diversity specific to a particular stress. EST analyses of expressed genes in *Metarhizium anisopliae* have shown large adaptation to environmental conditions, including changes in genes predicted to play roles in secondary metabolite production (*92-95*). Drug discovery efforts have traditionally relied on the mass screening of extracts to identify promising leads. In attempts to coax yields of natural products from fungi, fermentation biologists have inherently developed methodologies for evaluating the effect of nutritive and environmental conditions on production. Of course, this approach gives negative results if the genomic and chemical diversity of the organism is not expressed.

One of the greatest issues in bioprospecting is to ensure that cultures are assessed for metabolic potential under conditions that allow expression of metabolites. This process has been traditionally driven through empirical design, using multiple media formulations and different environmental parameters, such as solid/liquid, shaken/nonshaken, temperature, time of culture, humidity, and light/dark manipulations. Plackett-Burman designs have been used successfully to minimize the number of treatments needed to statistically evaluate the effect of a single variable in a multi-variable system (*96*), but this process is still labor-intensive. An understanding of the regulation of secondary metabolite production will be critical to manipulate these organisms (*97-100*).

The cues that organisms use to elicit secondary metabolites are poorly known, as are the roles that these natural products play in the organism's interactions within their environmental niches. One possible approach that arises from the genetic information to date is that these molecular tools may be used to evaluate conditions that will allow expression of particular genes. One means of evaluating the organisms' potential for chemical diversity may be to prescreen cultures for the expression of the gene fragment of interest to evaluate potential production of unique compounds of interest and to use this information to define new production conditions. Thus, it seems likely that when working with isolates of unknown potential in axenic liquid culture, standard formulations that maximize growth may not be useful to stimulate secondary metabolite production.

Acknowledgement

This work was partially supported by Biotechnology Research and Development Corporation grant no. 32-1-096 and the USDA, Agricultural Research service. Mention of a trademark, proprietary product, or vendor does not constitute a guarantee or warranty of the product by the U.S. Department of Agriculture and does not imply its approval to the exclusion of other products or vendors that may also be suitable.

References

1. Piggott, A.;Karuso, P. *Com. Chem. High Throughput Screening*. **2004**, *7*, 607-630.
2. Bull, A. T.; Goodfellow, M.;Slater, J. H. *Ann. Rev. Microbiol.* **1992**, *46*, 219-252.
3. Cole, R. J.;Cox, R. H. *Handbook of Toxic Fungal Metabolites*. Academic Press: New York, 1981; pp. 806-807.
4. Langley, D. *Mycologist*. **1997**, *11*.
5. Turner, W. B. *Fungal Metabolites*. Academic Press: London, 1971.
6. Roberts, D. W. In *Microbial Control of Pests and Plant Diseases 1970-1980*; Burges, H. D.; Ed; Academic Press: London, 1981; pp. 441-464.
7. Hawksworth, D. L. *Mycol. Res.* **2001**, *105*, 1422-1432.
8. Wildman, H. G. *Can. J. Bot.* **1995**, *73*, S907-S916.
9. Roberts, D. W.;Humber, R. A. In *Biology of Conidial Fungi*; Cole, G. T.; Kendrick, B.; Eds; Academic Press: New York, 1981; pp. 201-235.

10. Hajek, A. E.;St. Leger, R. J. *Ann. Rev. Entomol.* **1994**, *39,* 293-322.
11. Charnley, A. K. In *Biotechnology of Fungi for Improving Plant Growth*; Whipps, J. M. ;Lumsden, R. D.; Eds; Oxford University Press: London, 1989; pp. 86-125.
12. Roberts, D. W.; Gupta, S.;St. Leger, R. J. *Pesq. Agropec. Bras.* **1992**, *27,* 325-347.
13. Gupta, S.; Krasnoff, S. B.; Renwick, J. A. A.; Roberts, D. W.; Steiner, J. R.;Clardy, J. *J. Org. Chem.* **1993**, *58,* 1062-1067.
14. Gupta, S.; Krasnoff, S. B.; Roberts, D. W.; Renwick, J. A. A.; Brinen, L.;Clardy, J. *J. Org. Chem.* **1992**, *57,* 2306-2313.
15. Billich, A.;Zocher, R. *J. Biol. Chem.* **1987**, *262,* 17258-17259.
16. Bisset, J. *Can. J. Bot.* **1983**, *61,* 1311-1329.
17. Krasnoff, S. B.; Reategui, R. F.; Wagenaar, M. M.; Gloer, J. B.;Gibson, D. M. *J. Nat. Prod.* **2004**, *68,* 50-55.
18. Chu, M.; Mierzwa, R.; Truumees, I.; Gentile, F.; Patel, M.; Gullo, V.; Chan Tze, M.;Puar Mohindar, S. *Tetrahedron Lett.* **1993**, *34,* 7537-7540.
19. Hodge, K. T.; Krasnoff, S. B.;Humber, R. A. *Mycologia.* **1996**, *88,* 715-719.
20. Gupta, S.; Roberts, D. W.;Renwick, J. A. A. *J.Chem. soc. Perkin trans. I.* **1989**2347-2357.
21. Krasnoff, S. B.; Gibson, D. M.; Belofsky, G. N.; Gloer, K. B.;Gloer, J. B. *J. Nat. Prod.* **1996**, *59,* 485-489.
22. Samuels, R. I.; Charnley, A. K.;Reynolds, S. E. *Mycopathologia.* **1988**, *104,* 51-58.
23. Cerenius, L.; Thornqvist, P. O.; Vey, A.; Johansson, M. W.;Soderhall, K. *J. Insect Physiol.* **1990**, *36,* 785-790.
24. Dumas, D.; Ravallec, M.; Matha, V.;Vey, A. *J. Invert. Pathol.* **1996**, *67,* 137-146.
25. Huxham, I. M.; Lackie, A. M.;McCorkindale, N. J. *J. Insect Physiol.* **1989**, *35,* 97-106.
26. James, P. J.; Kershaw, M. J.; Reynolds, S. E.;Charnley, A. K. *J. Insect Physiol.* **1993**, *39,* 797-804.
27. Riederer, B.; Han, M.;Keller, U. *J. Biol. Chem.* **1996**, *271,* 27524-27530.
28. Panaccione, D. G. *Mycol. Res.* **1996**, *100,* 429-436.
29. Crueger, W.;Crueger, A. *Biotechnology: A Textbook of Industrial Microbiology*; 2nd ed. Academic Press: London, 1990.
30. O'Hagan, D. *Nat. Prod. Rep.* **1995**, *12,* 1-32.
31. Blanco, G.; Fu, H.; Menez, C.;Khosla, C. *Chem. Biol.* **1996**, *3,* 193-196.
32. Khosla, C.; Gokhale, R. S.; Jacobsen, J. R.;Cane, D. E. *Ann. Rev. Biochem.* **1999**, *68,* 219-253.
33. Hopwood, D. A. *Chem. Rev.* . **1997**, *97,* 2465-2497.
34. Bingle, L. E. H.; Simpson, T. J.;Lazarus, C. M. *Fungal Gen. Biol.* **1999**, *26,* 209-223.

35. Metsä-Ketelä, M.; Salo, V.; Halo, L.; Hautala, A.; Hakala, J.; Mäntsälä, P.;Ylihonko, K. *FEMS Microbiol. Lett.* **1999**, *180,* 1-6.
36. Nicholson, T. P.; Rudd, B. A. M.; Dawson, M.;Lazarus, C. M. *Chem. Biol.* **2001**, *8,* 157-178.
37. Beck, J.; Ripka, S.; Siegner, A.; Schiltz, E.;Schweizer, E. *Eur. J. Biochem.* **1990**, *192,* 487-498.
38. Yang, G.; Rose, M. S.; Turgeon, B. G.;Yoder, O. C. *Plant Cell.* **1996**, *8,* 2139-2150.
39. Tanako, Y.; Kubo, Y.; Shimizu, K.; Mise, K.; Okuno, T.;Furusawa, I. *Mol. Gen. Genet.* **1996**, *248,* 270-277.
40. Chang, P. K.; Cary, J. W.; Yu, J.; Bhatnagar, D.;Cleveland, T. E. *Mol. Gen. Genet.* **1995**, *248,* 270-277.
41. Brown, D. W.; Yu, J.-H.; Kelnar, H. S.; Fernandes, M.; Nesbitt, T. C.; Keller, N. P.; Adams, T. H.;Leonard, T. J. *Proc. Natl. Acad. Sci. U.S.A.* **1996**, *93,* 1418-1422.
42. Watanabe, C. M. H.; Wilson, D.; Linz, J. E.;Townsend, C. A. *Chem. Biol.* **1996**, *3,* 463-469.
43. Proctor, R. H.; Desjardins, A. E.; Plattner, R. D.;Hohn, T. M. *Fungal Gen. Biol.* **1999**, *27,* 100-112.
44. Hendrickson, L.; Davis, C. R.; Roach, C.; Nguyen, D. K.; Aldrich, T.; McAda, P. C.;Reeves, C. D. *Chem. Biol.* **1999**, *6,* 429-439.
45. Sauer, M.; Lu, P.; Sangari, R.; Kennedy, S.; Polishook, J.;Bills, G. *Mycol. Res.* **2002**, *196,* 460-470.
46. Schmitt, I.; Martin, M. P.; Kautz, S.;Lumbsch, H. T. *Phytochemistry (Amsterdam).* **2005**, *66,* 1241-1253.
47. Lee, T.; Yun, S.-H.; Hodge, K. T.; Humber, R. A.; Krasnoff, S. B.; Turgeon, B. G.; Yoder, O. C.;Gibson, D. M. *Appl. Microbiol. Biotechnol.* **2001**, *56,* 181-187.
48. Ayuso, A.; Clark, D.; Gonzalez, I.; Salazar, O.; Anderson, A.;Genilloud, O. *Appl. Microbiol. Biotechnol.* **2005**, *67,* 795-806.
49. Kroken, S.; Glass, N. L.; Taylor, J. W.; Yoder, O. C.;Turgeon, B. G. *Proc. Natl. Acad. Sci. U.S.A.* **2003**.
50. Minter, D. W.;Brady, B. L. *Transac. Brit. Mycol. Soc.* **1980**, *74,* 271-282.
51. Krasnoff, S. B.;Gupta, S. *J. Chem. Ecol.* **1994**, *20,* 293-302.
52. Evans, R. H.; Ellested, G. A.;Kunstmann, M. P. *Tetrahedron Lett.* **1969**, *1969,* 1791-1794.
53. Yamamoto, I.; Suide, H.; Henmi, T.;Yamano, T. *Takeda Ken. Ho.* **1970**, *29,* 1-10.
54. Yamano, T.; Hemmi, S.; Yamamot, I.;Tsubaki, K. 1971, Takeda Chemical Industries, Ltd.: Japan.
55. Mizuba, S.; Lee, K.;Jiu, J. *Can. J. Microbiol.* **1975**, *21,* 1781-1787.
56. Argoudelis, A. D.;Zieserl, J. F. *Tetrahedron Lett.* **1996**, *1966,* 1969-1973.
57. Bhattacharya, D.; Siddiqui, K. A.;Ali, E. *Ind. J. Mycol. Plant Pathol.* **1992**, *22,* 54-57.

58. Fukushima, T.; Tanaka, M.; Gohbara, M.;Fujimori, T. *Phytochemistry (Oxford)*. **1998**, *48,* 293-302.
59. Kim, J. C.; Choi, G. J.; Park, J. H.; Kim, H. T.;Cho, K. Y. *Pest Manage. Sci.* **2001**, *57,* 554-559.
60. Khambay, B. P. S.; Bourne, J. M.; Cameron, S.;Kerry, B. R. *Pest Manage. Sci.* **2000**, *56,* 1098-1099.
61. Wrigley, S. K.; Sadeghi, R.; Bahl, S.; Whiting, A. J.; Ainsworth, A. M.; Martin, S. M.; Katzer, W.; Ford, R.; Kau, D. A.; Robinson, N.; Hayes, M. A.; Elcock, C.; Mander, T.;Moore, M. *J. Antibiot.* **1999**, *52.*
62. Reynolds, H. T.;Currie, C. R. *Mycologia*. **2004**, *96,* 955-959.
63. Currie, C. R. *Nat. Biotechnol.* **1999**, *398,* 701-704.
64. Espada, A.;Dreyfuss, M. M. *J. Ind. Microbiol. Biotechnol.* **1997**, *19,* 7-11.
65. Verbist, L. *J. Antimicrob. Chemother.* **1990**, *25 (Suppl. B),* 1-6.
66. Riis, B.;Rattan, S. I. S. *Med. Sci. Res.* **1996**, *24,* 221-222.
67. Leadley, P. F.; Staunton, J.; Oliynyk, M.; Bisang, C.; Cortes, J.; Frost, E.; Hughes, T. Z. A.; Jones, M. A.; Kendrew, S. G.; Lester, J. B.; Long, P. F.; McArthur, H. A.; McCormick, E. L.; Oliynyk, Z.;Stark, C. B. W. *J. Ind. Microbiol. Biotechnol.* **2001**, *27,* 360-367.
68. Bills, G. F.; Platas, G.;Gams, W. *Mycol. Res.* **2004**, *108,* 1291-1300.
69. Isaka, M.; Tanticharoen, M.; Kongsaeree, P.;Thebtaranonth, Y. *J. Org. Chem.* **2001**, *66,* 4803-4808.
70. Wagenaar, M. M.; Gibson, D. M.;Clardy, J. *Org. Lett.* **2002**, *2,* 671-673.
71. McInnes, A. G.; Smith, D. G.; Wat, C. K.; Vining, L. C.;Wright, J. L. C. *J. Chem. Soc. Chem. Commun.* **1974**, *1974,* 281-282.
72. Wat, C. K.; McInnes, A. G.; Smith, D. G.; Wright, J. L. C.;Vining, L. C. *Can. J. Chem.* **1977**, *55,* 4090-4098.
73. Schmidt, K.; Guenther, W.; Stoyanova, S.; Schubert, B.; Li, Z.;Hamburger, M. *Org. Lett.* **2002**, *4,* 197-199.
74. Song, Z.; Cox, R. J.; Lazarus, C. M.;Simpson, T. J. *ChemBioChem*. **2004**, *5,* 1196-1203.
75. Schmidt, K.; Riese, U.; Li, Z.;Hamburger, M. *J. Nat. Prod.* **2003**, *66,* 378-383.
76. Casser, I.; Steffan, B.;Steglich, W. *Angew. Chem., Int. Ed.* **1987**, *26,* 586-587.
77. Daferner, M.; Anke, T.;Sterner, O. *Tetrahedron*. **2002**, *58,* 7781-7784.
78. Harrison, P. H. M.; Duspara, P. a.; Jenkins, S. I.; Kassam, S. A.; Liscombe, D. K.;Hughes, D. w. *J. Chem. Soc. Perkin Transac. I.* **2000**, *24,* 597-609.
79. Svensson, M.; Lundgren, L. N.; Woods, C.; Fatehi, J.;Stenlid, J. *Phytochemistry (Oxford)*. **2001**, *56,* 747-751.
80. Moffitt, M. C.;Neilan, B. A. *J. Mol. Evol.* **2003**, *56,* 446-457.
81. Grube, M.;Blaha, J. *Mycol. Res.* **2003**, *107,* 1419-1426.
82. Varga, J.; Rigo, K.; Kocsube, S.; Farkas, B.;Pal, K. *Res. Microbiol.* **2003**, *154,* 593-600.

83. Metz, J. G.; Roessler, P.; Facciotti, D.; Levering, C.; Dittrich, F.; Lassner, M.; Valentine, R.; Lardizabal, K.; Domergue, F.; Yamada, A.; Yazawa, K.; Knauf, V.;Browse, J. *Science.* **2001**, *293,* 290-293.
84. Sasaki, S.; Hashimoto, R.; Kiuchi, M.; Inoue, K.; Ikumoto, T.; Hirose, R.; Chiba, K.; Hoshino, Y.; Okumoto, T.;Fujita, T. *J. Antibiot.* **1994**, *47,* 420-433.
85. Miyake, Y.; Kozutsumi, Y.; Nakamura, S.; Fujita, T.;Kawasaki, T. *Biochem. Biophys. Res. Commun.* **1995**, *211,* 396-403.
86. Chen, J. K.; Lane, W. S.;Schreiber, S. L. *Chemistry and Biology (London).* **1999**, *6,* 221-235.
87. Schmid, G.; Guba, M.; Papyan, A.; Ischenko, I.; Bruekel, M.; Bruns, C. J.; Jauch, K. W.;Graeb, C. *Transplant. Proc.* **2005**, *37,* 110-111.
88. Kahan, B. D. *Transplant. Proc.* **2004**, *36,* 531S-543S.
89. Kiuchi, M.; Adachi, K.; Kohara, T.; Minoguchi, M.; Hanano, T.; Aoki, Y.; Mishina, t.; Arita, M.; Nasko, N.; Ohtsuki, M.; Hoshino, Y.; Teshima, K.; Chiba, K.; Sasaki, S.;Fujita, T. *J. Med. Chem.* **2000**, *43,* 2946-2961.
90. Tunlid, A.;Talbot, N. J. *Current Opinion in Microbiology.* **2002**, *5,* 513-519.
91. Lautru, S.; Deeth, R. J.; Bailey, L. M.;Challis, G. L. *Nat. Chem. Biol.* **2005**, *1,* 265-269.
92. Freimoser, F. M.; Hu, g.;St. Leger, R. J. *Microbiol.* **2005**, *151,* 361-371.
93. Freimoser, F. M.; Screen, S.; Hu, G.;St. Leger, R. J. *Microbiol.* **2003**, *149,* 1893-1900.
94. Wang, C.; Hu, G.;St. Leger, R. J. *Fungal Gen. Biol.* **2005**, *42,* 704-718.
95. Wang, C.;St. Leger, R. J. *Eukaryotic Cell.* **2005**, *4,* 937-947.
96. Monaghan, R.;Koupal, L. R. In *Novel Microbial Products for Medicine and Agriculture*; Demain, A. L.; Somkuti, G. A.; Hunter-Cevera, J. C.;Rossmore, H. W.; Eds; Elsevier: Amsterdam, 1989, pp 25-32.
97. Bok, J. W.;Keller, N. P. *Eukaryotic Cell.* **2004**, *3,* 527-535.
98. Calvo, A. M.; Wilson, R. A.; Bok, J. W.;Keller, N. P. *Microbiol. Mol. Biol. Rev.* **2002**, *66,* 447-459.
99. Smid, E. J.; Molenaar, D.; Hugenholtz, J.; deVos, W. M.;Teusink, B. *Curr. Opin. Biotechnol.* **2005**, *16,* 190-197.
100. Askenazi, M.; Driggers, E. M.; Holtzman, D. A.; Norman, T. C.; Iverson, S.; Zimmer, D. P.; Boers, M.-E.; Blomquist, P. R.; Martinez, E. J.; Monreal, A. W.; Feibelman, T. P.; Mayorga, M. E.; Maxon, M. E.; Sykes, K.; Tobin, J. V.; Cordero, E.; Salama, S. R.; Trueheart, J.; Royer, J. C.;Madden, K. T. *Nat. Biotechnol.* **2003**, *21,* 150-156.

Chapter 5

Polyketide Biosynthesis in Fungi

Kenneth C. Ehrlich

Southern Regional Research Center, Agricultural Research Service, U.S. Department of Agriculture, 1100 Robert E. Lee Boulevard, New Orleans, LA 70124

Elaborate arrays of polyketide metabolites are produced by filamentous fungi. Polyketides are produced by iterative condensation of malonylCoA by a specialized type of fatty acid synthase that often lacks some or all of the reducing domains normally found in such enzymes. Fungal polyketide synthases (PKSs) have a single polypeptide chain with multiple independently acting catalytic domains. The structure of the resulting polyketide depends on the type of coenzymeA (CoA) starter unit upon which the polyketide is built, whether or not reducing domains are present in the PKS, how the polyketide chain is terminated, and the types of oxidative enzymes that perform subsequent modifications. A contiguous set of genes encoding the regulatory and metabolic enzymes is usually required for the biosynthesis of polyketides. The most extensively studied biosynthetic gene cluster is for the carcinogenic mycotoxin, aflatoxin (AF). AF is formed from a hexanoyl CoA precursor by PKS-catalyzed condensation of seven malonyl CoA units and a series of oxidative steps performed by six cytochrome P450 monooxygenases, nine oxidoreductases, and three non-oxidative enzymes. Since there is no evidence that these enzymes are produced sequentially, the subsequent oxidations of the polyketide precursor are most likely dictated by the chemical susceptibility of the individual precursor metabolites formed at each step.

Filamentous fungi produce diverse natural products by condensation of acetate and malonate (1, 2). Such products are often highly pigmented due to the presence of multiple conjugated aromatic rings, but also can resemble linear fatty acid-like hydrocarbons. Many polyketides (PK) are considered to be mycotoxins because of their toxicity to bacteria, animals, or plants (3). Some mycotoxins, such as the aflatoxins, are highly mutagenic because they are able to intercalate and alkylate DNA (4). Others, such as melanins, are known virulence factors in plants and cause a variety of toxic effects (5). Melanins are ubiquitous PKs, that are produced by the fungus for both protection and invasion. PK sizes range from diketides to dodecaketides, for example 6-methylsalicylic acid formed by condensation of acetyl CoA with malonyl CoA is one of the smallest, while T-toxin and fumonisins are among the largest and, like fatty acids, are formed from repeated condensation reactions. There are many theories of why fungi make PKs (6-10). Among these are: PKs are a defense response by fungi to stress, they provide protection from UV damage, they are by-products of primary metabolism, they are virulence factors, they increase fungal fitness (11, 12), they are involved in fungal developmental processes that require small molecule signalling, and they provide protection from predators for reproductive structures such as conidia and sclerotia.

The fungal polyketide synthase

Fungal polyketides (PK) are products of iterative condensation of malonyl CoA units by Type I polyketide synthases (PKS) that contain modular catalytic units within a single polypeptide (13). These enzymes are specialized fatty acid synthases (FASs) and many lack some or all of the FAS-type reducing domains. All Type I PKSs have ketoacylsynthase (KS) and acyl transferase (AT) domains and sites for attachment of acyl carrier protein (ACP), also called phosphopantetheine. In addition, some have catalytic sites that are typically present in all FASs, namely ketoreductase (KR), enoyl reductase (ER) and dehydratase (DH) sites. These additional sites are present only in reducing PKSs. Certain reducing PKSs have methyl transferase (MT) or adenylation (AMP) domains that serve to further modify the polyketide. Almost all non-reducing PKSs have a specialized domain, usually close to their C-termini, that functions as a Claisen-like cyclase (CLC) (14). Domain structures for some typical fungal PKSs are shown schematically in Figure 1. Complexity of the resulting PK depends on the number of malonyl CoA units condensed, the subsequent modification processes, and whether or not the PK is built onto a preformed fatty acid CoA "starter unit."

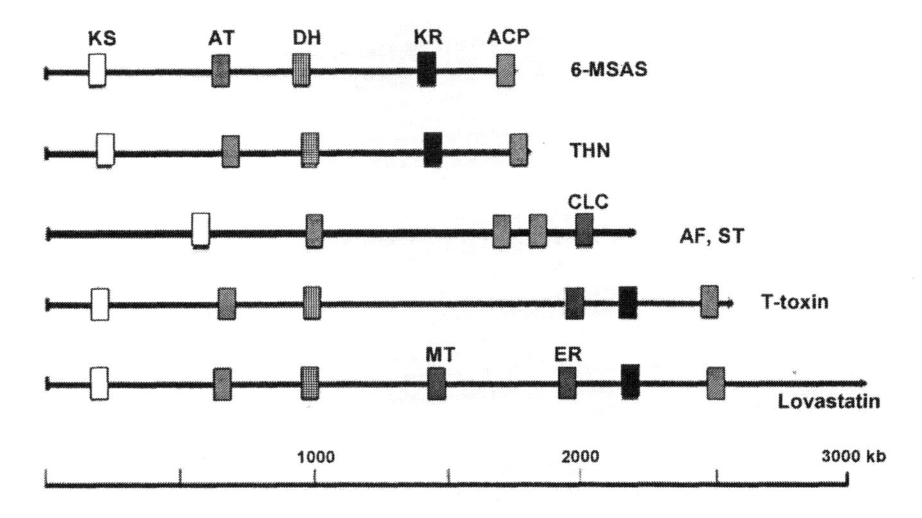

Figure 1. Schematic representation of domain architecture in fungal Type I PKSs based on reported gene sequences. KS. ketoacyl, AT, acyltransferase; DH, dehydratase; KR, ketoreductase; ER, enoyl reductase; MT, methyltransferase; CLC, Claisen-like cyclase; MSAS, methylsalicylic acid synthase; THN, tetrahydronaphthalene; AF, ST, aflatoxin, sterigmatocystin

Evolution of PKS genes

Comparison of the phylogenetic distribution of functional domains in PKSs has provided insight into the genealogy of the Type I PKSs (10) (Figure 2). The comparisons show that PKSs group into three main clades including one for reducing PKSs, another for non-reducing PKSs, and a third for bacterial-type PKSs. Further subdivisions of these clades involve gains or losses of particular PKS domains, for example, the gain of the CLC domain in non-reducing PKSs, gains or losses of domains characteristic of non-ribosomal peptide synthetases (NRPS), and losses of a methyl transferase (MT) domain found in bacterial PKSs, fungal 6-methylsalicylic acid synthases (MSASs) and NRPSs. The evidence from this genealogical study suggests that all of the different types of PKS genes existed in fungal species for more than 300 million years. Although the fungal PKS genes in the clade that also contains bacterial- type (Type II) PKS genes probably were acquired from bacteria by horizontal gene transfer, PKSs in the other fungal clades most likely evolved by selective gene

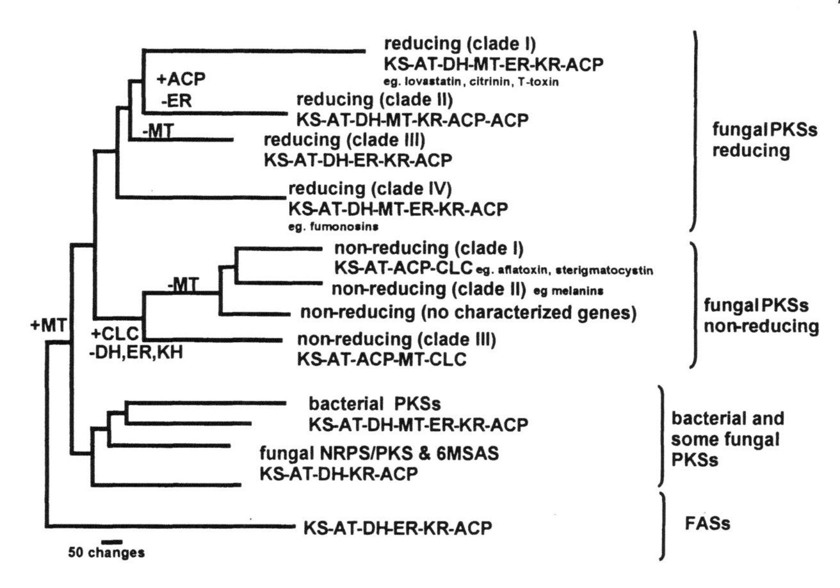

Figure 2. Genealogy of Type I PKSs showing major clades and catalytic domains. Gains (+) and losses (-) are shown above the branches with the types of domains gained or lost. Domain abbreviations are given in Figure 1.

duplications or gene losses that allowed domain swapping and adaptive translocations (10, 15).

Whole genome analyses show that individual fungi may contain genes for more than 30 PKSs (10). For most of these, no specific product has been identified. Linear reduced PKs, for example T-toxin and PM-toxin have been frequently identified as plant virulence factors, while nonreducing PKSs more typically produce aromatic, pigmented PKs. Therefore, the different types of PKSs that produce these compounds may have served different evolutionary functions. If the radiation of these types of fungi occurred from 300 to 700 million years ago (16), it is likely the fungi had to interact with insects and early green plants (algae) (17). Fungi form symbiotic as well as competitive relationships with both of these types of organisms and, to flourish and to propagate, they may have required a diverse array of both pigmented and protectorant metabolites, for which PKs were ideally suited.

Determination of chain length by fungal polyketide synthases

Determination of PK chain length is different in the iterative Type I fungal PKSs than it is in bacterial Type II PKSs. In bacterial PKSs a separate protein called a chain length factor forms a complex that acts as a decarboxylase to prevent further chain extension after the modular PKS completes synthesis of the polyketide chain. Watanabe and Ebizuka (18) found that the CLC domain both catalyzes the condensation reactions to stabilize the polyketide as an aromatic ring and also hydrolyzes the polyketomethylene intermediate from the ACP domain after the cavity of the PKS is filled. As an example of this mode of chain length determination, *Colletotrichum lagenarium* PKS1, in the absence of a functional CLC domain, produces mainly a hexaketide isocoumarin, while with a functional CLC domain, produces mainly pentaketide products.

Polyketide gene clusters

In general, the gene encoding a PKS is part of a dedicated biosynthetic gene cluster, whose components contain most or all of the enzymes necessary for formation of a particular PK metabolite (19). The product formed by a non-reducing PKS is inherently unstable unless it aromatizes or forms stabilized ring compounds by other means such as lactonization or esterification. Usually formation of such compounds requires further oxidation by associated oxidative modifying enzymes. For reasons that are still not completely understood, the genes encoding these modifying enzymes, and the genes encoding the transcription regulatory factors and transporters necessary for metabolite biosynthesis are clustered with the PKS (20). Gene clusters for PK biosynthesis have been identified for lovastatin (21), sterigmatocystins (22), aflatoxins (23), dothistromin (24), and fumonisins (25). Cluster organization could have a role in the maintenance of the genes for PK biosynthesis after the selective pressures that fostered formation of the cluster are lost (6, 26, 27). It is also possible that the cluster organization provides a region of "active" chromatin where the genes are highly and coordinately expressed under a common inducing stimulus (28-30). Support for this latter hypothesis comes from the observation that when a gene within the AF cluster is moved to a distant locus, its expression is 500-fold lower than when it is within the cluster (28).

The best studied polyketide biosynthesis gene cluster is the aflatoxin (AF) cluster (Figure 3). Expression of most of the genes in the cluster is regulated by a yeast Gal4-like Cys_6Zn_2 binuclear zinc transcription factor, AflR (31) Sites for AflR are in the promoter regions of most of the AF biosynthesis genes (Figure 3). Similar types of transcription factors have been shown to regulate expression of genes in other fungal gene clusters, including those involved in primary or secondary metabolism (20). Globally acting transcription factors, BrlA, AreA and PacC, which regulate transcription of genes in response to developmental stimuli, nitrogen source and pH, respectively, also mediate the expression of genes in the AF biosynthesis cluster. What regulates expression of AflR is still uncertain, but proteins involved in mediating responses of AF-producing

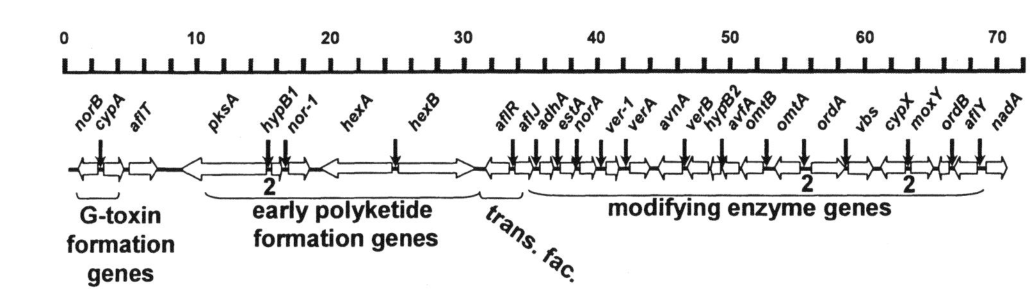

Figure 3. Schematic diagram of the aflatoxin gene cluster showing approximate locations of coding regions. The vertical arrows indicate the presence of AflR-binding sites in the promoter regions of genes and the numbers indicate the presence of two sites in the region.

Aspergillus species to environmental stress, carbon source, lipids, and certain amino acids all are known to affect AF production by modulating transcription of AF biosynthesis genes (32-38).

Polyketide formation in aflatoxin biosynthesis

Aflatoxins (AFs) and their precursors, the sterigmatocystins (STs) (6) are bisfurans produced by several species of *Aspergillus*. AF biosynthesis requires the expression of at least 25 genes and involves 23 enzymatic steps (Figure 3) (30). Following formation of the condensed polyketide, noranthrone (Figure 4), a series of 14 oxidative, two methyltransferase steps, and one esterase step are required to make AF. The 66 kb biosynthetic gene cluster encodes all of the oxidative enzymes required: including six cytochrome P450 monooxygenases and several types of oxidoreductases, some unique to the AF biosynthesis pathway.

Polyketide formation in AF biosynthesis utilizes a hexanoyl CoA starter unit that is synthesized by aid of two specialized FASs (39) (Figure 4A). The genes for these FASs are transcribed from a common promoter region (40). The alpha subunit FAS has separate ACP and PP domains as well as KS and KR domains while the beta subunit has KS, AT, ACP and thioesterase (TE) domains (41).

Two subunit FASs are also present in the palmitate synthase from *Saccharomyces cerevisiae*. These proteins form a multi-subunit complex, $\alpha_6\beta_6$ and share active sites across the two subunits. The FAS proteins involved in primary metabolism in *A. nidulans* are shorter than HexA and HexB, used for formation of the hexanoyl CoA starter unit. Recent evidence suggests that the PKS is part of this protein complex (Figure 4B). The FAS/PKS complex (also called the NorS complex) has the stoichiometry $\alpha_2\beta_2\gamma_2$. Presumably hexanoylCoA never becomes a free unit and is transferred directly to the PKS by an internal trans thioesterification process. This explains why added precursor hexanoylCoA feeds poorly into a strain of *Aspergillus* in which HexA was inactivated. Furthermore, release of hexanoylCoA from the complex has not been found.

Anthrone formation and oxidation

The initial product formed by the NorS complex is predicted to be an anthrone (41). The enzyme required for conversion of the anthrone to the first stable intermediate in the AF biosynthesis pathway, norsolorinic acid, has not yet been identified. In *A. terreus*, emodin is formed from the anthrone by incorporation of molecular oxygen in a reaction catalyzed by the enzyme, emodinanthrone oxygenase (42). Putative enzymes encoded in the AF cluster

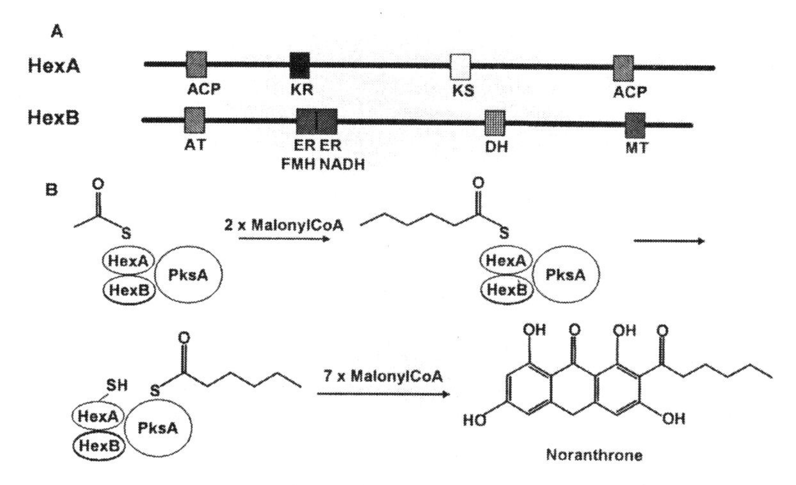

Figure 4. AF pathway FASs and their role in formation of norsolorinic acid anthrone. A. Domain structure of the FASs, HexA and HexB. B. Proposed complex formation for the initial steps in AF biosynthesis. Steps include loading of the HexA/HexB complex, synthesis of hexanoyl CoA, transfer of hexanoylCoA to PksA, and iterative condensation to form the anthrone.

similar to such an oxidase that may carry out this reaction may be encoded by the genes, *hypB1* or *hypB2*. The predicted enzymes have structures and conserved catalytic Trp residues similar to those of emodinanthrone oxidase (42). A recent study found that knockout of the homolog (*stcM*) of this gene in the *A. nidulans* sterigmatocystin gene cluster led to a marked decrease in ST formation without buildup of precursor metabolites, a result consistent with such a function (Keller et al., unpublished results). It is known that anthrones are oxidized by molecular oxygen in the absence of an enzyme, so it is not surprising that ST biosynthesis was not completely absent in the knockout culture.

Further oxidation steps

Further oxidative reactions of norsolorinic acid are carried out in three stages (Figure 6) by enzymes encoded by genes in the AF gene cluster (23). The oxidative enzymes, for which conversion steps have been assigned, include six cytochrome P450 monooxygenases (CypA, AvnA, VerA, VerB, OrdA, CypX), two Baeyer-Villiger type oxidases (MoxY and AflY), and three NADH-dependent reductases (Nor-1, AvfA, OrdB). Recently, preliminary evidence has been obtained suggesting that the AF cluster gene *norA*, which is predicted to encode a ketoreductase, may be involved in the decarboxylation/dehydration step in the conversion of versicolorin A to demethylST (Cary and Ehrlich, unpublished data). Non-oxidative enzymes required for the conversion steps are: two O-methyltransferases (OmtA and OmtB) and an esterase (EstA). In the first stage the hydrocarbon moiety derived from hexanoylCoA undergoes a series of oxidations that ultimately give rise to the bisfuran, versicolorin A. At least ten enzymes are needed for the steps shown in Figure 5A. Most of these steps and the enzymes involved have been confirmed by gene knockout and complementation studies (30). In the second stage, the anthraquinone, versicolorin A, is oxidized and rearranged to form the xanthone, sterigmatocystin, in a set of reactions for which no stable intermediate has yet been isolated. Postulated intermediates are shown in the conversion scheme (Fig 5B). The sequence of steps has been deduced by metabolite feeding studies and expected reactions for the types of enzymes involved. In the third stage, sterigmatocystin is converted to the aflatoxins (shown in Figure 5C for formation of aflatoxins B1 and G1). As with formation of demethylST from versicolorin A (43), the individual steps shown have been deduced mainly by determining the most likely catalytic functions for the different types of enzymes.

The biosynthesis of the aflatoxins has provided insight into the genes and enzymes required for biosynthesis of other closely related polyketides, for example, metabolites related to anthraquinones, where the anthraquinone has undergone further oxidation and/or reduction of the A-ring, Baeyer-Villiger oxidations of the B-ring, and rearrangements to xanthones. Examples of these types of reactions are cytochrome P450 catalyzed oxidation of chrysophanol (44) and emodin to produce islandicin and secalonic acids in *Penicillium islandicum* and *P. oxalicum*, respectively, and tajixanthone and related compounds by *A. versicolor*, and geodin by *A. terreus* (43). The bisfuran pine pathogenic compound, dothistromin, is a metabolite related to versicolorin A (24).

Role of PKs in fungal adaptation

PKs probably enabled fungi to survive and prosper as free-living organisms in the diverse environments in which they needed to adapt (45, 46). Often it is only possible to guess at the adaptive roles of some of the metabolites. However, there is a commonality to their biosynthesis that suggests that the PK biosynthesis gene clusters may have evolved in order to provide a basic need in many different types of fungi irrespective of the specific niche to which they had to adapt. Our previous studies suggested that the aflatoxin biosynthesis cluster has been maintained for more than 100 to 300 million years (My) based on the calculated time of divergence of section *Nidulantes* and section *Flavi* isolates (40). We speculate that prior to this time both types of fungi had a basal gene cluster that consisted of only the genes for a PKS and a few other enzymes

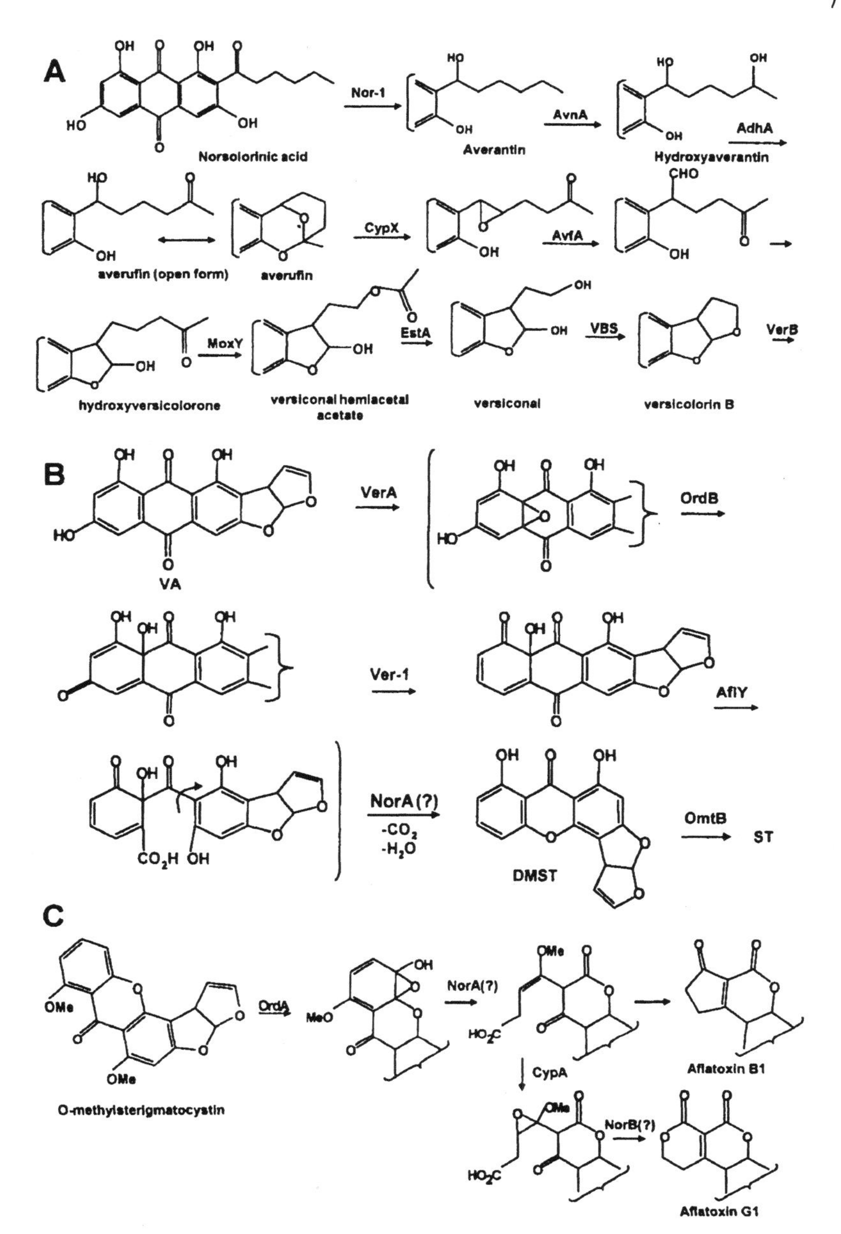

Figure 5. Sequence of enzymatic conversion steps in the biosynthesis of aflatoxins. A. Steps from norsolorinic acid to versicolorins. B. Steps from versicolorin A to sterigmatomcystin. C. Steps from O-methylsterigmatocystin to the aflatoxins. Some of the intermediates shown are hypothetical. A question mark indicates that the enzyme suggested for the conversion step has not been proven.

capable of stabilizing the nascent PK. Such a basal cluster would have allowed fungi to synthesize aromatic, highly pigmented, molecules that probably, like plant pigments, attracted insects to foster spore dispersal. The medium-chain FAS genes, HexA and HexB could have evolved from yeast FASs to allow fungi to survive in anaerobic environments, but in the presence of the PKS allowed biosynthesis of colorful anthraquinones. As the fungi dispersed into less hospitable niches, a need may have arisen for them to be more competitive with associated microflora or plants. These conditions may have selected for the duplication and movement of suitable genes, which originally served other functions, into the proximity of this primordial cluster. This permitted the elaboration of a greater variety of metabolites, some of which were toxic to the competing microflora. It also may have allowed the fungi to become plant pathogens.

Where did the genes come from to convert the bisfurans to compounds that are even more highly toxic in *Aspergillus* species. We suggest that genes involved in fungal virulence were moved as a block to the initial cluster at some time in the formation of the AF and ST biosynthesis clusters. The proteins encoded by the genes *ordB* and *ver-1* are similar to genes involved in appessorium development, a hardened mycelial structure needed for penetration of the cell walls of host organisms. In the ST cluster of *A. nidulans* all of the genes necessary for the conversion of versicolorin A to ST are adjacent, suggesting that these genes may have been "captured" by the cluster as a discreet unit. Also in both types of cluster these genes are at the proximal end of the cluster, again indicating that they form a biosynthetic block.. In line with the AF cluster being "cobbled" together from discreet gene units, the gene(s) necessary for aflatoxin G formation are at the extreme distal end of the cluster and therefore, also probably were acquired by the cluster after formation of the initial cluster unit.

While these conclusions are highly speculative, they help account for the great variety of metabolites made by closely related species of fungi. Our previous studies found that the genes in the AF biosynthesis cluster encoding oxidative steps were much more highly conserved than were the genes encoding the polyketide formation steps and the transcription regulatory proteins (40, 47). These results are consistent with higher selection pressures being placed on the oxidative enzyme-encoding genes than on the basal cluster genes.

References

1. Turner, W. B.; Aldridge, D. C. *Fungal Metabolites II*. Academic Press: London, U.K., 1983.
2. *Handbook of secondary fungal metabolites, 3-volume set.* Elsevier: Academic Press, 2003.

3. Bhatnagar, D.; Ehrlich, K. C.; Chang, P.-K., *Mycotoxins*. In *Encyclopedia of Life Sciences*, Eds.; Nature Publishing Company: London, 2000; pp 564-573.
4. Eaton, D.; Gallagher, E. *Annu. Rev. Pharmacol. Toxicol.* **1994**, *34*, 135-172.
5. Henson, J. M.; Butler, M. J.; Day, A. W. *Annu. Rev. Phytopathol.* **1999**, *37*, 447-471.
6. Bhatnagar, D.; Ehrlich, K. C.; Cleveland, T. E. *Appl. Microbiol. Biotechnol.* **2003**, *61*, 83-93.
7. Jarvis, B. B.; Miller, J. D., *Natural products, complexity and evolution.* In *Phytochemical Diversity and Redundancy in Ecological Interactions*, Romeo, J. T.; Saunders, J. A.; Barbosa, P., Eds.; Plenum Press: New York, 1996; pp 265-295.
8. Lillehoj, E. B., *Aflatoxin: an ecologically elicited genetic/activation signal.* In *Mycotoxins and Animal Foods*, Smith, J. E.; Henderson, R. S., Eds.; CRC Press: Boca Raton, FL, 1991; pp 2-30.
9. Demain, A. L., *Adv. Biochem. Eng. Biotechnol.* **2000**, *69*, 1-39.
10. Kroken, S.; Glass, N. L.; Taylor, J. W.; Yoder, O. C.; Turgeon, B. G. *Proc Natl Acad Sci U S A* **2003**, *100*, 15670-15675.
11. Pringle, A.; Taylor, J. *Trends Microbiol.* **2002**, *10*, 474-481.
12. Wilkinson, H.; Ramaswamy, A.; Sim, S. C.; Keller, N. P. *Mycologia* **2004**, *96*, 1190-1198.
13. Shen, B. *Curr. Opin. Chem. Biol.* **2003**, *7*, 285-295.
14. Fujii, I.; Watanabe, A.; Sankawa, U.; Ebizuka, Y. *Chem. Biol.* **2001**, *8*, 189-197.
15. O'Donnell, K. O.; Kistler, H. C.; Tacke, B. K.; Casper, H. H. *Proc. Natl. Acad. Sci. U. S. A.* **2000**, *97*, 7905-7910.
16. Kasuga, T.; White, T. J.; Taylor, J. W. *Mol. Biol. Evol.* **2002**, *19*, 2318-2324.
17. Lutzoni, F.; Pagel, M.; Reeb, V. *Nature* **2001**, *411*, 937-940.
18. Watanabe, A.; Ebizuka, Y. *Chem. Biol.* **2004**, *11*, 1101-1106.
19. Keller, N. P.; Hohn, T. M. *Fungal Genet. Biol.* **1997**, *21*, 17-29.
20. Cary, J. W.; Chang, P.-K.; Bhatnagar, D., *Clustered metabolic pathway genes in filamentous fungi.* In *Agriculture and Food Production*, Khachatourians, G. G.; Arora, D. K., Eds.; Elsevier, B.V.: 2001; pp 165-198.
21. Kennedy, J.; Auclair, K.; Kendrew, S. G.; Park, C.; Vederas, J. C.; Hutchinson, C. R. *Science* **1999**, *284*, 1368-1372.
22. Brown, D. W.; Yu, J. H.; Kelkar, H. S.; Fernandes, M.; Nesbitt, T. C.; Keller, N. P.; Adams, T. H.; Leonard, T. L. *Proc. Natl. Acad. Sci. U.S.A.* **1996**, *93*, 1418-1422.
23. Yu, J.; Bhatnagar, D.; Cleveland, T. E. *FEBS Lett.* **2004**, *564*, 126-130.
24. Bradshaw, R. E.; Bhatnagar, D.; Ganley, R. J.; Gillman, C. J.; Monahan, B. J.; Seconi, J. M. *Appl. Environ. Microbiol.* **2002**, *68*, 2885-2892.

25. Proctor, R. H.; Brown, D. W.; Plattner, R. D.; Desjardins, A. E. *Fungal Genet. Biol.* **2003**, *38*, 237-249.
26. Walton, J. D. *Fungal Genet. Biol.* **2000**, *30*, 167-171.
27. Sidhu, G. S. *Eur. J. Plant Pathol.* **2002**, *108*, 705-711.
28. Liang, S. H.; Wu, T. S.; Lee, R.; Chu, F. S.; Linz, J. E. *Appl. Environ. Microbiol.* **1997**, *63*, 1058-1065.
29. Yu, J.; Bhatnagar, D.; Cleveland, T. E. *FEBS Lett.* **2004**, *564*, 126-130.
30. Yu, J.; Chang, P.-K.; Ehrlich, K. C.; Cary, J. W.; Bhatnagar, D.; Cleveland, T. E.; Payne, G. A.; Linz, J. E.; Woloshuk, C. P.; Bennett, J. W. *Appl. Environ. Microbiol.* **2004**, *70*, 1253-1262.
31. Ehrlich, K. C.; Montalbano, B. G.; Cary, J. W. *Gene* **1999**, *230*, 249-257.
32. Bok, J.-W.; Keller, N. P. *Eukaryot. Cell.* **2004**, *3*, 527-535.
33. Burow, G. B.; Nesbitt, J. D.; Keller, N. P. *Mol. Plant Microbe Int.* **1997**, *10*, 380-387.
34. Calvo, A. M.; Bok, J.-W.; Brooks, W.; Keller, N. P. *Appl. Environ. Microbiol.* **2004**, *70*, 4733-4739.
35. Hicks, J. K.; Yu, J. H.; Keller, N. P.; Adams, T. H. *EMBO J.* **1997**, *16*, 4916-4923.
36. Keller, N. P.; Adams, T. H. *Proceedings: 19th Fungal Genetics Conference, March 18-23* **1997**, 1.
37. Keller, N. P.; Nesbitt, C.; Sarr, B.; Phillips, T. D.; Burow, G. B. *Phytopathol.* **1997**, *87*, 643-648.
38. Tag, A.; Hicks, J.; Garifullina, G.; Ake, C.; Phillips, T. D.; Beremand, M.; Keller, N. *Mol. Microbiol.* **2000**, *38*, 658-665.
39. Watanabe, C. M.; Townsend, C. A. *Chem. Biol.* **2002**, *9*, 981-988.
40. Ehrlich, K. C.; Yu, J.; Cotty, P. J. *J. Appl. Microbiol.* **2005**, *99*, 518-527.
41. Hitchman, T. S.; Schmidt, E. W.; Trail, F.; Rarick, M. D.; Linz, J. E.; Townsend, C. A. *Bioorganic Chem.* **2001**, *29*, 293-307.
42. Fujii, I.; Chen, Z. G.; Ebizuka, Y.; Sankawa, U. *Biochem. Int.* **1991**, *25*, 1043-1049.
43. Henry, K. M.; Townsend, C. A. *J. Am. Chem. Soc.* **2005**, *127*, 3724-3733.
44. Ahmed, S. A.; Bardshiri, E.; Simpson, T. J. *Chem. Commun.* **1987**, 883-884.
45. Schmidt, S.; Sunyaev, S.; Bork, P.; Dandekar, T. *Trends Biochem. Sci.* **2003**, *28*, 336-341.
46. Jarvis, B. B. *An. Acad. Bras. Cienc.* **1995**, *67*, 329-345.
47. Ehrlich, K. C.; Montalbano, B. G.; Cotty, P. J. *Fungal Genet. Biol.* **2003**, *38*, 63-74.

Chapter 6

Biochemical and Molecular Analysis of the Biosynthesis of Fumonisins

Liangcheng Du[1], Fengan Yu[1], Xiangcheng Zhu[1], Kathia Zaleta-Rivera[1], Ravi S. Bojja[1,3], Yousong Ding[1,4], Han Yi[1,4], and Qiaomei Wang[2]

[1]Department of Chemistry, University of Nebraska, Lincoln, NE 68588
[2]Department of Horticulture, Zhejiang University, Hangzhou, China
[3]Current address: Max Planck Institute for Molecular Physiology, Dortmund, Germany 44202
[4]Current address: Life Sciences Institute, University of Michigan, 210 Washtenaw Avenue, Ann Arbor, MI 48109

Fumonisins are mycotoxins produced by several widespread fungal pathogens of corn. Ingestion of fumonisin contaminated food and feeds causes several fatal diseases in livestock and is associated with esophageal cancer and neural tube defects in humans. In the recent years, fumonisin research has become one of the most active areas in fungal secondary metabolism. The gene cluster (*FUM*) required for fumonisin biosynthesis in *Fusarium verticillioides* has been cloned. The *FUM* genes are predicted to encode various fascinating enzymes, including a seven-domain iterative modular polyketide synthase (PKS), the first example of a fusion protein between the two subunits of serine palmitoyltransferase, a naturally fused cytochrome P450 monooxygenase and reductase, and a "stand-alone" nonribosomal peptide synthetase complex. This review highlights the progress in the studies of the molecular mechanism for fumonisin biosynthesis.

Fumonisins are a group of mycotoxins produced by a number of filamentous fungi, including *Fusarium verticillioides* (synonym *F. moniliforme*, teleomorph *Gibberella moniliformis*, synonym *G. fujikuroi* mating population A), which is a common fungal contaminant of corn and maize-derived products worldwide (*1*-*4*). The fungus is found in all corn kernels and causes ear, stalk, and seedling rots (*5*). Fumonisins cause several fatal animal diseases, including leukoencephalomalacia in horses, pulmonary edema in swine, and cancer in rats and mice (*1, 6, 7*). In addition, fumonisins are associated with human esophageal cancer (*8, 9*) and neural tube defects (*10*). In 2002, the Joint FAO/WHO Expert Committee on Food Additives released a provisional maximum tolerable daily intake of 2 mg/kg body weight for fumonisins (*11*).

The mode of fumonisin action is believed to occur through disrupting the biosynthesis of ceramide and sphingolipids (*12, 13*). Sphinglipids are ubiquitous constituents of membranes of eukaryotic cells and crucial signaling molecules involved in numerous cellular processes. Fumonisins are structurally similar to sphingoids, such as sphinganine, sphingosine, and phytosphingosine, which are the long-chain base backbones of sphingolipids. Thus, fumonisins can specifically inhibit sphinganine *N*-acyltransferase (ceramide synthase) (*14*).

In spite of the importance of fumonisins, detailed molecular mechanism for their biosynthesis has just started to emerge in recent years. An understanding of the molecular mechanism is critical to the development of rational approaches to mycotoxin elimination in agriculture and food industry. This review intends to highlight the progress in this area.

Biosynthetic Origin of Fumonisins

Fumonisins were first isolated from a strain of *F. verticillioides* on moldy maize by Benzuidenhout et al. in 1988 (*15*). So far, twenty eight fumonisin analogs are known (*16*). They are classified into four groups, the A, B, C, and P-series fumonisins. The B-series fumonisins are the most abundant analogs produced by wild-type strains, with fumonisin B_1 (FB_1) accounting for approximately 70% of the total content (*2*). FB_1 is also believed to be the most toxic analog (*3, 9*). It is a tricarballylic ester formed between propane-1,2,3-tricarboxylic acid and the hydroxyl at C-14 and C-15 of 2*S*-amino-12*S*,16*R*-dimethyl-3*S*,5*R*,10*R*,14*S*,15*R*-pentahydroxy-eicosane (Figure 1) (*15, 17, 18*).

The biosynthetic origin of fumonisins has been studied by several groups. C-3 to C-20 of the carbon backbone are derived from acetate, and C-1 and C-2, as well as the C-2 amino, are derived from alanine (*17, 19*). The two methyl groups at C-12 and C-16 are derived from methionine (*20*). Therefore, the biosynthesis of the carbon backbone of fumonisins is likely to involve a polyketide mechanism (*21*). The origin of the hydroxyl on the carbon backbone has also

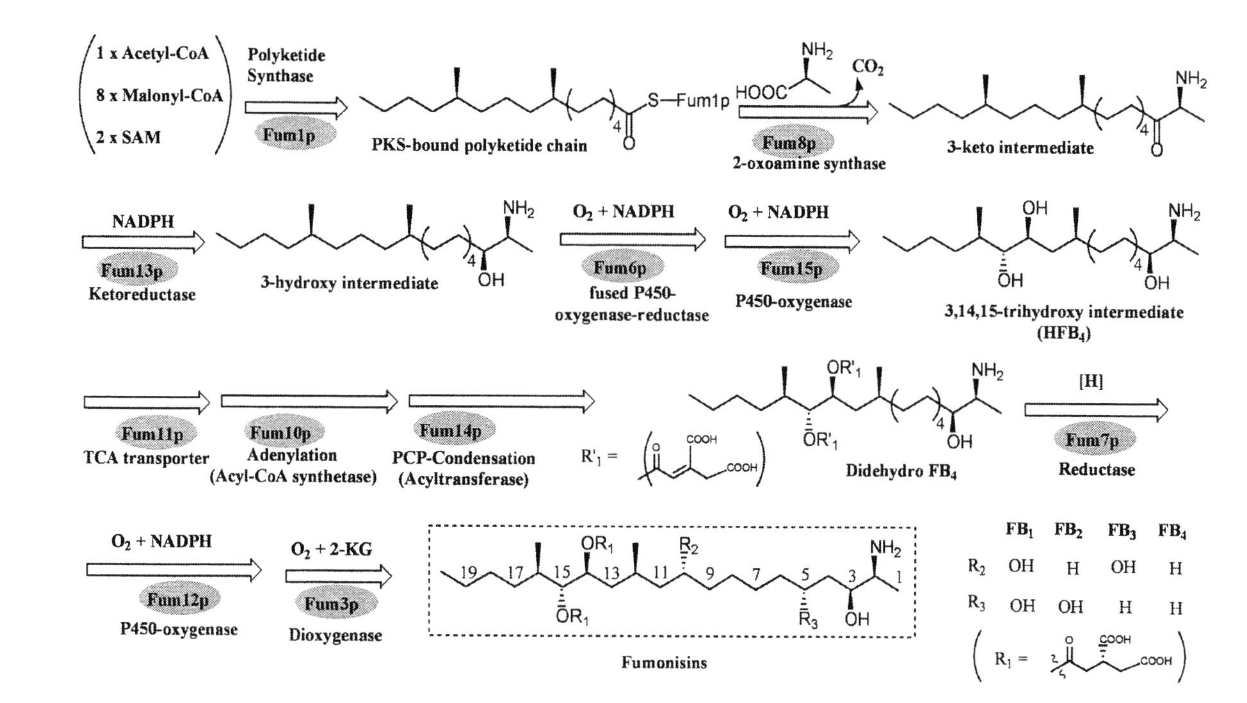

Figure 1. A proposed pathway for the biosynthesis of fumonisins.

been studied (*22*). The hydroxyl at C-5, C-10, C-14, and C-15 of FB_1 are derived from molecular oxygen, whereas the hydroxyl at C-3 is from acetate. Finally, the origin of the two tricarballylic esters is not totally clear, but most likely from the citric acid cycle (*17*).

Molecular Mechanism for Biosynthesis

Biosynthetic Gene Cluster and Biosynthetic Pathway

A 15-gene cluster (*FUM*) required for fumonisin biosynthesis in *F. verticillioides* has been cloned (*21, 23, 24*). The *FUM* genes were predicted to encode at least five different types of enzymes, some of which exhibit highly intriguing features (see below). Recently, several libraries of expressed sequence tags (EST) of *F. verticillioides* have been sequenced, and results showed that the majority of the *FUM* genes have alternative spliced transcripts (*25*). A previously unidentified gene, *FUM20*, was also found by the EST sequencing project. No function has been proposed for this gene.

A biosynthetic pathway for fumonisins has been proposed (Figure 1) (*26, 27*). The pathway can be divided into four stages. The first stage is the assembly of the dimethylated carbon backbone (C-3 to C-20) from acetyl-CoA, malonyl-CoA, and *S*-adenosylmethionine (SAM) via a polyketide biosynthetic mechanism. *FUM1* encodes a polyketide synthase (Fum1p), which is the logical candidate for the first step. The second stage is the release of the enzyme-bound carbon chain from Fum1p via the decarboxylative condensation of alanine with the polyketide chain. This produces a 20-carbon-long intermediate with a 3-keto group (Figure 1). *FUM8* was predicted to encode a serine palmitoyltransferase (SPT) homolog (*24*), which is the first committed enzyme in sphingolipid pathway (*28*). The SPT homolog is likely to catalyze the second step. The third stage is the oxidoreductions of the 3-keto intermediate, including the reduction of the 3-keto to a 3-hydroxyl by Fum13p (*29, 30*), the hydroxylation at carbon C-14, C-15, and C-10 by P450 monooxygenases (Fum6p, Fum12p, and Fum15p). The final stage is the formation of tricarballylic esters at C-14 and C-15. Biochemical data have indicated that the nonribosomal peptide synthetase complex, Fum10p-Fum14p-Fum7p, is responsible for the esterification (Figure 1) (*31, 32*). The biosynthesis of fumonisins is completed when a final hydroxyl is added at C-5 by Fum3p (*33, 34*).

Characterization of Biosynthetic Genes and Enzymes

FUM1—*Polyketide Backbone Synthesis*

FUM1 (or *FUM5*) has been proposed to synthesize the carbon backbone (C-3 to C-20) of fumonisins (*23, 26*). Only one polyketide synthase (PKS) gene is found in the *FUM* cluster, and a deletion of this gene in *F. verticillioides* led to the elimination of fumonisin production (*21*). Very little is known about the mechanism by which Fum1p controls the carbon backbone structure. In bacteria, especially *Streptomyces*, the molecular mechanism for polyketide biosynthesis is relatively well understood (*35-37*). Although many biologically active, structurally complex polyketides have been isolated from filamentous fungi, the biosynthetic mechanism for these polyketides, especially the reduced polyketides like fumonisins, remain largely unknown (*38*).

FUM1 represents a typical iterative modular PKS and could serve as a model system for the study of fungal reduced polyketides. Fum1p contains seven domains, β-ketoacyl synthase (KS), acyltransferase (AT), dehydratase (DH), methyltransferase (MT), β-ketoacyl reductase (KR), enoylreductase (ER), and acyl carrier protein (ACP) (Figure 2) (*23*). The domain manipulation strategy has been successfully used to elucidate the biosynthetic mechanism for bacterial modular PKSs (*35-37*). However, this strategy is not directly applicable to fungal modular PKSs, because only a single set of domains is present in the fungal enzymes. We exploited this strategy by using a group of fungal polyketide synthases that share an identical domain architecture but synthesize different carbon chains (*39, 40*). We have mutated the active site of the MT domain of *FUM1* and shown that the mutant strain produced demethylated intermediates of fumonisin biosynthesis (*39*). The results not only confirmed the role of MT domain but also demonstrated that a genetic system has been developed for manipulating the single-modular PKS in this filamentous fungus. More recently, we replaced the KS domain of *FUM1* with the KS domain of *PKS1* for T-toxins in *Cochliobolus heterostrophus* (Figure 2) (*40, 41*). The KS-replaced mutant produced fumonisins, rather than T-toxin-like metabolites. The results show that the KS domain of fungal PKSs is functionally exchangeable and that the KS domain alone does not control the structure of polyketide products. This is the first successful domain manipulation in polyketide synthases for fungal reduced polyketides. We also replaced the entire *FUM1* in *F. verticillioides* with *PKS1*. The mutant did not produce fumonisins or T-toxin analogs, suggesting that the interaction between PKS and downstream enzymes in the biosynthetic pathway may play an important role in the structural determination of the products.

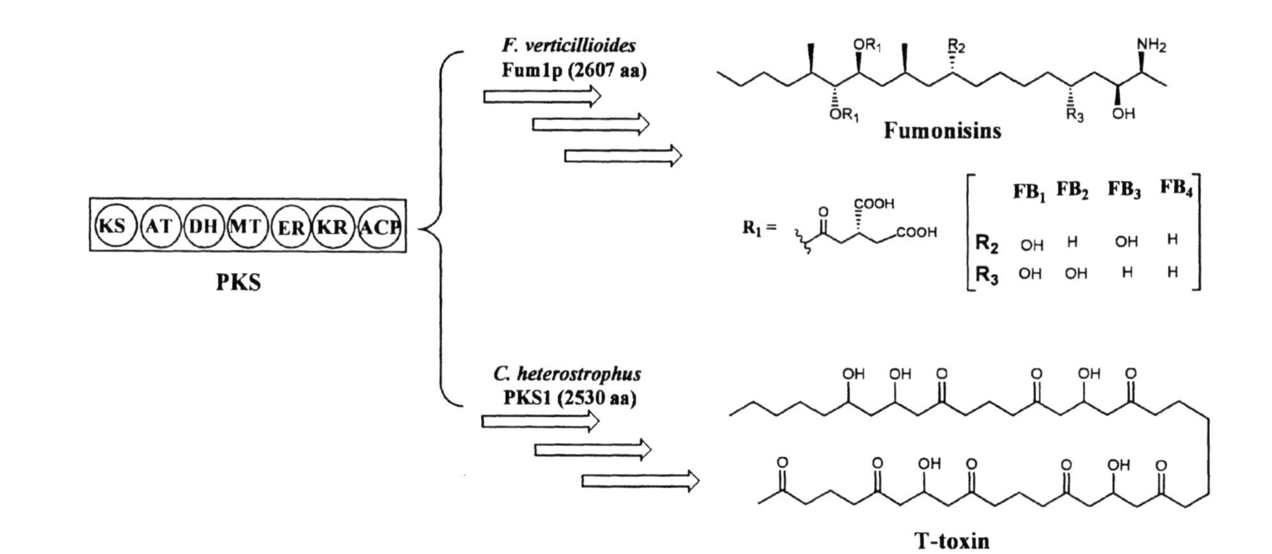

Figure 2. The domain organization of fumonisin Fum1p and T-toxin PKS1 and the chemical structure of the metabolites.

FUM8 —Amino Acid Transfer and Polyketide Chain Release

FUM8 was predicted to encode a pyridoxal phosphate-dependent α-oxoamine synthase to catalyze the decarboxylative condensation between alanine and acyl-*S*-Fum1p (Figure 1) (*23, 26*). Members of this group of enzymes include SPT involved in sphigolipid biosynthesis (*42*), 5-aminolevulinate synthase (ALAS) in heme biosynthesis (*43*), and 8-amino-7-oxononanoate synthase (AONS) in biotin biosynthesis (*44*). While ALAS and AONS are homodimers, SPT is a heterodimer composing of two separate subunits, Lcb1p and Lcb2p, which are encoded by the *LCB1* and *LCB2*, respectively. Lcb1p and Lcb2p are homologous to each other, and Lcb2p contains the pyridoxal phosphate-binding lysine residue and therefore is the catalytic subunit of SPT (*42*). However, both proteins are required for SPT activity. Unique among this family of enzymes, Fum8p is a natural fusion protein of a Lcb2p homolog and a Lcb1p homolog (*24*). The *N*-terminal half (Lcb2p domain) of Fum8p is similar to human Lcb2p (37.1%) and yeast Lcb2p (32.8%), and the *C*-terminal half (Lcb1p domain) to human Lcb1p (24.3%) and yeast Lcb2p (23.4%). The motifs that are conserved in this family of enzymes are present in Fum8p (Figure 3).

Fum8p provides an ideal model system to study SPT-type enzymes, because it is the only known example of a single-protein SPT found in eukaryotes. All known eukaryotic SPTs are membrane-associated heterodimers of Lcb1p and Lcb2p (*28*). This complexity has hampered the efforts to study the structure and function, because Lcb2p is unstable unless it is associated with Lcb1p and it has been difficult to simultaneously overproduce both subunits. The feature of Fum8p ensures the co-production of both SPT subunits. More interestingly, it was reported that mutations in human *LCB1* (*SPTLC1*) cause hereditary sensory neuropathy type I (HSN1) in humans (*45, 46*). HSN1 is the most common hereditary disorder of peripheral sensory neuron. No medical treatments exist that can cure this neural disorder. Fum8p provides an opportunity to study the structure and function of this group of enzymes, which could potentially lead to a better understanding of the molecular mechanism underlying HSN1.

We have expressed *FUM8* in *E. coli* and yeast to study the activity of Fum8p in vitro. In addition to the Lcb1p and Lcb2p domains, Fum8p contains a putative "membrane-associate" (MA) domain of about 65 residues at its *N*-terminus (Figure 4A). *FUM8* with the MA domain was refractory to heterologous expression. However, the removal of MA domain led to the expression of *FUM8*, as well as the Lcb2p and Lcb1p domains, in *E. coli* and/or yeast (Ding and Du, unpublished data). Moreover, when the Lcb2p and Lcb1p domains were separately expressed in *E. coli*, no soluble protein was obtained. However, a small fraction (estimated ~ 5%) of Fum8p was obtained as soluble protein when the two-domain were expressed as a fusion protein. Three mutants (K344G, K344E, K344L) at the active site of Fum8p Lcb2p domain have also

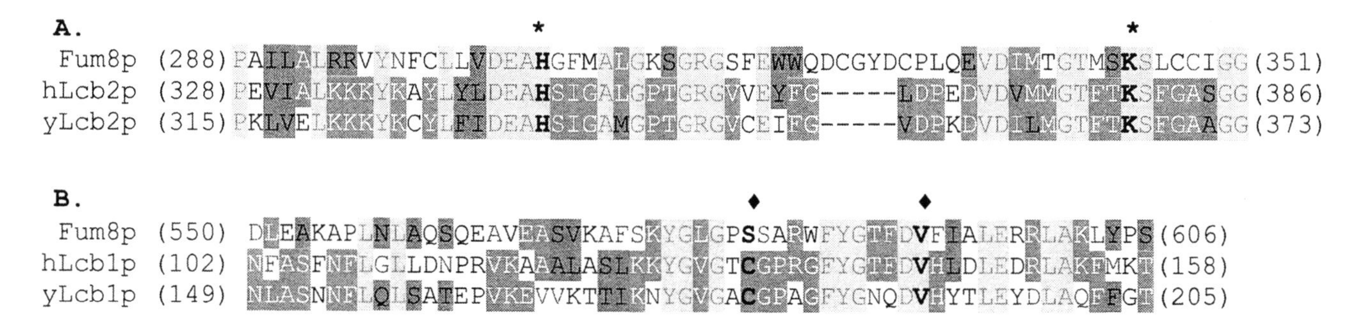

Figure 3. Sequence alignment of two conserved regions in serine palmitoyltransferases (SPTs). (A). A region from the N-terminus (Lcb2p domain) of Fum8p (839 residues in full length) aligned with Lcb2p subunit of human (hLcb2p, 562 residues) and yeast (yLcb2p, 561 residues). The asterisks mark the positions of the active sites of Lcb2p. Lys-366 of yLcb2p and Lys-379 of hLcb2p are the PLP-binding site, whereas His-334 of yLcb2p and His-347 of hLcb2p the hydrogen-bonding site with 3'-O of PLP (42). (B). A region from the C-terminus (Lcb1p domain) of Fum8p aligned with Lcb1p subunit of human (hLcb1p, 473 residues) and yeast (yLcb1p, 558 residues). The diamonds mark the positions of the mutations in hLcb1p, C133W, C133Y, and V144D, which cause the disease of hereditary sensory neuropathy type I (45). Note that Fum8p has a serine residue at the cysteine-133 position of hLcb1p.

been generated by site-directed mutagenesis (Figure 4A). When *E. coli* cells containing *FUM8-3* gene were fed with 2-^{14}C-alanine and stearoyl-CoA, a ^{14}C-labeled product that closely co-migrated with sphinganine on a TLC plate was produced (Figure 4B) (Ding and Du, unpublished). The product was not produced in the control experiments. Since sphinganine is a condensation product of serine and palmitoyl-CoA (Figure 5C) (*42*), the ^{14}C-labeled products derived from 2-^{14}C-alanine and stearoyl-CoA could be a compound closely related to the first condensation products in the fumonisin pathway (see Figure 1 and Figure 4B).

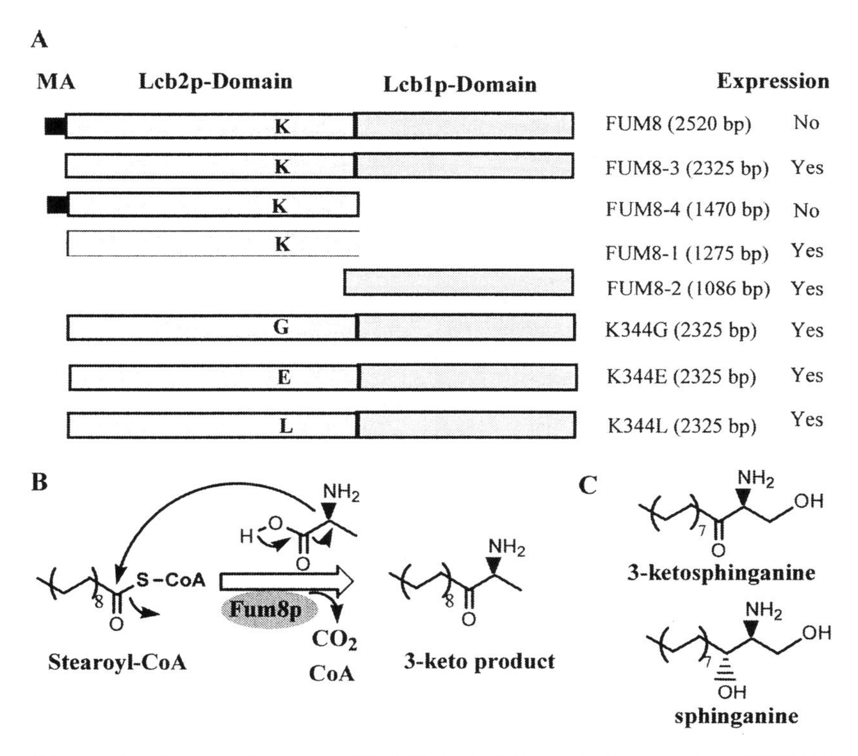

Figure 4. Characterization of FUM8. (A). Schematic illustration of FUM8 expression constructs. MA, "membrane-associate" domain. ThePLP-binding lysine residue is indicated, and the mutants of the lysine residue are indicated by replacing K with G, E, or L. (B). An in vitro test of Fum8p expressed in E. coli, using ^{14}C-alanine and stearoyl-CoA as substrates to produce a 3-keto product, which is analogous to 3-ketosphinganine. (C). Chemical structure of 3-ketosphinganine and sphinganine.

FUM6, FUM9, FUM12, FUM13, FUM15 —Oxidoreduction of the Backbone

Six genes, *FUM6*, *FUM7*, *FUM9* (*FUM3*), *FUM12*, *FUM13*, and *FUM15*, were predicted to encode redox enzymes (*23*). Among them, *FUM7* was suggested to be involved in the formation of tricarballylic esters (see below), whereas the other five genes are involved in the oxidoreduction of fumonisin backbone.

FUM6, *FUM12*, and *FUM15* show similarity to genes of cytochrome P450 monooxygenases (*23*). Interestingly, *FUM6* is predicted to encode a natural fusion between a heme-P450 monooxygenase and a FMN/FAD-containing reductase (*24*), representing the first self-sufficient P450 enzyme involved in fungal polyketide biosynthesis. Examples of P450 fusion enzymes include P450BM-3, a fatty acid hydroxylase isolated from *Bacillus megaterium* (*47, 48*), and P450foxy, a fatty acid ω-terminal hydroxylase isolated from *F. oxysporum*a (*49*). A deletion of *FUM6* in *F. verticillioides* led to the elimination of fumonisin production (*24*). Interestingly, the production of fumonisins was restored in the *FUM6* deletion mutant when it was co-cultured with the *FUM1* deletion mutant or the *FUM8* deletion mutant (*26*). No fumonisin was produced when the *FUM1* mutant and the *FUM8* mutant were co-cultured. The results suggest that a biosynthetic intermediate(s) was produced by the *FUM6* mutant and taken up by the *FUM1* mutant or the *FUM8* mutant, which converted the intermediate(s) into fumonisins. Further analysis of the co-cultures showed that metabolites having the general carbon skeleton of fumonisins with 1-4 hydroxyl groups were accumulated in the cultures (*26*). The results suggest that the oxidation of the carbon backbone of fumonisins takes place after the polyketide chain release from Fum1p, which is catalyzed by Fum8p. In spite of the studies, the precise role of *FUM6*, *FUM12*, and *FUM15* in fumonisin biosynthesis has not been established but is most likely related to the three hydroxylations at C-10, C-14, and C-15 (Figure 1).

FUM9 (*FUM3*) deletion mutants accumulated FB_3 and FB_4, but not FB_1 and FB_2 (*33*), indicating that the function of *FUM3* is for C-5 hydroxylation. This has been confirmed by in vitro characterization of the heterologously expressed Fum3p (*34*). The purified Fum3p converted FB_3 to FB_1, in the presence of 2-ketoglutarate, Fe^{++}, ascorbic acid, and catalase, which are the characteristics of 2-ketoglutarate-dependent dioxygenases.

FUM13 shows a sequence similarity to short-chain dehydrogenase/reductase genes. The deletion of this gene in *F. verticillioides* resulted in production of a low level of normal B-series fumonisins as well as the 3-keto form of FB_3 and FB_4 (*29*). The results suggest that *FUM13* may be responsible for the 3-ketoreduction of fumonisins and that an additional 3-ketoreductase gene may be present in the fungus. The production of normal fumonisins in the mutants made it difficult to define the function of *FUM13*. We expressed *FUM13* in *E. coli* and purified Fum13p (*30*). Using the 3-keto FB_3 as substrate, we showed that

Fum13p was able to convert 3-keto FB_3 to FB_3 in the presence of NADPH. The results demonstrate that *FUM13* gene encodes an NADPH-dependent ketoreductase responsible for the 3-ketoreduction in fumonisin biosynthesis. This was further supported by functional complementation of a yeast *TSC10* mutant strain by *FUM13* (*30*). *TSC10* encodes 3-ketosphinganine reductase in the sphingolipid pathway (*50*). The results support the in vivo function of *FUM13* in 3-ketoreduction.

FUM7, FUM10, FUM11, FUM14 —Tricarballylic Esterifications

Fumonisins contain two tricarballylic esters at C-14 and C-15. Four genes, *FUM7*, *FUM10*, *FUM11*, and *FUM14*, have been proposed to be involved in the esterifications (*31, 32*). The deletion mutants of *FUM10* and *FUM14* produced the hydrolyzed forms of FB_3 and FB_4 (called HFB_3 and HFB_4), which lack the tricarballylic esters (*31, 32*). Deletion of *FUM7* resulted in the accumulation of a fumonisin-like metabolite, probably having a carbon-carbon double-bond in the tricarboxylic side-chains (*32*). Interestingly, deletion of *FUM11* resulted in the accumulation of the half-hydrolyzed forms of FB_3 and FB_4 (called $PHFB_3$ and $PHFB_4$) (*32*). However, it is difficult to define the function of the individual genes based on the metabolites accumulated in the mutants.

These *FUM* genes, except *FUM11* that encodes a tricarxylate transporter, may encode a "stand-alone" complex of nonribosomal peptide synthetase (NRPS) (*31*). The *FUM10*-encoded acyl-CoA synthetase is analogous to the adenylation domain of NRPS (*51*). The *FUM14*-encoded protein contains two domains of NRPS, peptidyl carrier protein (PCP) and condensation domain (C). The *FUM7*-encoded dehydrogenase can be regarded as a reductase domain of NRPS (*52, 53*). We have purified Fum7p, Fum10p, and Fum14p by expressing the corresponding *FUM* genes in *E. coli* or yeast. Among them, Fum14p has been characterized (*31*). Fum14p was able to convert HFB_3 and HFB_4 to FB_3 and FB_4, respectively, when incubated with tricarballylic thioester of *N*-acetylcysteamine. In addition, the condensation domain was able to convert HFB_1 to FB_1. These data provide direct evidence for the role of Fum14p in the esterification of fumonisins. More interestingly, the results are the first example of an NRPS condensation domain catalyzing a C-O bond (ester) formation, instead of the typical C-N bond (amide) formation in nonribosomal peptides.

Based on the paradigm of NRPS-catalyzed reactions, we proposed a mechanism for the biosynthesis of the tricarballylic esters of fumonisins (Figure 5). The first step is the ATP-dependent activation of a tricarboxylate substrate by Fum10p to form an acyl-AMP, which is subsequently transferred to the PCP domain of Fum14p. The exact substrate for Fum10p is not known, but is likely to be an intermediate of the tricaroxylatic acid cycle (*17*). The data obtained from the *FUM7* deletion mutant suggested that the substrate of Fum10p could be a tricarboxylate containing a double bond (*32*), such as aconitic acid. The second step is the C domain-catalyzed condensation between acyl-S-PCP of Fum14p and HFB_3 or HFB_4 to produce a didehydro-intermediate of FB_3 or FB_4, respectively. The last step is the reduction of the double bond in the didehydro-intermediate by Fum7p to produce FB_3 and FB_4 (Figure 5). A hydroxylation at carbon-5 will convert FB_3 and FB_4 to FB_1 and FB_2, respectively (*33, 34*).

FUM17, FUM18, and FUM19 —Resistance Mechanism

Three *FUM* genes, *FUM17*, *FUM18*, and *FUM19*, were proposed to be involved in the self-protection (resistance) of the fungus from the mycotoxins (*23*). Both *FUM17* and *FUM18* show sequence similarities to genes for longevity assurance factors (ceramide synthases) found in yeast and mammals (*54*). Since fumonisins are specific inhibitors of ceramide synthase, it is logical to propose that *FUM17* and *FUM18* are related to the resistance mechanism in *F. verticillioides*. They may provide "extra" copies of ceramide synthase genes to maintain a proper level of ceramides and sphingolipids that are essential for the fungus life. However, a simultaneous deteletion of both *FUM17* and *FUM18* did not affect the fumonisin production in the mutant (*23*), suggesting that this fungus may possess other resistance mechanisms. One such mechanism could be to sequester the mycotoxins from the ceramide synthases by pumping the toxins out of the cells. *FUM19* shows a sequence similarity to ABC transporter (*23*), which may serve as an efflux pump for the mycotoxins. A deletion of *FUM19* did slightly reduce, but did not eliminate, the production of fumonisins in the mutant (*23*). We recently deleted all three genes simultaneously and found that the mutant still produced fumonisins with a level similar to the wild type strain (Yu and Du, unpublished). Thus, the exact resistance mechanism for fumonisin production in *F. verticillioides* is still not clear.

Conclusion

Throughout the world, corn is infected by mycotoxigenic fungi that produce toxic secondary metabolites. *F. verticillioides* is one of the most common fungal pathogens of corn. The fungal infection can occur in the field before harvest or during the storage stage after harvest. The understanding of molecular mechanisms for their biosynthesis is of economical significance, because this knowledge could lead to rational approaches toward toxin elimination/reduction in agriculture and food industry. Although significant progress in this area has been made in the past six years, the detailed molecular mechanism for the individual steps of fumonisin biosynthesis remains to be studied. Among them,

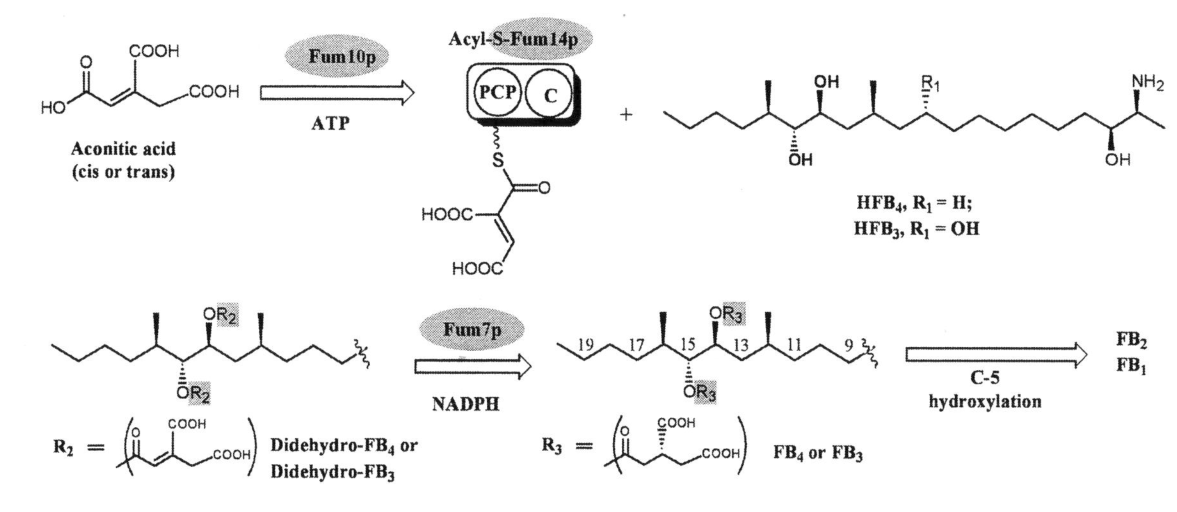

Figure 5. A proposed mechanism for the NRPS-catalzyed tricarballylic ester formation of fumonisins. C, condensation domain of nonribosomal peptide synthetase; PCP, peptidyl carrier protein.

the mechanism by which the fungal PKS controls the structure of fumonisin carbon chain is the least understood. With the genetic and biochemical tools available to study *F. verticillioides*, we are now in a good position to tackle these individual steps to reveal molecular insights in fumonisin biosynthesis.

Acknowledgement

Research in the Du laboratory has been supported by Elsa Pardee Foundation, Layman Award from UNL, NSF EPSCOR II, and NSFC (#30428023).

References

1. Marasas, W. F. *Environ. Health Perspect. 109 Suppl.* **2001**, *2*, 239-243.
2. Nelson, P. E.; Desjardins, A. E.; Plattner, R. D. *Annu. Rev. Phytopathol.* **1993**, *31*, 233-252.
3. Thiel, P. G.; Marasas, W. F.; Sydenham, E. W.; Shephard, G. S.; Gelderblom, W. C. *Mycopathologia* **1992**, *117*, 3-9.
4. Bullerman, L. B. *Adv. Exp. Med. Biol.* **1996**, *392*, 27-38.
5. Miller, J. D. *Environ. Health Perspect. 109 Suppl.* **2001**, *2*, 321-324.
6. Gelderblom, W. C.; Jaskiewicz, K.; Marasas, W. F.; Thiel, P. G.; Horak, R. M.; Vleggaar, R.; Kriek, N. P. *Appl. Environ. Microbiol.* **1988**, *54*, 1806-1811.
7. Marasas, W. F.; Kellerman, T. S.; Gelderblom, W. C.; Coetzer, J. A.; Thiel, P. G.; van der Lugt, J. J. *Onderstepoort. J. Vet. Res.* **1988**, *55*, 197-203.
8. Wang, H.; Wei, H.; Ma, J.; Luo, X. *J. Environ. Pathol. Toxicol. Oncol.* **2000**, *19*, 139-141.
9. Marasas, W. F. *Nat. Toxins* **1995**, *3*, 193-198.
10. Hendricks, K. *Epidemiology* **1999**, *10*, 198-200.
11. WHO Techn Report 906; **2002**; p 16-27.
12. Merrill, A. H., Jr.; Sullards, M. C.; Wang, E.; Voss, K. A.; Riley, R. T. *Environ. Health Perspect. 109 Suppl.* **2001**, *2*, 283-289.
13. Riley, R. T.; Enongene, E.; Voss, K. A.; Norred, W. P.; Meredith, F. I.; Sharma, R. P.; Spitsbergen, J.; Williams, D. E.; Carlson, D. B.; Merrill, A. H., Jr. *Environ. Health Perspect. 109 Suppl.* **2001**, *2*, 301-308.
14. Wang, E.; Norred, W. P.; Bacon, C. W.; Riley, R. T.; Merrill, A. H., Jr. *J. Biol. Chem.* **1991**, *266*, 14486-14490.
15. Bezuidenhout, S. C.; Gelderblom, W. C. A.; Gorst-Allman, C. P.; Horak, R. M.; Marasas, W. F. O.; Spiteller, G.; Vleggaar, R. *J. Chem. Soc. Chem. Commun.* **1988**, 743-745.
16. Rheeder, J. P.; Marasas, W. F.; Vismer, H. F. *Appl. Environ. Microbiol.* **2002**, *68*, 2101-2105.

17. Blackwell, B. A.; Edwards, O. E.; Fruchier, A.; ApSimon, J. W.; Miller, J. D. *Adv. Exp. Med. Biol.* **1996**, *392*, 75-91.
18. Shi, Y.; Peng, L. F.; Kishi, Y. *J. Org. Chem.* **1997**, *62*, 5666-5667.
19. Branham, B. E.; Plattner, R. D. *Mycopathologia* **1993**, *124*, 99-104.
20. Plattner, R. D.; Shackelford, D. D. *Mycopathologia* **1992**, *117*, 17-22.
21. Proctor, R. H.; Desjardins, A. E.; Plattner, R. D.; Hohn, T. M. *Fungal Genet. Biol.* **1999**, *27*, 100-112.
22. Caldas, E. D.; Sadilkova, K.; Ward, B. L.; Jones, A. D.; Winter, C. K.; Gilchrist, D. G. *J. Agric. Food Chem.* **1998**, *46*, 4734-4743.
23. Proctor, R. H.; Brown, D. W.; Plattner, R. D.; Desjardins, A. E. *Fungal Genet. Biol.* **2003**, *38*, 237-249.
24. Seo, J. A.; Proctor, R. H.; Plattner, R. D. *Fungal Genet. Biol.* **2001**, *34*, 155-165.
25. Brown, D. W.; Cheung, F.; Proctor, R. H.; Butchko, R. A.; Zheng, L.; Lee, Y.; Utterback, T.; Smith, S.; Feldblyum, T.; Glenn, A. E.; Plattner, R. D.; Kendra, D. F.; Town, C. D.; Whitelaw, C. A. *Fungal Genet. Biol.* **2005**, *42*, 848-861.
26. Bojja, R. S.; Cerny, R. L.; Proctor, R. H.; Du, L. *J. Agric. Food Chem.* **2004**, *52*, 2855-2860.
27. Wang, Q.; Wang, J.; Yu, F.; Zhu, X.; Zaleta-Rivera, K.; Du, L. *Progress in Natural Science* **2005**, *in press*.
28. Hanada, K. *Biochim. Biophys. Acta* **2003**, *1632*, 16-30.
29. Butchko, R. A.; Plattner, R. D.; Proctor, R. H. *J. Agric. Food Chem.* **2003**, *51*, 3000-3006.
30. Yi, H.; Bojja, R. S.; Fu, J.; Du, L. *J. Agric. Food Chem.* **2005**, *53*, 5456-5460.
31. Zaleta-Rivera, K.; Xu, C.; Yu, F.; Butchko, R. A.; Proctor, R. H.; Hidalgo-Lara, M. E.; Raza, A.; Dussault, P. H.; Du, L. **2005**, *Submitted.*
32. Butchko, R. A.; Plattner, R. D.; Proctor, R. H. *Aflatoxin/fumonisin elimination and fungal genomics workshop*, Savannah, Georgia, **2003**.
33. Butchko, R. A.; Plattner, R. D.; Proctor, R. H. *Appl. Environ. Microbiol.* **2003**, *69*, 6935-6937.
34. Ding, Y.; Bojja, R. S.; Du, L. *Appl. Environ. Microbiol.* **2004**, *70*, 1931-1934.
35. Staunton, J.; Weissman, K. J. *Nat. Prod. Rep.* **2001**, *18*, 380-416.
36. Khosla, C.; Gokhale, R. S.; Jacobsen, J. R.; Cane, D. E. *Annu. Rev. Biochem.* **1999**, *68*, 219-253.
37. Shen, B. *Biosynthesis: Aromatic Polyketides, Isoprenoids, Alkaloids* **2000**, *209*, 1-51.
38. Hutchinson, C. R.; Kennedy, J.; Park, C.; Kendrew, S.; Auclair, K.; Vederas, J. *Antonie Van Leeuwenhoek* **2000**, *78*, 287-295.
39. Yu, F.; Zhu, X.; Du, L. *FEMS Microbiol Lett* **2005**, *248*, 257-264.
40. Zhu, X.; Yu, F.; Bojja, R. S.; Zaleta-Rivera, K.; Du, L. **2005**, *Submitted.*

41. Yang, G.; Rose, M. S.; Turgeon, B. G.; Yoder, O. C. *Plant Cell* **1996**, *8*, 2139-2150.
42. Gable, K.; Han, G.; Monaghan, E.; Bacikova, D.; Natarajan, M.; Williams, R.; Dunn, T. M. *J. Biol. Chem.* **2002**, *277*, 10194-10200.
43. Ferreira, G. C.; Vajapey, U.; Hafez, O.; Hunter, G. A.; Barber, M. J. *Protein Sci.* **1995**, *4*, 1001-1006.
44. Webster, S. P.; Alexeev, D.; Campopiano, D. J.; Watt, R. M.; Alexeeva, M.; Sawyer, L.; Baxter, R. L. *Biochemistry* **2000**, *39*, 516-528.
45. Dawkins, J. L.; Hulme, D. J.; Brahmbhatt, S. B.; Auer-Grumbach, M.; Nicholson, G. A. *Nat. Genet.* **2001**, *27*, 309-312.
46. Bejaoui, K.; Wu, C.; Scheffler, M. D.; Haan, G.; Ashby, P.; Wu, L.; de Jong, P.; Brown, R. H., Jr. *Nat. Genet.* **2001**, *27*, 261-262.
47. Narhi, L. O.; Fulco, A. J. *J. Biol. Chem.* **1986**, *261*, 7160-7169.
48. Fulco, A. J. *Annu. Rev. Pharmacol. Toxicol.* **1991**, *31*, 177-203.
49. Nakayama, N.; Takemae, A.; Shoun, H. *J. Biochem. (Tokyo)* **1996**, *119*, 435-440.
50. Beeler, T.; Bacikova, D.; Gable, K.; Hopkins, L.; Johnson, C.; Slife, H.; Dunn, T. *J. Biol. Chem.* **1998**, *273*, 30688-30694.
51. Marahiel, M. A.; Stachelhaus, T.; Mootz, H. D. *Chem. Rev.* **1997**, *97*, 2651-2674.
52. Reimmann, C.; Patel, H. M.; Serino, L.; Barone, M.; Walsh, C. T.; Haas, D. *J. Bacteriol.* **2001**, *183*, 813-820.
53. Patel, H. M.; Walsh, C. T. *Biochemistry* **2001**, *40*, 9023-9031.
54. Guillas, I.; Kirchman, P. A.; Chuard, R.; Pfefferli, M.; Jiang, J. C.; Jazwinski, S. M.; Conzelmann, A. *EMBO J.* **2001**, *20*, 2655-2665.

Chapter 7

Benzoic Acid-Specific Type III Polyketide Synthases

L. Beerhues[1], B. Liu[1,2], T. Raeth[1], T. Klundt[1], T. Beuerle[1], and M. Bocola[3]

[1]Institute of Pharmaceutical Biology, Technical University of Braunschweig, Mendelssohnstrasse 1, D–38106 Braunschweig, Germany
[2]Key Laboratory of Photosynthesis and Environmental Molecular Physiology, Institute of Botany, Chinese Academy of Sciences, Nanxincun 20, Haidian District, Beijing 100093, China
[3]Max Planck Institute of Coal Research, D–45470 Mülheim/Ruhr, Germany

Benzophenone synthase (BPS) and biphenyl synthase (BIS) catalyze the formation of the same linear tetraketide from benzoyl-CoA and three molecules of malonyl-CoA. However, BPS cyclizes this intermediate via intramolecular C6→C1 Claisen condensation, whereas BIS uses intramolecular C2→C7 aldol condensation. Benzophenone derivatives include polyprenylated polycyclic compounds with high pharmaceutical potential. Biphenyl derivatives are the phytoalexins of the economically important Maloideae.

Type III polyketide synthases (PKSs) generate a diverse array of natural products by condensing multiple acetyl units from malonyl-CoA to specific starter substrates (*1,2*). The homodimeric enzymes orchestrate a series of acyltransferase, decarboxylation, condensation, cyclization, and aromatization reactions at two functionally independent active sites. Due to their ability to vary either the starter molecule or the type of cyclization and the number of condensations, they are, along with terpene synthases, one of the major generators of carbon skeleton diversity in plant secondary metabolites (*3*). Among the starter substrates used, benzoyl-CoA is a rare starter molecule. It is utilized by bacterial type I PKSs to form soraphen A, enterocin and the wailupemycins (*4*). in plants, benzoyl-CoA is the starter unit for two type III PKSs, benzophenone synthase (BPS) and biphenyl synthase (BIS), both of which were cloned in our laboratory.

Benzophenone and Xanthone Biosynthesis

BPS catalyzes the formation of 2,4,6-trihydroxybenzophenone (*5*), i.e. the C_{13} skeleton of benzophenones and xanthones (Figure 1). Subsequent 3'-hydroxylation to give 2,3,4,6-tetrahydroxybenzophenone is catalyzed by a cytochrome P450 monooxygenase (*6*) that was detected in cell cultures of *Hypericum androsaemum* (Clusiaceae).

Complex Benzophenone Derivatives

Simple benzophenones can undergo stepwise prenylation using prenyl donors such as dimethylallyl diphosphate (*7*). Accompanying cyclizations of the prenyl substituents lead to the formation of bridged polycyclic compounds that are widely distributed in Clusiaceae (=Guttiferae). The combination of challenging chemical structure and intriguing pharmacological activity makes the complex constituents attractive molecules for biotechnological research. For example, garcinol (Figure 1) is a potent inhibitor of histone acetyltransferases both *in vitro* and *in vivo* and induces apoptosis (*8*). Sampsoniones that are characterized by unique caged tetracyclic skeletons include cytotoxic compounds whose mechanisms of action and target structures are under study (*9,10*). Further polyprenylated polycyclic benzophenone derivatives exhibit antibacterial and HIV-inhibitory activities (*11,12*).

In *Centaurium erythraea* (Gentianaceae) cell cultures, BPS prefers 3-hydroxybenzoyl-CoA as starter substrate, yielding immediately 2,3',4,6-tetrahydroxybenzophenone (*13*). 3-Hydroxybenzoate:CoA ligase was purified to apparent homogeneity (*14*). 3-Hydroxybenzoic acid is derived from an intermediate of the shikimate pathway, as shown by feeding experiments with

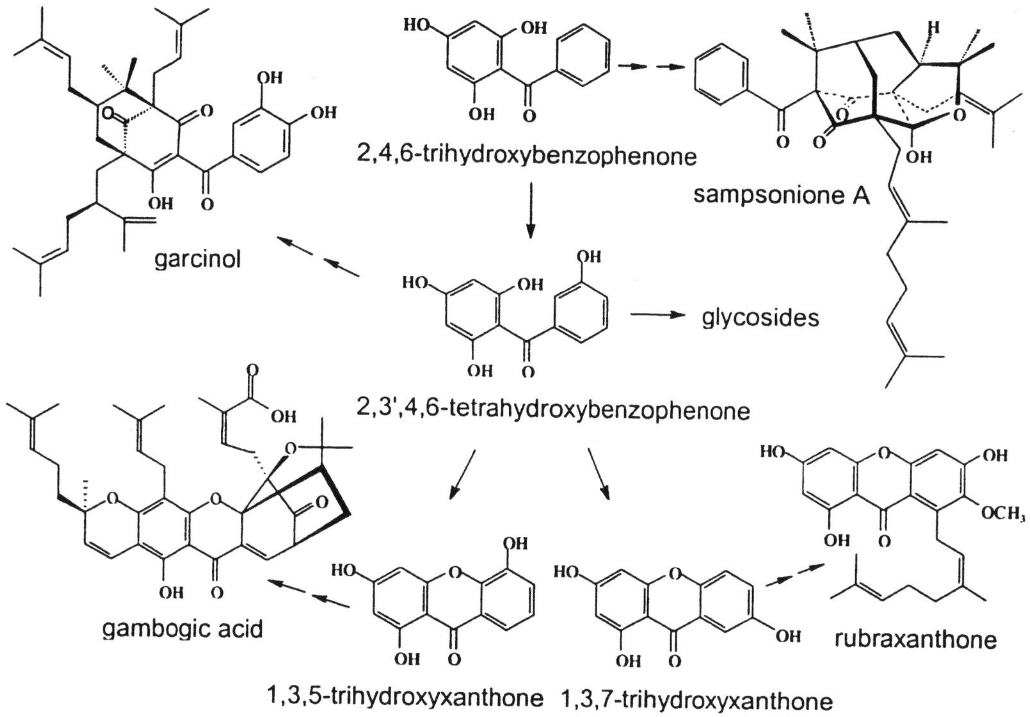

Figure 1. Benzophenone and xanthone biosynthesis and examples of pharmacologically active derivatives. (Adapted with permission from reference 5. Copyright 2003 Blackwell Publishing Ltd)

radiolabelled compounds (*15*). This result was confirmed by a retrobiosynthetic NMR study with *Swertia chirata* root cultures (*16*). In contrast to Clusiaceae, Gentianaceae accumulate glycosylated rather than prenylated constituents (*17*).

Xanthone Formation

2,3',4,6-Tetrahydroxybenzophenone is not only a prenyl acceptor but also the immediate precursor of xanthone biosynthesis (Figure 1). It undergoes regioselective oxidative phenol couplings that occur either *ortho* or *para* to the 3' hydroxy group and are catalyzed by xanthone synthases (*18*). These cytochrome P450 enzymes were proposed to be P450 oxidases. The underlying reaction mechanism is likely to involve two one-electron oxidation steps. The first one-electron transfer and a deprotonation generate a phenoxy radical whose electrophilic attack at C-2' or C-6' leads to the cyclization of the benzophenone. The intermediate hydroxycyclohexadienyl radical is transformed by the loss of a further electron and proton to 1,3,5- and 1,3,7-trihydroxyxanthones. These regioselective ring closures were detected in cell cultures of *C. erythraea* and *H. androsaemum* that accumulate primarily 5- and 7-oxygenated xanthones, respectively (*19,20*). In both cell cultures, the first modifying step at the xanthone level is 6-hydroxylation by P450 monoxygenases (*21*).

The isomeric products of the xanthone synthases are the precursors of the majority of plant xanthones (*17*). This class of secondary metabolites also includes prenylated compounds with high pharmaceutical potential. Gambogic acid is a bridged polycyclic derivative that induces apoptosis independent of the cell cycle through a novel mechanism of caspase activation (*22*). Rubraxanthone inhibits methicillin-resistant strains of *Staphylococcus aureus* (*23*). Its activity compares to that of the antibiotic vancomycin. Bioactive complex benzophenone derivatives may serve as potential lead compounds for designing drugs. Synthetic approaches to these fascinating structures have recently been developed (*24,25*).

The ecological function of polyprenylated benzoyl- and acylphloroglucinol derivatives was studied in *H. calycinum* flowers (*26*). The constituents accumulate in those parts of petals that are exposed to the outside in the unopened bud. They are, besides flavonoids, a new class of products that serve as floral UV pigments. The constituents also accumulate in anthers and ovarian walls and provide protection against herbivores. Thus, phloroglucinol derivatives fulfill both attractive and defensive functions in *H. calycinum* flowers.

Benzophenone Synthase

BPS catalyzes the formation of the carbon skeleton of benzophenone derivatives. The enzyme was cloned from *H. androsaemum* cell cultures and functionally expressed in *E. coli* (*5*). The ~43 kDa protein shared 53-63% amino acid sequence identity with other members of the type III PKS superfamily. BPS preferred benzoyl-CoA as starter substrate (Table I).

BPS catalyzed the stepwise condensation of benzoyl-CoA with three molecules of malonyl-CoA to give a tetraketide intermediate that was cyclized by intramolecular Claisen condensation into 2,4,6-trihydroxybenzophenone (Figure 2). The enzyme was inactive with CoA-linked cinnamic acids such as 4-coumaroyl-CoA, the preferred starter substrate for chalcone synthase (CHS). BPS and CHS from *H. androsaemum* cell cultures shared 60.1% amino acid sequence identity. CHS is ubiquitous in higher plants and the prototype enzyme of the type III PKS superfamily (*1,2*). It uses the same reaction mechanism like BPS to form 2',4,4',6'-tetrahydroxychalcone, the precursor of flavonoids (Figure 2).

Flavonoids serve as flower pigments, UV protectants, phytoalexins and signal molecules (*1,2*). Unlike BPS, CHS exhibits a broad substrate specificity (Table I). *H. androsaemum* CHS and the enzymes from other sources (*27*) accepted CoA-linked benzoic acids as minor substrates and formed the respective benzophenones. An efficient but unphysiological starter unit is benzoyl-CoA for 2-pyrone synthase (2-PS) that performs only two decarboxylative condensations with malonyl-CoA to produce 6-phenyl-4-hydroxy-2-pyrone (*28*). In cell cultures of *H. androsaemum,* benzoic acid originates from cinnamic acid by side-chain degradation (*29*). The underlying mechanism is CoA-dependent and non-*β*-oxidative. The complete sequence of enzymes involved was detected.

Table I. Substrate Specificities of BPS and CHS from *H. androsaemum* Cell Cultures and BIS from S. *aucuparia* Cell Cultures

	Enzyme activity ('% of max. each)		
Starter Substrate	*BPS*	*CHS*	*BJS*
Benzoyl-CoA	100	22	100
3-Hydroxybenzoyl-CoA	19	22	12
4-Hydroxybenzoyl-CoA	2	0	30
2-Hydroxybenzoyl-CoA	0	0	43
4-Coumaroyl-CoA	0	100	0
Cinnamoyl-CoA	0	88	0
3-Coumaroyl-CoA	0	34	0
2-Coumaroyl-CoA	0	0	0
Acetyl-CoA	0	0	0

SOURCE: Reproduced with permission from reference 5. Copyright 2003 Blackwell Publishing Ltd.

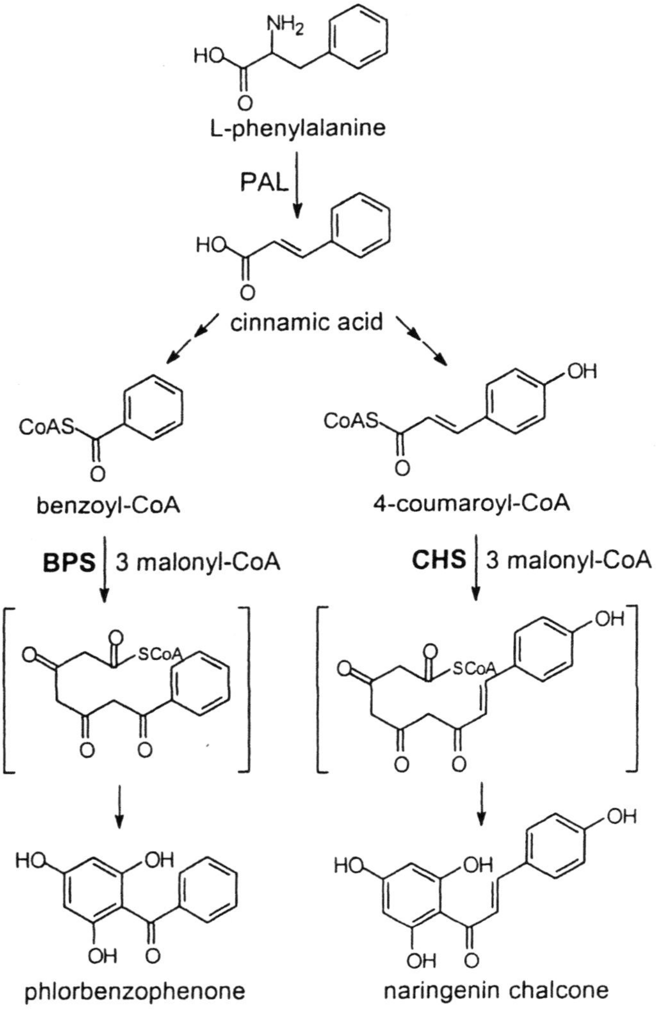
NH2
HO
O
L-phenylalanine
PAL
HO
O
cinnamic acid
CoAS
O
benzoyl-CoA
OH
CoAS
O
4-coumaroyl-CoA
BPS
3 malonyl-CoA
CHS
3 malonyl-CoA
O
O
SCoA
O
O
OH
O
O
SCoA
O
O
HO
OH
OH
O
phlorbenzophenone
OH
HO
OH
OH
O
naringenin chalcone

Figure 2. Reactions catalyzed by BPS and CHS from H. androsaemum cell cultures. Both starter substrates originate from cinnamic acid, itself supplied by phenylalanine ammonia-lyase (PAL). (Reproduced with permission from reference 5. Copyright 2003 Blackwell Publishing Ltd.)

The crystal structures of CHS, 2-PS and stilbene synthase (STS) were determined and provide a framework for understanding substrate and product specificities (*30,31,32*). They defined a number of amino acids shaping the geometry of the initiation/elongation cavity that binds the starter molecule and accommodates the growing polyketide chain. Some of these residues varied between BPS and CHS and were exchanged for each other by site-directed mutagenesis (*5*). A CHS triple mutant (M9: L263M/F265Y/S338G) exhibited strongly reduced activity with 4-coumaroyl-CoA and preferred benzoyl-CoA over the activated cinnamic acid. The BPS mutants either resembled functionally the wild-type enzyme or lacked the ability for benzophenone and chalcone formation. BPS was modeled on the basis of the crystal structure of alfalfa CHS (Figure 3). This homology model is guiding the generation of further site-directed mutants.

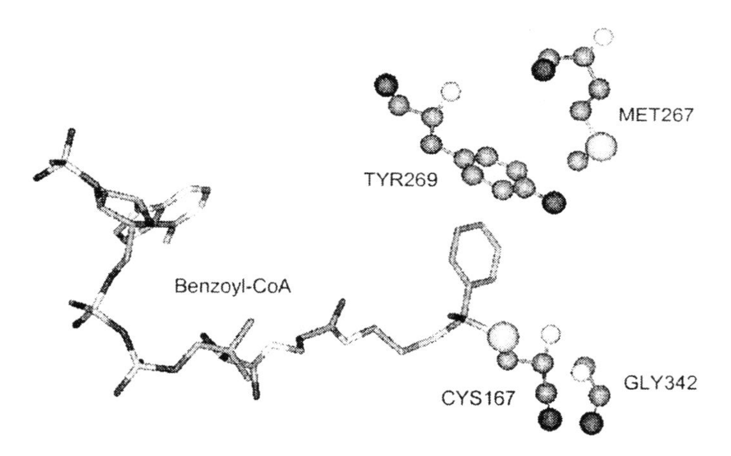

Figure 3. Homology model of the active site cavity of BPS. The CoA-linked starter unit is loaded onto the catalytic cysteine. The other three amino acids shown were introduced by site-directed mutagenesis into the CHS mutant M9.

Biphenyl and Dibenzofuran Phytoalexins

Phytoalexins are low-molecular-mass antimicrobial compounds that are formed *de novo* after microbial infection (*33*). They provide one strategy of the multi-component pathogen defence program. Many of the over 200,000 plant secondary metabolites are either constitutive or inducible antimicrobial agents.

Defence Compounds of the Maloideae

The phytoalexins of the economically important Rosaceae, especially of the subfamily Maloideae, are biphenyls and dibenzofurans (*34*). Maloideae include some of the best-known ornamental and edible-fruit plants in the temperate regions of the world. Valuable fruits are apples, pears and stone fruits such as peach, apricot, cherry, plum and almond. Furthermore, strawberries, blackberries and raspberries belong to the subfamily.

The ability to form biphenyl and dibenzofuran phytoalexins is confined to the Maloideae (*34*). Following fungal infection, a number of Maloideae species accumulated the defence compounds in the sapwood, but not in the leaves (*34, 35*). An exception was *Sorbus aucuparia* leaves that produced the biphenyl aucuparin in response to biotic and abiotic elicitation (*36*). Similarly, cell cultures of *S. aucuparia* responded to yeast extract treatment with the accumulation of aucuparin (*37*).

Interesting observations of the induction of the phytoalexins were also made with cell cultures of two *Malus domestica* cultivars (*38,39*). While cultured cells of a scab-susceptible cultivar (Macintosh) failed to produce phytoalexins, cell cultures of a scab-resistant cultivar (Liberty) produced phytoalexins in response to yeast extract treatment. Biphenyls and dibenzofurans were formed simultaneously and had similar patterns of substituents, suggesting their biogenic relationship (Figure 4). In an earlier study, no rosaceous species was found to form both classes of phytoalexins simultaneously (*34*). The antifungal activity of biphenyls and dibenzofurans was demonstrated (*40,41*). The phytoalexins inhibited both spore germination and mycelial growth at concentrations that are thought to be present at localized infection sites.

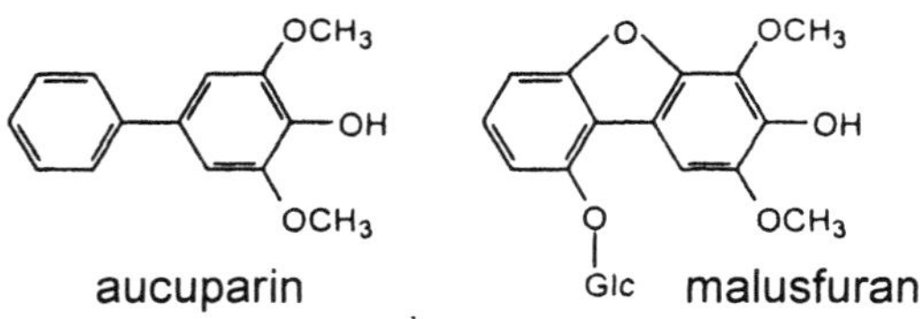

Figure 4. Major phytoalexins from cell cultures of a scab-resistant apple cultivar (39).

Biphenyl Synthase

The carbon skeleton of the phytoalexins of the Maloideae is formed by BIS that was cloned from yeast-extract-treated cell cultures of *S. aucuparia* and functionally expressed in *E. coli* (*42*) Addition of the elicitor to the cell cultures resulted in a rapid, strong and transient induction of the enzyme at the transcriptional level. BIS preferred benzoyl-CoA as starter substrate (Table 1). It catalyzed the iterative condensation of the starter moiety with three acetyl units from the decarboxylation of malonyl-CoA. BIS and BPS form identical linear tetraketides (Figure 5). However, while BPS cyclizes this intermediate via an intramolecular C6→C1 Claisen condensation, BIS catalyzes an intramolecular C2→C7 aldol condensation and decarboxylative elimination of the terminal carboxyl group to give 3,5-dihydroxybiphenyl. BIS and BPS share 54% amino acid sequence identity.

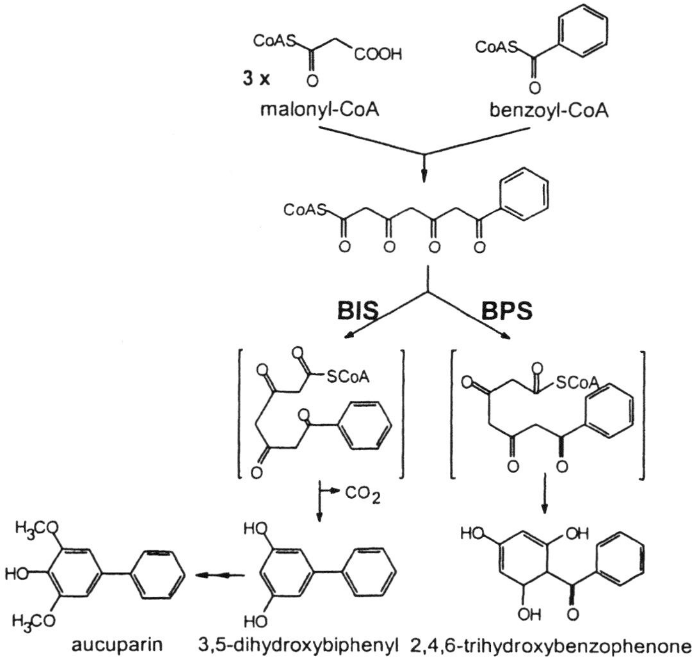

Figure 5. Reactions of BIS and BPS. Aucuparin accumulates in elicitor-treated Sorbus aucuparia cell cultures. (Adapted with permission from reference 37. Copyright 2003 Springer- Verlag.)

The alternative intramolecular cyclizations are also catalyzed by STS and CHS which, however, use 4-coumaroyl-CoA as starter molecule, leading to the formation of stilbenes and flavonoids, respectively *(1,2)*. CoA esters of cinnamic acids are accepted by neither BPS nor BIS (Table 1). Recently, the crystal structure of STS from *Pinus sylvestris* has been resolved and the mutagenic conversion of alfalfa CHS into pine STS identified the thioesterase-like "aldol switch" that is responsible for the aldol cyclization specificity in STS (*32*). Interestingly, all "aldol switch" amino acids differ from the corresponding residues in BIS, indicating the lack of a conserved set of "aldol switch" amino acids (*42*).

A phylogenetic tree was constructed by selecting one CHS each from various families of dicots, monocots, gymnosperms and ferns (*42*). In addition, all plant type III PKSs that functionally differ from CHS were included (Figure 6).

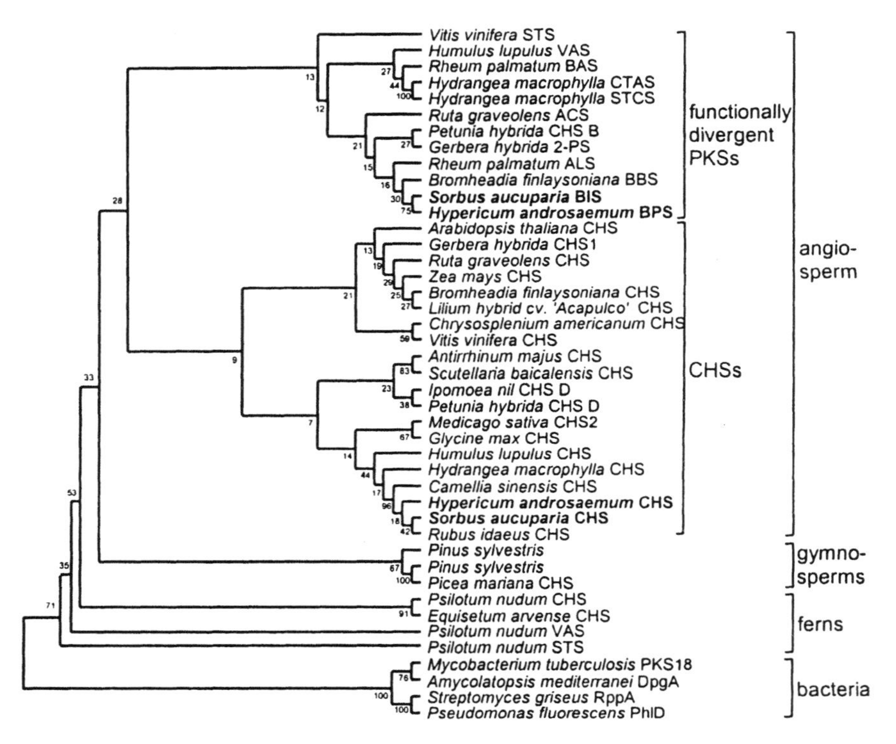

Figure 6. Phylogenetic tree of type III PKSs. (Reproduced with permission from reference 42. Copyright 2006 American Society of Plant Biologists.)

The outgroup was the consensus sequence of four bacterial type III PKSs. The bacterial enzymes extend the array of the polyketide scaffolds. They share only ~25% amino acid sequence identity with plant PKSs and each other (*43*). The phylogenetic tree reflected the systematic grouping of the higher plants. The PKSs from angiosperms fall into two clusters. One cluster comprises CHSs including the enzymes from *H. androsaemum* and *S. aucuparia* cell cultures.

The other cluster is formed by the functionally divergent enzymes including BIS and BPS. The two groups originate from an ancient duplication of the ancestral gene. One gene retained the CHS function that is indispensable for plants due to the important functions of flavonoids. The other gene underwent impressive functional diversification in various taxa. Among the new gene products are BIS and BPS that group together closely. This indicates that these enzymes result from a recent functional divergence of their common ancestor.

References

1. Schröder, J. *Comprehensive Natural Products Chemistry;* Sankawa, U., Ed.; Elsevier Science: Amsterdam, 1999; Vol. 1, pp. 749-771.
2. Austin, M.B.; Noel, J.P. *Nat. Prod. Rep.* **2003**, *20*, 79.
3. Samappito, S.; Page, J.E.; Schmidt, J.; De-Eknamkul, W.; Kutchan, T.M. *Phytochemistry* **2003**, *62,* 313.
4. Moore, B.S.; Hertweck, C. *Nat. Prod. Rep.* **2002**, *19,* 70.
5. Liu, B.; Falkenstein-Paul, H.; Schmidt, W.; Beerhues, L. *Plant J.* **2003**, *34,* 847.
6. Schmidt, W.; Beerhues, L. *FEBS Lett.* **1997**, *420,* 143.
7. Hu, L.H.; Sim, KY. *Tetrahedron* **2000**, *56,* 1379.
8. Balasubramanyam, K.; Altaf, M.; Varier, R.A.; Swaminathian, V.; Ravindran, A.; Sadhale, P.P.; Kundu, T.K. *J. Biol. Chem.* **2004**, *279,* 33716.
9. Hu, L.H.; Sim, K.Y. *Org. Lett.* **1999**, *23,* 879.
10. Matsumoto, K.; Akao, Y.; Kobayashi, E.; Ito, T.; Ohguchi, K.; Tanaka, T.; Iinuma, M.; Nozawa, Y. *Biol. Pharm. Bull.* **2003**, *26*, 569.
11. Cuesta Rubio, O.; Cuellar Cuellar, A.; Rojas, N.; Velez Castro, H.; Rastrelli, L.; Aquino, R. *J. Nat. Prod.* **1999**, *62,* 1013.
12. Fuller, R.W.; Blunt, J.W.; Boswell, J.L.; Cardellina II, J.H.; Boyd, M.R. *J. Nat. Prod.* **1999**, *62,* 130.
13. Beerhues, L. *FEBS Lett.* **1996**, *383,* 264.
14. Barillas, W.; Beerhues, L. *Biol. Chem.* **2000**, *381*, 155.
15. Abd El-Mawla, A.M.A.; Schmidt, W.; Beerhues, L. *Planta* **2001**, *212,* 288.
16. Wang, C.Z.; Maier, U.H.; Keil, M.; Zenk, M.H.; Bacher, A.; Rohdich, F.; Eisenreich, W. *Eur. J. Biochem.* **2003**, *270,* 2950.
17. Bennett, G.J.; Lee, H.H. *Phytochemistry* **1989**, *28,* 967.
18. Peters, S.; Schmidt, W.; Beerhues, L. *Planta* **1998**, *204,* 64.
19. Beerhues, L.; Berger, U. *Planta* **1995**, *197,* 608.
20. Schmidt, W.; Abd El-Mawla, A.M.A.; Wolfender, J.L.; Hostettmann, K.; Beerhues, L. *Planta Med.* **2000**, *66,* 380.
21. Schmidt, W.; Peters, S.; Beerhues, L. *Phytochemistry* **2000**, *53,* 427.
22. Zhang, H.Z.; Kasibhatla, S.; Wang, Y.; Herich, J.; Guastella, J.; Tseng, B.; Drewe, J.; Cai, S.X. *Bioorg. Med. Chem.* **2004**, *12,* 309.

23. Iinuma, M.; Tosa, H.; Tanaka, T.; Asai, F.; Kobayashi, Y.; Shimano, R.; Miyauchi, K.I. *J. Pharm. Pharmacol.* **1996**, *48,* 861.
24. Tisdale, E.J.; Slobodov, I.; Theodorakis, E.A. *PNAS* **2004**, *101,* 12030.
25. Ciochina, R.; Grossman, R.B. *Org. Lett.* **2003**, *5,* 4619.
26. Gronquist, M.; Bezzerides, A.; Attygalle, A.; Meinwald, J.; Eisner, M.; Eisner, T. *PNAS* **2001**, *98,* 13745.
27. Morita, H.; Takahashi, Y.; Noguchi, H.; Abe, I. *Biochem. Biophys. Res. Comm.* **2000**, *279,* 190.
28. Eckermann, S.; Schröder, C.; Schmidt, J.; Strack, D.; Edrada, R.A.; Helanutta, Y.; Elomaa, P.; Kotilainen, M.; Kilpelainen, I.; Proksch, P.; Teen, T.H.; Schröder, J. *Nature* **1998**, *396,* 387.
29. Abd El-Mawla, A.M.A.; Beerhues, L. *Planta* **2002**, *214,* 727.
30. Ferrer, J.L.; Jez, J.M.; Bowman, M.E.; Dixon, R.A.; Noel, J.P. *Nat. Struct. Biol.* **1999**, *6*, 775.
31. Jez, J.M.; Austin, M.B.; Ferrer, J.L.; Bowman, M.E.; Schröder, J.; Noel, J.P. *Chem. Biol.* **2000**, *7,* 919.
32. Austin, M.B.; Bowman, M.E.; Ferrer, J.L.; Schröder, J.; and Noel, J.P. *Chem. Biol.* **2004**, *11,* 1179.
33. Dixon, R.A. *Nature* **2001**, *411,* 843.
34. Kokubun, T.; Harborne, J.B. *Phytochemistry* **1995**, *40,* 1649.
35. Hrazdina, *C. Phytochemistry* **2003**, *64*, 485.
36. Kokubun, T.; Harborne, J.B. *Z. Naturforsch.* **1994**, *49c,* 628.
37. Liu, B.; Beuerle, T.; Klundt, T.; Beerhues, L. *Planta* **2004**, *218,* 492.
38. Hrazdina, C.; Borejsza-Wysocki, W.; Lester, C. *Phytopathology* **1997**, *87,* 868.
39. Borejsza-Wysocki, W.; Lester, C.; Attygalle, A.B.; Hrazdina, C. *Phytochemistry* **1999**, *50,* 231.
40. Kokubun, T.; Harborne, J.B.; Eagles, J.; Waterman, P.C. *Phytochemistry* **1995**, *40,* 57.
41. Hrazdina, C.; Borejsza-Wysocki, W.; Lester, C. *Phytopathology* **1997**, *87,* 868.
42. Liu, B.; Raeth, T.; Beuerle, T.; Beerhues, L. *Submitted.*
43. Austin, M.B.; Izumikawa, M.; Bowman, M.E.; Udwary, D.W.; Ferrer, J.L.; Moore, B.S.; Noel, J.P. *J. Biol. Chem* **2004**, *279,* 45162.

Chapter 8

Engineered Biosynthesis of Plant Polyketides: Chain Length Control in Novel Type III Polyketide Synthases

Ikuro Abe

School of Pharmaceutical Sciences, University of Shizuoka, 52–1 Yada, Shizuoka 422–8526, Japan

A growing number of functionally divergent type III polyketide synthases (PKSs), the chalcone synthase (CHS) superfamily enzymes, have been cloned and characterized, which include recently obtained pentaketide chromone synthase (PCS) and octaketide synthase (OKS) from aloe (*Aloe arborescens*). Recombinant PCS expressed in *Escherichia coli* catalyzed successive condensations of malonyl-CoA to produce a pentaketide, 5,7-dihydroxy-2-methylchromone, while recombinant OKS yielded octaketides, SEK4 and SEK4b, the longest polyketides produced by the structurally simple type III PKS. PCS and OKS share 91% amino acid sequence identity, and maintain the conserved Cys-His-Asn catalytic triad. The most characteristic feature is that the CHS active-site residue 197 (numbering in *Medicago sativa* CHS) is uniquely replaced with Met in PCS and Gly in OKS, respectively. Site-directed mutagenesis revealed that the steric bulk of the chemically inert residue lining the active-site cavity determines the polyketide chain length and the product specificity.

The CHS superfamily of type III PKSs produce a variety of plant secondary metabolites (*e.g.* chalcones, stilbenes, benzophenones, biphenyls, acridones, phloroglucinols, resorcinols, pyrones, and chromones) with remarkable structural diversity and biological activities. The type III PKSs are structurally and mechanistically distinct from the modular type I and the dissociated type II PKSs of bacterial origin; the simple homodimer of 40-45 kDa proteins carry out complete series of decarboxylation, condensation, and cylization/aromatization reactions with a single active-site by directly utilizing CoA-linked substrates without the involvement of the phosphopantetheine-armed acyl carrier proteins (*1*). The functional diversity of the type III PKSs derives from the differences of their selection of starter substrate, number of polyketide chain extensions, and mechanisms of cyclization/aromatization reactions. For example, CHS, the pivotal enzyme for the flavonoid biosynthesis, selects 4-coumaroyl-CoA as a starter and performs three condensations with malonyl-CoA to produce a tetraketide, 4,2',4',6'-tetrahydroxychalcone (naringenin chalcone) with a new aromatic ring system (Figure 1A) (*1*). On the other hand, 2-pyrone synthase (2PS) selects acetyl-CoA as a starter, and carries out only two condensations with malonyl-CoA to produce a triketide, triacetic acid lactone (TAL) (Figure 1B) (*2*). Recent crystallographic and site-directed mutagenesis studies on plant (*2-4*) and bacterial (*5,6*) type III PKSs have revealed that the enzymes share a common three-dimensional overall fold with a conserved Cys-His-Asn catalytic triad, and a common active-site machinery of the polyketide formation reactions which proceeds through starter molecule loading at the active-site Cys, malonyl-CoA decarboxylation, polyketide chain elongation, and subsequent cyclization and aromatization of the enzyme bound intermediate.

Aloe (*Aloe arborescens*) is a medicinal plant rich in aromatic polyketides such as pharmaceutically important aloenin (hexaketide), aloesin (heptaketide), and barbaloin (octaketide) (Figure 2A). Therefore, in addition to regular CHSs involved in the flavonoid biosynthesis, presence of functionally distinct multiple PKS enzymes that catalyze the initial key reactions in the biosynthesis of the secondary metabolites was expected. The cDNAs encoding the novel plant-specific type III PKSs; PCS (*7a*) and OKS (*7b*) (the GenBank^{TM} accession no. AY823626 and AY567707, respectively), were cloned and sequenced from young roots of *A. arborescens* by RT-PCR using inosine-containing degenerate primers based on the conserved sequences of known CHSs. The deduced amino acid sequences (both coding *M*r 44 kDa protein with 403 amino acids) showed 50-60% identity to those of other type III PKSs of plant origin (Figure 3); OKS shares 91% identity (368/403) with PCS, 60% identity (240/403) with alfalfa (*Medicago sativa*) CHS (*3a*), and 54% identity (216/403) with a heptaketide-producing aloesone synthase (ALS) recently obtained from rhubarb (*Rheum palmatum*) (*4c*). In contrast, it shows only 23% identity (93/403) with a bacterial Type III PKS, 1,3,6,8-tetrahydroxynaphthalene synthase (THNS or RppA) from *Streptomyces griseus* (*5a*).

A. arborescens PCS and OKS were heterologously expressed in *Escherichia coli* BL21(DE3)pLysS (pET vector). The purified enzymes gave a single band with *Mr* 44 kDa on SDS-PAGE, while the native PCS and OKS appeared to be a homodimer since they had *Mr* 88 kDa as determined by gel-filtration. The recombinant PCS did not produce chalcone, but instead efficiently accepted malonyl-CoA as a sole substrate to yield a pentaketide, 5,7-dihydroxy-2-methylchromone, as a single product (Figure 1D) (*7a*). The aromatic pentaketide is a biosynthetic precursor of khellin and visnagin, the well known anti-asthmatic furochromones in *Ammi visnaga* (*8*). On the other hand, despite the 91% amino acid sequence identity with PCS, the recombinant OKS did not produce the pentaketide, but instead efficiently produced a 1:4 mixture of SEK4 and SEK4b (Figure 1E, 4A), the longest polyketides generated by the structurally simple type III PKS (*7b*). The octaketides SEK4/SEK4b are the products of the minimal type II PKS for actinorhodin (*act* from *Streptomyces coelicolor,* a heterodimeric complex of ketosynthase and chain length factor) (*9*). Since the aloe plant does not produce SEK4/SEK4b, but produces significant amount of anthrones and anthraquinones (octaketides) (Figure 2A), it is tempting to speculate that the enzyme is originally involved in the biosynthesis of anthrones and anthraquinones in the plant. However, maybe because of misfolding of the protein in the *E. coli* expression system, or because of absence of interactions with tailoring enzymes such as yet unidentified ketoreductase (*8*), the recombinant OKS just yielded SEK4/SEK4b as shunt products as in the case of the minimal type II PKS (Figure 2C). The physiological role of OKS in the medicinal plant still remains to be elucidated.

Interestingly, both PCS and OKS accepted acetyl-CoA, resulting from decarboxylation of malonyl-CoA, as a starter substrate (*7*), but not so efficiently as in the case of *R. palmatum* ALS (*4c*). This was confirmed by the ^{14}C incorporation rate from [1-^{14}C]acetyl-CoA in the presence of cold malonyl-CoA, while the yield from [2-^{14}C]malonyl-CoA was almost at the same level either in the presence or absence of cold acetyl-CoA. In contrast, the yield of the heptaketide by *R. palmatum* ALS was two- to three-fold higher in the presence of acetyl-CoA. Theoretical ^{14}C specific incorporation from [1-^{14}C]acetyl-CoA should be 20% (1/5) in PCS and 12.5% (1/8) in OKS if acetyl-CoA serves as a starter of the polyketide forming reactions, which was largely consistent with the experimental results (*7*).

The recombinant PCS showed the K_M = 71.0 μM and k_{cat} = 445 x 10^{-3} min^{-1}, with a broad pH optimum within a range of 6.0-8.0, while OKS showed the K_M = 95.0 μM and k_{cat} = 94.0 x 10^{-3} min^{-1} for malonyl-CoA (SEK4b forming activity) with a pH optimum at 7.5 (7). Like other type III PKSs (3h,10), both PCS and OKS exhibited unusually broad substrate tolerance; the enzymes also accepted aromatic (4-coumaroyl, cinnamoyl, and benzoyl) and aliphatic (n-hexanoyl, n-octanoyl, and n-decanoyl) CoA esters as a starter, and carried out sequential condensations with malonyl-CoA to produce triketide and tetraketide α-pyrones

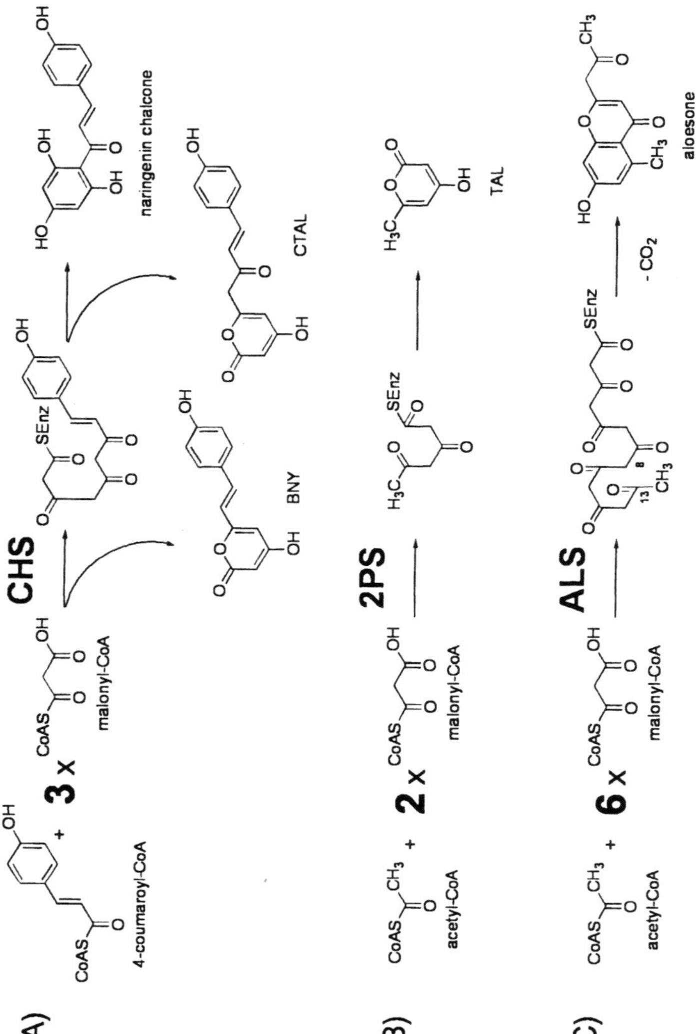

(A) 4-coumaroyl-CoA + 3 x malonyl-CoA — CHS → naringenin chalcone; BNY; CTAL

(B) acetyl-CoA + 2 x malonyl-CoA — 2PS → TAL

(C) acetyl-CoA + 6 x malonyl-CoA — ALS → aloesone ($- CO_2$)

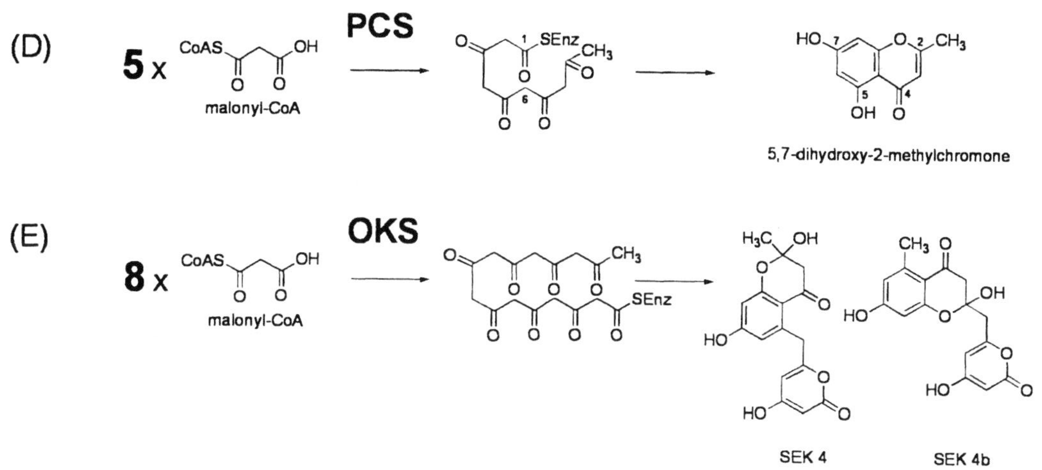

(D) 5 x malonyl-CoA —PCS→ 5,7-dihydroxy-2-methylchromone

(E) 8 x malonyl-CoA —OKS→ SEK 4, SEK 4b

Figure 1. Proposed mechanism for the formation of (A) naringenin chalcone from 4-coumaroyl-CoA and three molecules of malonyl-CoA by CHS, (B) triacetic lactone from acetyl-CoA and two molecules of malonyl-CoA by 2PS, , (C) aloesone from acetyl-CoA and six molecules of malonyl-CoA by ALS, (D) 5,7-dihydroxy-2-methylchromone from five molecules of malonyl-CoA by PCS, and (E) SEK4 and SEK4b from eight molecules of malonyl-CoA by OKS. Bis-noryangonin (BNY) and 4-coumaroyltriacetic acid lactone (CTAL) are derailment by-products of the CHS reactions in vitro *when the reaction mixtures are acidified before extraction. In* A. arborescens *PCS and OKS, acetyl-CoA, resulting from decarboxylation of malonyl-CoA, is also accepted as a starter but not so efficiently as in the case of* R. palmatum *ALS.*

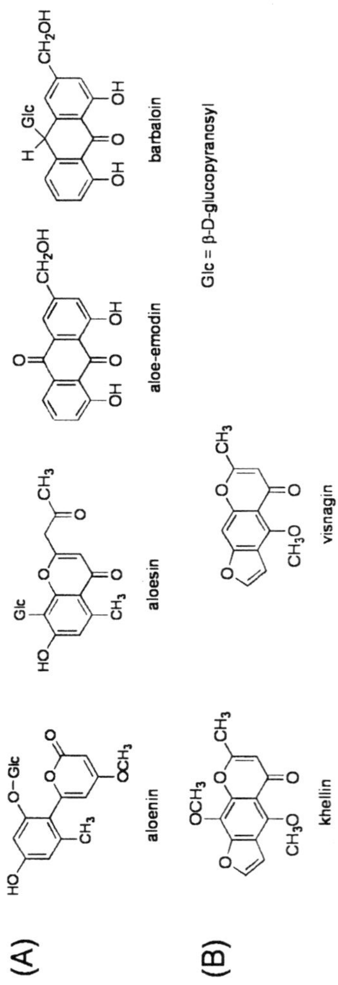

(A)

aloenin

aloesin

aloe-emodin

barbaloin

(B)

khellin

visnagin

Glc = β-D-glucopyranosyl

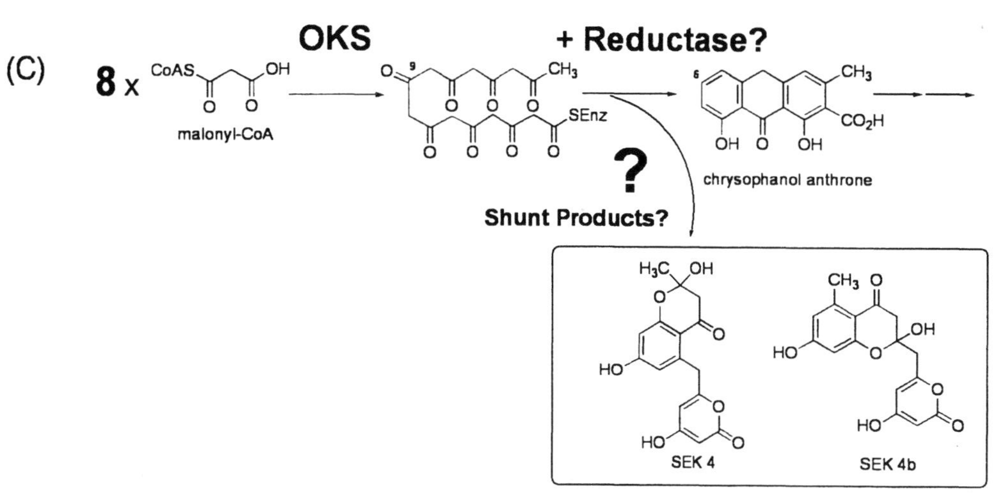

Figure 2. (A) Structures of aromatic polyketides produced by A. arborescens. *(B) Khellin and visnagin in* Ammi visnaga. (C) *A hypothetical scheme for the involvement of OKS and as yet unidentified ketoreductase in the biosynthesis of anthrones and anthraquinones. In the absence of interactions with the reductase, OKS just affords SEK4/SEK4b as shunt products. (Adapted from reference 7b. Copyright 2005 American Chemical Society.)*

(*7*). In particular, it is noteworthy that OKS readily accepted *n*-eicosanoyl (C_{20}) CoA, much longer than the previously reported *Scutellaria baicalensis* CHS that accepted only up to the C_{12} ester (*10f*), suggesting difference of the active-site structure between the two enzymes.

Comparison of the primary sequences revealed that the catalytic triad (Cys164, His303, and Asn336) and most of the CHS active-site residues (Met137, Gly211, Gly216, Phe215, Phe265, and Pro375) (numbering in *M. sativa* CHS) (*3*) are well conserved in *A. arborescens* PCS and OKS (Figure 3). One of the most characteristic features is that both PCS and OKS lack CHS's conserved residues, Thr197, Gly256, and Ser338. The residues lining the active-site cavity (*3a*) are uniquely replaced in PCS (T197M/G256L/S338V) and in OKS (T197G/G256L/S338V). The three residues are sterically altered in a number of functionally divergent type III PKSs including the heptaketide-producing *R. palmatum* ALS (T197A/G256L/S338T) (*4c)*, and the triketide-producing 2PS (T197L/G256L/ S338I) from daisy (*Gerbera hybrida*) (*2*). In *G. hybrida* 2PS, it has been proposed that the three residues control starter substrate selectivity and polyketide chain length by steric modulation of the active-site cavity (*2b*). Indeed, a CHS triple mutantation (T197L/G256L/ S338I) yielded an enzyme that was functionally identical to 2PS (*2b*).

As mentioned above, the pentaketide-producing PCS and the octaketide-producing OKS share 91% sequence identity, and the CHS's active-site residue Thr197 is uniquely replaced with Met207 in PCS (T197M/G256L/S338V) and with less bulky Gly207 in OKS (T197G/G256L/S338V). In order to investigate the structure function relationship of the two enzymes, we constructed a PCS mutant in which Met207 was substituted with Gly as in OKS. As a result, there was a dramatic change in the enzyme activity; PCS M207G mutant efficiently produced a 1:4 mixture of SEK4 and SEK4b instead of the pentaketide (formation of only trace amount of 5,7-dihydroxy-2-methylchromone was detected by LC-MS) (*7a*). Surprisingly, the pentaketide-forming PCS was transformed into an octaketide synthase by the single amino acid substitution.

On the other hand, when Gly207 in OKS was substituted with bulky Met as in PCS, OKS G207M mutant completely lost the octaketide-forming activity, but instead efficiently afforded an unnatural pentaketide, 2,7-dihydroxy-5-methylchromone, as a single product (Figure 4B and 5) (*7b*). The octaketide-producing OKS was thus converted to a pentaketide synthase by the point mutation. Interestingly, the pentaketide product is a regio isomer of the PCS's 5,7-dihydroxy-2-methylchromone, which is formed by a C-1/C-6 Claisen-type cyclization (Figure 1D). While in the OKS mutant, a C-4/C-9 aldol-type cyclization yielded 2,7-dihydroxy-5-methylchromone (Figure 5A). Despite the 91% sequence identity, OKS and PCS are not exactly functionally interconvertible by the single amino acid replacement, suggesting further subtle structural differences between the two enzymes.

These results suggested that the chemically inert single residue determines the polyketide chain length and the product specificity by steric modulation of the active-site cavity. To further test the hypothesis, we constructed a series of OKS mutants (G207A, G207T, G207L, G207F, and G207W) (*7b*). First, when Gly207 was replaced with Ala as in the case of heptaketide-producing ALS (*4c*), G207A mutant indeed yielded the heptaketide aloesone from seven molecules of malonyl-CoA in addition to SEK4/SEK4b (Figure 4C and 5). Interestingly, the heptaketide is a biosynthetic precursor of aloesin (Figure 2A), the anti-inflammatory agent of the aloe plant (*4c*). Next, when Gly207 was substituted with Thr as in the case of CHS (*3*), OKS G207T mutant completely lost the octaketide-forming activity, but efficiently yielded a hexaketide, 6-(2,4-dihydroxy-6-methylphenyl)-4-methoxy-2-pyrone, from six molecules of malonyl-CoA along with the pentaketide 2,7-dihydroxy-5-methylchromone (Figure 4D and 5). Notably, the hexaketide is a precursor of aloenin (Figure 2A), the anti-histamic agent of the medicinal plant (*11*). Moreover, other bulky substitutions G207L and G207F yielded 2,7-dihydroxy-5-methylchromone with a trace amount of SEK4/SEK4b (Figure 5). Finally, when Gly207 was substituted with the most bulky Trp, OKS G207W mutant just afforded a tetraketide, tetracetic acid lactone, along with TAL, without formation of a new aromatic ring system (Figure 4E and 5).

It was thus demonstrated that the single residue 207 in *A. arborescens* PCS and OKS (corresponding to Thr197 in *M. sativa* CHS) determines the polyketide chain length and the product specificity depending on the steric bulk of the side chain. The small-to-large substitutions in place of Gly207 in OKS resulted in loss of the octaketide-producing activity and the concomitant formation of shorter chain length polyketides (from triketide to heptaketide) (Figure 5). It is conceivable that the site-directed mutagenesis caused steric contraction of the active-site cavity in which the polyketide chain elongation reactions take place, resulting in the shortening of the product chain length. Recently, an analogous result for the chain length control has been reported for the minimal type II PKS (*12*). Here it should be noted that the functional diversity of the type III PKSs was evolved from the simple steric modulation of the active-site cavity with maintaining the Cys-His-Asn catalytic triad. Presumably, OKS and the G207 mutants catalyze the chain initiation and elongation, and possibly initiate the first aromatic ring formation reaction at the methyl end of the polyketide intermediate (Figure 5). The partially cyclized polyketide intermediates would be then released from the active-site and undergo subsequent spontaneous lactone ring formations, thereby completing the construction of the fused ring systems.

Very recently, X-ray crystal structures of *A. arborescens* PCS, both the pentaketide-producing wild-type enzyme and the octaketide-producing PCS M207G mutant, have been solved at 1.6 Å resolution (*13*). The crystal structures revealed that PCS and CHS share the common three-dimensional overall fold, including the CoA binding tunnel and the Cys-His-Asn catalytic triad. On the

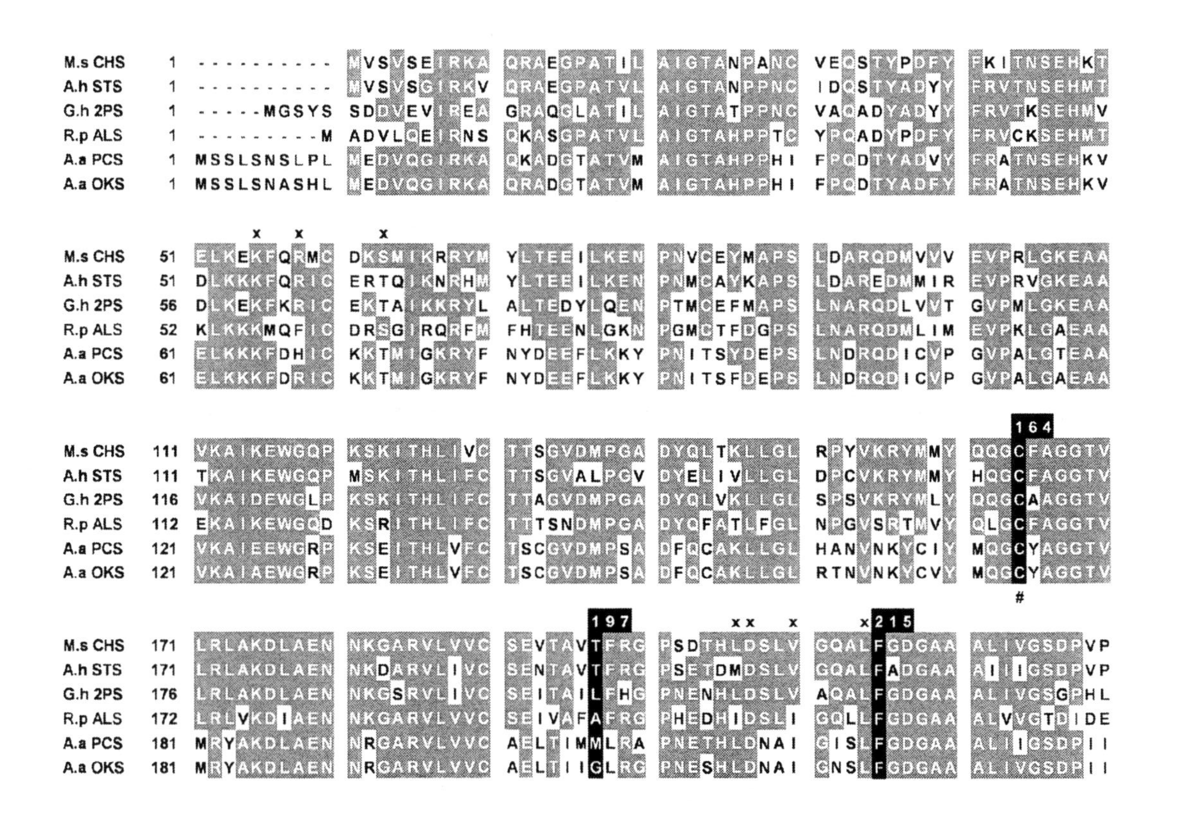
M.s CHS
A.h STS
G.h 2PS
R.p ALS
A.a PCS
A.a OKS
164
197
215

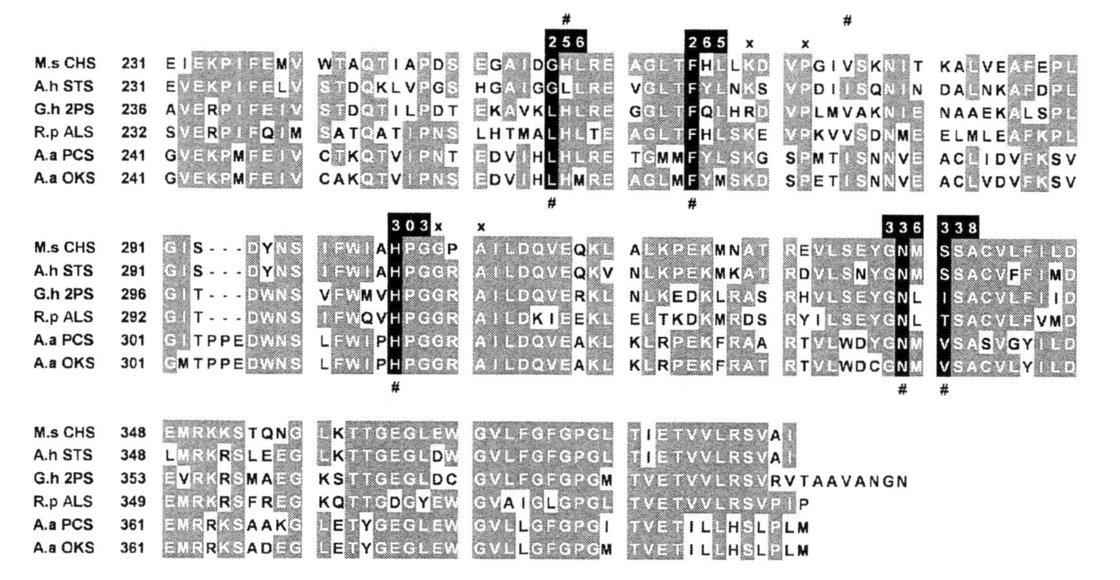

Figure 3. Comparison of the primary sequences of A. arborescens PCS/OKS and other CHS-superfamily type III PKSs. M.s CHS, M. sativa CHS; A.h STS, Arachis hypogaea stilbene synthase; G.h 2PS, G. hybrida 2PS; R.p ALS, R. palmatum ALS; A.a PCS, A. arborescens PCS; A.a OKS, A. arborescens OKS. The catalytic triad (Cys164, His303, and Asn336), and the residues lining the active-site cavity (Thr197, Phe215, Gly256, F265, and Ser338) (numbering in M. sativa CHS) are marked with #. Residues for the CoA binding are marked with x.

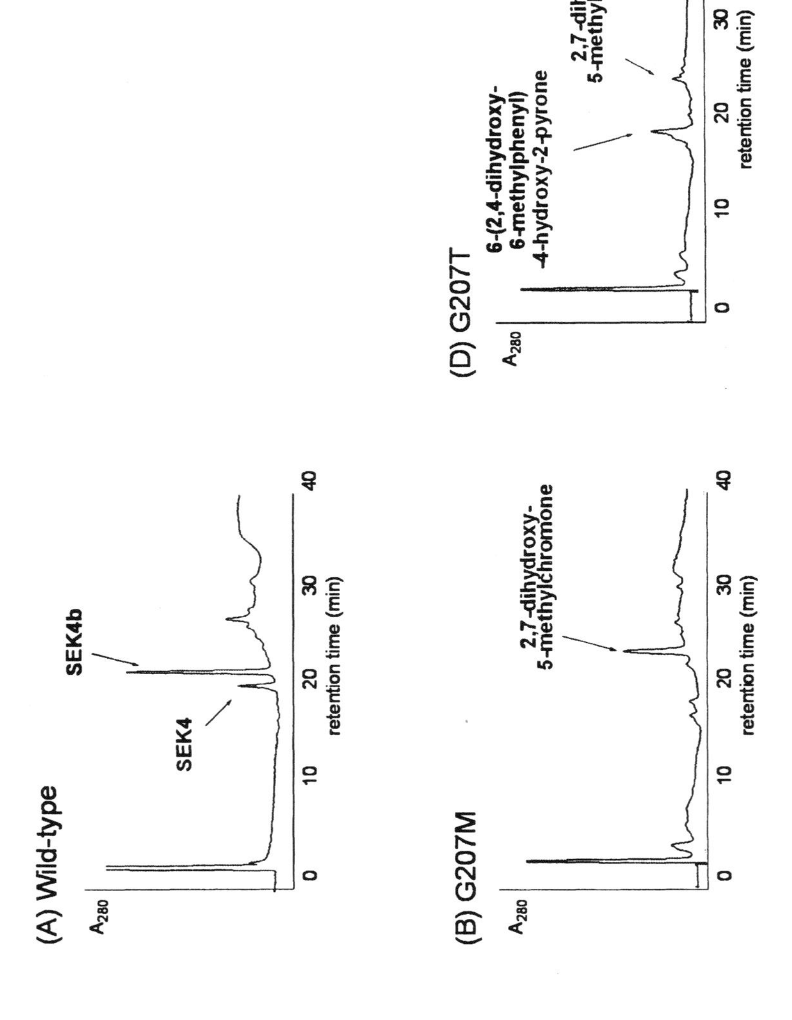
(A) Wild-type
A_{280}
SEK4b
SEK4
0
10
20
30
40
retention time (min)
(B) G207M
A_{280}
2,7-dihydroxy-
5-methylchromone
0
10
20
30
40
retention time (min)
(D) G207T
A_{280}
6-(2,4-dihydroxy-
6-methylphenyl)
-4-hydroxy-2-pyrone
2,7-dihydroxy-
5-methylchromone
0
10
20
30
40
retention time (min)

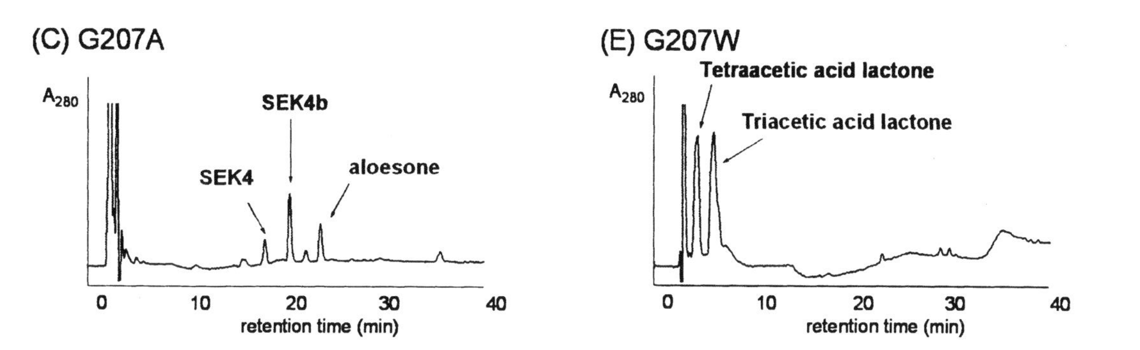

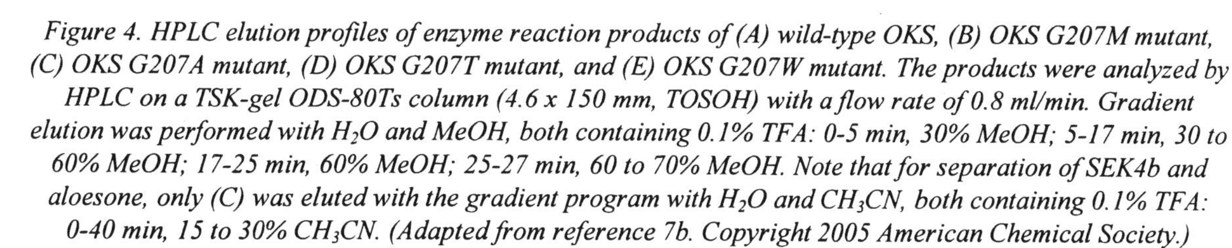

Figure 4. HPLC elution profiles of enzyme reaction products of (A) wild-type OKS, (B) OKS G207M mutant, (C) OKS G207A mutant, (D) OKS G207T mutant, and (E) OKS G207W mutant. The products were analyzed by HPLC on a TSK-gel ODS-80Ts column (4.6 x 150 mm, TOSOH) with a flow rate of 0.8 ml/min. Gradient elution was performed with H_2O and MeOH, both containing 0.1% TFA: 0-5 min, 30% MeOH; 5-17 min, 30 to 60% MeOH; 17-25 min, 60% MeOH; 25-27 min, 60 to 70% MeOH. Note that for separation of SEK4b and aloesone, only (C) was eluted with the gradient program with H_2O and CH_3CN, both containing 0.1% TFA: 0-40 min, 15 to 30% CH_3CN. (Adapted from reference 7b. Copyright 2005 American Chemical Society.)

(A)

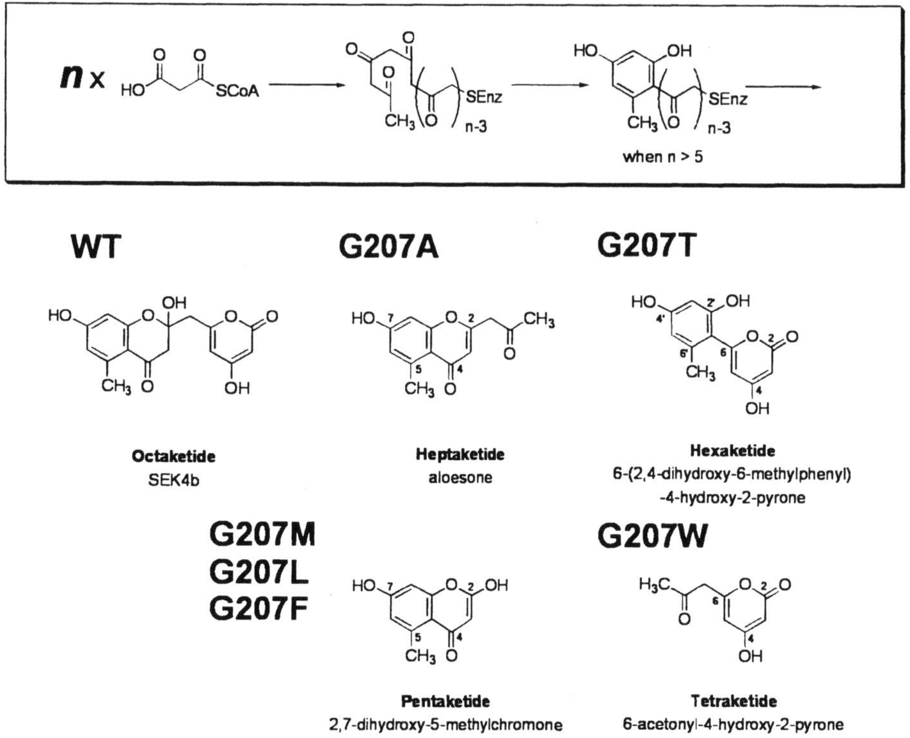

n x → → → when n > 5

WT

Octaketide
SEK4b

G207A

Heptaketide
aloesone

G207T

Hexaketide
6-(2,4-dihydroxy-6-methylphenyl)
-4-hydroxy-2-pyrone

G207M
G207L
G207F

Pentaketide
2,7-dihydroxy-5-methylchromone

G207W

Tetraketide
6-acetonyl-4-hydroxy-2-pyrone

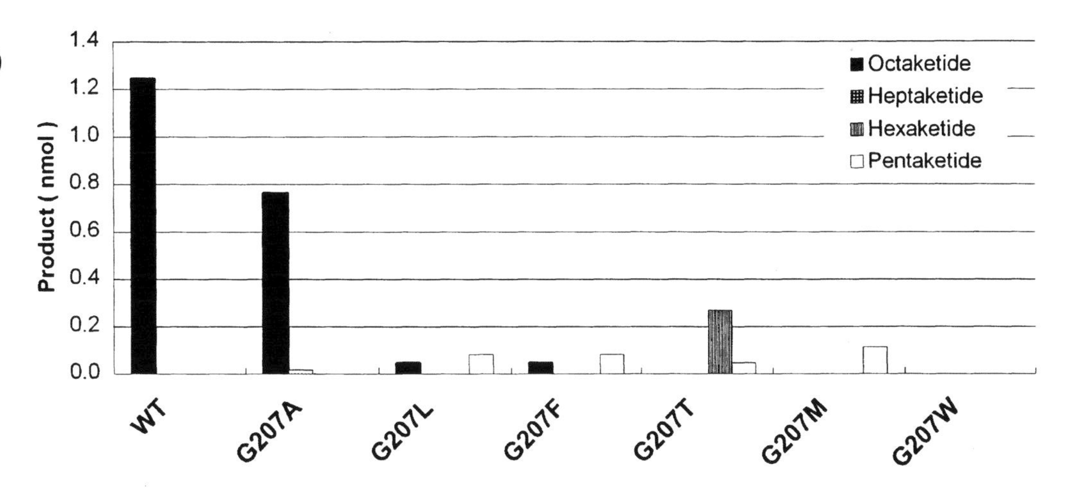

Figure 5. (A) Proposed mechanism for the formation of aromatic polyketides by OKS and its mutants. The enzymes catalyze chain initiation/elongation, and possibly initiating the first aromatic ring formation reaction at the methyl end of the polyketide intermediate. The partially cyclized intermediate are then released from the active-site and undergo spontaneous cyclizations, which leads to formation of the fused ring systems. (B) Distribution pattern of aromatic polyketides produced by OKS and its mutants. (Reproduced from reference 7b. Copyright 2005 American Chemical Society.)

other hand, the residue 207 lining the active-site cavity indeed occupies a crucial position for the polyketide chain elongation reactions. The residue is located at the entrance of a novel buried pocket that extends into the "floor" of the active-site cavity (Figure 6). The large-to-small M207G substitution widely opens the gate to the buried pocket, thus expanding a putative polyketide chain elongation tunnel, which lead to formation of the longer octaketides SEK4/SEK4b instead of the pentaketide. Interestingly, a similar active-site architecture has been also recently reported for a bacterial pentaketide-producing THNS from *S. coelicolor* (*5e*) but shares only ca. 20% amino acid sequence identity with the plant type III PKSs.

In addition to the above mentioned "downward expanding" active-site residue 197, the "horizontally restricting" G256L substitution (*5e*) (numbering in *M. sativa* CHS) in *A. arborescens* PCS and OKS would control the starter

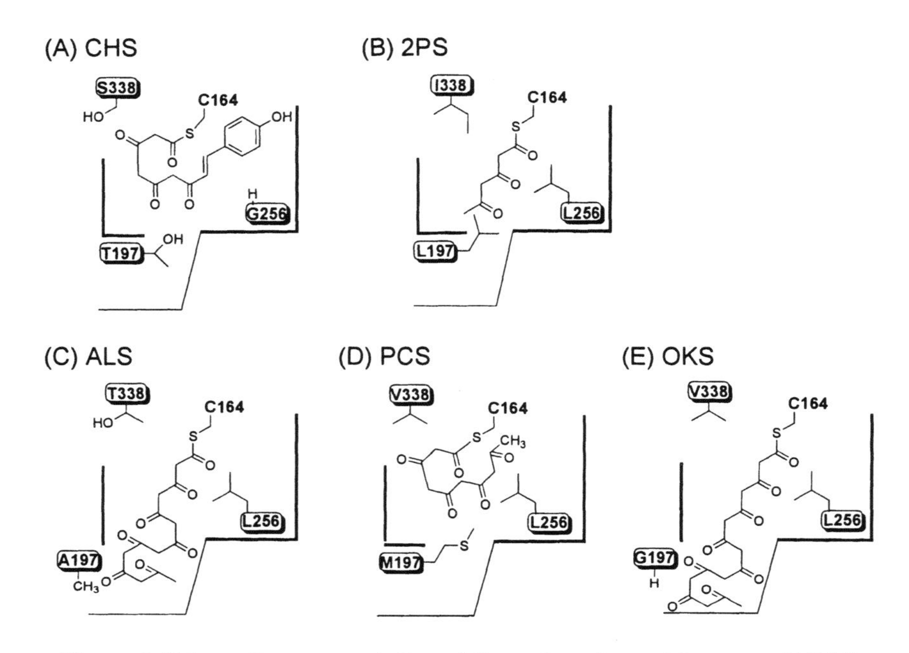

Figure 6. Schematic representation of the active-site architecture of (A) M. sativa *CHS, (B)* G. hybrida *2PS, (C)* R. palmatum *ALS, and (D)* A. arborescens *PCS, and (E)* A. arborescens *OKS (numbering in* M. sativa *CHS). The "horizontally restricting" G256L substitution controls the starter substrate selectivity, while the "downward expanding" substitution of T197A (ALS) and T197G (OKS) open a gate to a novel buried pocket that extends into the "floor" of the active-site cavity. On the other hand, the residue 338 located in proximity of the catalytic Cys164 at the "ceiling" of the active-site cavity guides the growing polyketide chain to extend into the buried pocket.*

substrate selectivity as in the case of the previously reported triketide-producing 2PS (*2b*) and the heptaketide-producing ALS (*14*). The small-to-large replacement relative to CHS would facilitate the enzyme to utilize the smaller malonyl-CoA (acetyl-CoA) starter instead of the bulky 4-coumaroyl-CoA while providing adequate volume for the longer chain polyketides formation (Figure 6). It has been recently demonstrated that when Leu256 of ALS was substituted with less bulky Gly, ALS L256G mutant became to accept 4-coumaroyl-CoA as a starter to efficiently produce the tetraketide CTAL, whereas wild-type ALS does not accept the coumaroyl starter at all (*14*). Finally, the residue 338 (numbering in *M. sativa* CHS) located in proximity of the catalytic Cys164 at the "ceiling" of the active-site cavity is also likely to play a crucial role in the polyketide chain elongation reactions (Figure 6) (*14,15*). Interestingly, the CHS active-site Ser338 is replaced with Val in both PCS and OKS. It is likely that the residue 338 guides the growing polyketide chain to extend into the buried pocket, thus expanding a putative polyketide elongation tunnel, thereby leading to formation of the longer polyketides. Very surprisingly, it was recently demonstrated that *S. baicalensis* CHS S338V mutant did produce a trace amount of SEK4/SEK4b in addition to the major product TAL (*15*). Moreover, the SEK4/SEK4b-forming activity was dramatically increased in an OKS-like CHS triple mutant (T197G/G256L/S338V). It is quite remarkable that even CHS could be engineered to produce octaketides by a single amino acid substitution.

In conclusion, *A. arborescens* PCS and OKS are novel plant-specific type III PKSs that produce pentaketide 5,7-dihydroxy-2-methylchromone and octaketides SEK4/SEK4b, respectively. In the two enzymes, the CHS active-site residues are uniquely replaced with T197M/G256L/S338V in PCS and T197G/G256L/S338V in OKS, which controls the substrate and the product specificities of the enzymes. Most importantly, a single residue 197 determines the polyketide chain length and the product specificity. The functional diversity of the type III PKSs was shown to be evolved from the simple steric modulation of the chemically inert residue lining the active-site cavity. These findings revolutionized our concept for the catalytic potential of the structurally simple type III PKSs, and suggest novel strategies for the engineered biosynthesis of pharmaceutically important plant polyketides.

References

1. For recent reviews, see: (a) Schröder, J. in *Comprehensive Natural Products Chemistry*, Vol. 2, pp. 749-771, Elsevier, Oxford, 1999. (b) Austin, M. B.; Noel, J. P. *Nat. Prod. Rep.* **2003**, *20*, 79-110. (c) Staunton, J.; Weissman, K. J. *Nat. Prod. Rep.* **2001**, *18*, 380-416.

2. (a) Eckermann, S.; Schröder, G.; Schmidt, J.; Strack, D.; Edrada, R. A.; Helariutta, Y.; Elomaa, P.; Kotilainen, M.; Kilpeläinen, I.; Proksch, P.; Teeri, T. H.; Schröder, J. *Nature* **1998**, *396*, 387-390. (b) Jez, J. M.; Austin, M. B.; Ferrer, J.; Bowman, M. E.; Schröder, J.; Noel, J. P. *Chem. Biol.* **2000**, *7*, 919-930.
3. (a) Ferrer, J. L.; Jez, J. M.; Bowman, M. E.; Dixon, R. A.; Noel, J. P. *Nat. Struct. Biol.* **1999**, *6*, 775-784. (b) Jez, J. M.; Ferrer, J. L.; Bowman, M. E.; Dixon, R. A.; Noel, J. P. *Biochemistry* **2000**, *39*, 890-902. (c) Jez, J. M.; Noel, J. P. *J. Biol. Chem.* **2000**, *275*, 39640-39646. (d) Jez, J. M.; Bowman, M. E.; Noel, J. P. *Biochemistry* **2001**, *40*, 14829-14838. (e) Tropf, S.; Kärcher, B.; Schröder, G.; Schröder, J. *J. Biol. Chem.* **1995**, *270*, 7922-7928. (f) Suh, D. Y.; Fukuma, K.; Kagami, J.; Yamazaki, Y.; Shibuya, M.; Ebizuka, Y.; Sankawa, U. *Biochem. J.* **2000**, *350*, 229-235. (g) Suh, D. Y.; Kagami, J.; Fukuma, K.; Sankawa, U. *Biochem. Biophys. Res. Commun.* **2000**, *275*, 725-730. (h) Jez, J. M.; Bowman, M. E.; Noel, J. P. *Proc. Natl. Acad. Sci. U S A* **2002**, *99*, 5319-5324. (i) Austin, M. B.; Bowman, M. E.; Ferrer, J.-L.; Schröder, J.; Noel, J. P. *Chem. Biol.* **2004**, *11*, 1179-1194.
4. (a) Abe, I.; Takahashi, Y.; Morita, H.; Noguchi, H. *Eur. J. Biochem.* **2001**, *268*, 3354-3359. (b) Abe, I.; Sano, Y.; Takahashi, Y.; Noguchi, H. *J. Biol. Chem.* **2003**, *278,* 25218-25226. (c) Abe, I.; Utsumi, Y.; Oguro, S.; Noguchi, H. *FEBS Lett.* **2004**, *562*, 171-176. (d) Abe, I.; Watanabe, T.; Noguchi, H. *Proc. Japan Acad., Ser. B.* **2005**, *81*, 434-440.
5. (a) Funa, N.; Ohnishi, Y.; Fujii, I.; Shibuya, M.; Ebizuka, Y.; Horinouchi, S. *Nature* **1999**, *400*, 897-899. (b) Funa, N.; Ohnishi, Y.; Ebizuka, Y.; Horinouchi, S. *J. Biol. Chem.* **2002**, *277*, 4628-4635. (c) Funa, N.; Ohnishi, Y.; Ebizuka, Y.; Horinouchi, S. *Biochem. J.* **2002**, *367*, 781-789. (d) Izumikawa, M.; Shipley, P. R.; Hopke, J. N.; O'Hare T.; Xiang, L.; Noel, J. P.; Moore, B. S. *J. Ind. Microbiol. Biotechnol.* **2003**, *30*, 510-515. (e) Austin, M. B.; Izumikawa, M.; Bowman, M. E.; Udwary, D. W.; Ferrer, J.-L.; Moore, B. S.; Noel, J. P. *J. Biol. Chem.* **2004**, *279*, 45162-45174.
6. (a) Saxena, P.; Yadav, G.; Mohanty, D.; Gokhale, R. S. *J. Biol. Chem.* **2004**, *278*, 44780-44790. (b) Sankaranarayanan, R.; Saxena, P.; Marathe, U.; Gokhale, R. S.; Shanmugam, V. M.; Rukmini, R. *Nat. Struct. Mol. Biol.* **2004**, *11*, 894-900. (c) Pfeifer, V.; Nicholson, G. J.; Ries, J.; Recktenwald, J.; Schefer, A. B.; Shawky, R. M.; Schröder, J.; Wohlleben, W.; Pelzer, S. *J. Biol. Chem.* **2001**, *276*, 38370-38377. (d) Tseng, C. C.; McLoughlin, S. M.; Kelleher, N. L.; Walsh, C. T. *Biochemistry* **2004**, *43*, 970-980.
7. (a) Abe, I.; Utsumi, Y.; Oguro, S.; Morita, H.; Sano, Y.; Noguchi, H. *J. Am. Chem. Soc.* **2005**, *127*, 1362-1363. (b) Abe, I.; Oguro, S.; Utsumi, Y.; Sano, Y.; Noguchi, H. *J. Am. Chem. Soc.* **2005**, *127*, 12709-12716.
8. Dewick, P. M. in *Medicinal Natural Products, A Biosynthetic Approach*, 2nd ed., Wiley, West Sussex, 2002.

9. (a) Fu, H.; Ebert-Khosla, S.; Hopwood, D. A.; Khosla, C. *J. Am. Chem. Soc.* **1994**, *116*, 4166-4170. (b) Fu, H.; Hopwood, D. A.; Khosla, C. *Chem. Biol.* **1994**, *1*, 205-210.
10. (a) Abe, I.; Morita, H.; Nomura, A.; Noguchi, H. *J. Am. Chem. Soc.* **2000**, *122*, 11242-11243. (b) Morita, H.; Takahashi, Y.; Noguchi, H.; Abe, I. *Biochem. Biophys. Res. Commun.* **2000**, *279*, 190-195. (c) Morita, H.; Noguchi, H.; Schröder, J.; Abe, I. *Eur. J. Biochem.* **2001**, *268*, 3759-3766. (d) Abe, I.; Takahashi, Y.; Noguchi, H. *Org. Lett.* **2002**, *4*, 3623-3626. (e) Abe, I.; Takahashi, Y.; Lou, W.; Noguchi, H. *Org. Lett.* **2003**, *5*, 1277-1280. (f) Abe, I.; Watanabe, T.; Noguchi, H. *Phytochemistry* **2004**, *65*, 2447-2453. (g) Oguro, S.; Akashi, T.; Ayabe, S.; Noguchi, H.; Abe, I. *Biochem. Biophys. Res. Commun.* **2004**, *325*, 561-567.
11. Suga, T. ; Hirata, T. *Bull. Chem. Soc. Jpn* **1978**, *51*, 872-877.
12. (a) Tang, Y.; Tsai, S.-C.; Khosla, C. *J. Am. Chem. Soc.* **2003**, *125*, 12708-12709. (b) Keatinge-Clay, A. T.; Maltby, D. A.; Medzihradszky, K. F.; Khosla, C.; Stroud, R. M. *Nat. Struct. Mol. Biol.* **2004**, *11*, 888-893.
13. Morita, H.; Kondo, M.; Oguro, S.; Noguchi, H.; Sugio, S.; Kohno, T.; Abe, I. manuscript in preparation.
14. Abe, I.; Watanabe, T.; Lou, W.; Noguchi, H. *FEBS J.* **2006**, *273*, 208-218.
15. Abe, I.; Watanabe, T.; Morita, H.; Kohno, T.; Noguchi, H. *Org. Lett.* **2006**, *8*, in press.

Chapter 9

Expression and Function of Aromatic Polyketide Synthase Genes in Raspberries (*Rubus idaeus* sp.)

Geza Hrazdina and Desen Zheng

Department of Food Science and Technology, Cornell University, NYSAES, Geneva, NY 14456

The genus *Rubus* contains a family of type III polyketide synthases which show differences in structure and function. We have cloned and characterized five aromatic polyketide synthase genes from raspberries. All five genes contain an intron of varying length and have 1173 bp coding sequences, with the exception of one gene that consists of 1149 bp. Four of the five genes encode proteins with 391 amino acid residues with a calculated protein mass of 42 kDa, while one gene coded for a shorter protein consisting of 383 amino acids. Sequence comparison of the five polyketide synthase genes showed high similarity both at the DNA and protein levels. Differences in the coding region were found mainly in the flanking sequences. Analysis of the reaction products showed that PKS1 and PKS5 were chalcone synthases, PKS2 that differs in six amino acids from PKS1 is silent, PKS3 is a *p*-coumarate triacetic acid lactone synthase (CTAS) and PKS4 is a benzalacetone synthase (BAS). The structural variations and the architecture of these PKS genes and enzymes is discussed.

Introduction

Type III polyketide synthases (PKS) are acyltransferases that transfer acyl groups from the corresponding CoA thioesters to form mono,-oligo- or polyketides. Unlike the type I and II polyketide synthases that occur widely in most higher organisms, fungi and bacteria, type III PKS's are found mainly in plants. One of the major differences between the three types is that type I and type II PKS's use acetyl-CoA as a starter substrate and add 2-carbon units from malonyl-CoA with the help of an acyl carrier protein (ACP). For reviews on the reaction mechanism of these proteins see articles by Asturias et al, (2005) and Hitchman et al., (1998). While type I and type II PKS's use ACP to transfer the C_2 units to the starter molecule or intermediate condensation product, type III PKS's load the starter substrate, usually an aromatic thioester, and malonyl-CoA for the chain extensions directly on the cysteine-phenylalanine pairs. The prototype of type III PKS's is chalcone synthase (CHS, E.C. 2.3.1.74) that condenses three acetate units deriving from malonyl-CoA with one molecule of *p*-coumaryl-CoA to form the C_{15} aromatic unit, naringenin chalcone. CHS was the first type III polyketide synthase reported (Kreuzaler and Hahlbrock, 1972), characterized (Hrazdina et al., 1976) and its gene cloned (Reimold et al., 1983).

Since these original reports many CHS genes and proteins have been reported in plants. As of the date of writing this chapter, 1221 reports appeared in the literature on CHS biochemistry and genetics, including x-ray crystallographic structure analysis (Jez et al., 2001). Presently 14 plant specific polyketide synthase type III enzymes/genes are known. Their names and reaction products are shown in Figure 1.

PKS Genes in Raspberries

The five raspberry PKS type III genes were cloned from raspberry genomic DNA by PCR based approaches (Zheng et al., 2001; Zheng and Hrazdina manuscript in preparation). Based on two highly conserved regions in plant type III PKS's, forward and reverse primers were designed and using RNA isolated from raspberry cell suspension cultures, followed by PCR amplification of the DNA fragments. Full sequence PKS genes were obtained by 3'and 5' extension of the fragments.

All five PKS genes showed high sequence homology, including the location of a single intron at Cys65. The size of the intron, however, varied in the individual genes (Figure 2). Intron length in PKS1 was 383 bp, in PKS2 and 3 it was 384 bp, in PKS4 was 711 bp and in PKS5 577 bp. PKS genes 1,2,3 and 5 coded for proteins having 391 amino acids, while PKS4 coded for a shorter protein composed of 383 amino acids. The identity between the five genes ranged from 85.1% to 99.4% (Table 1).

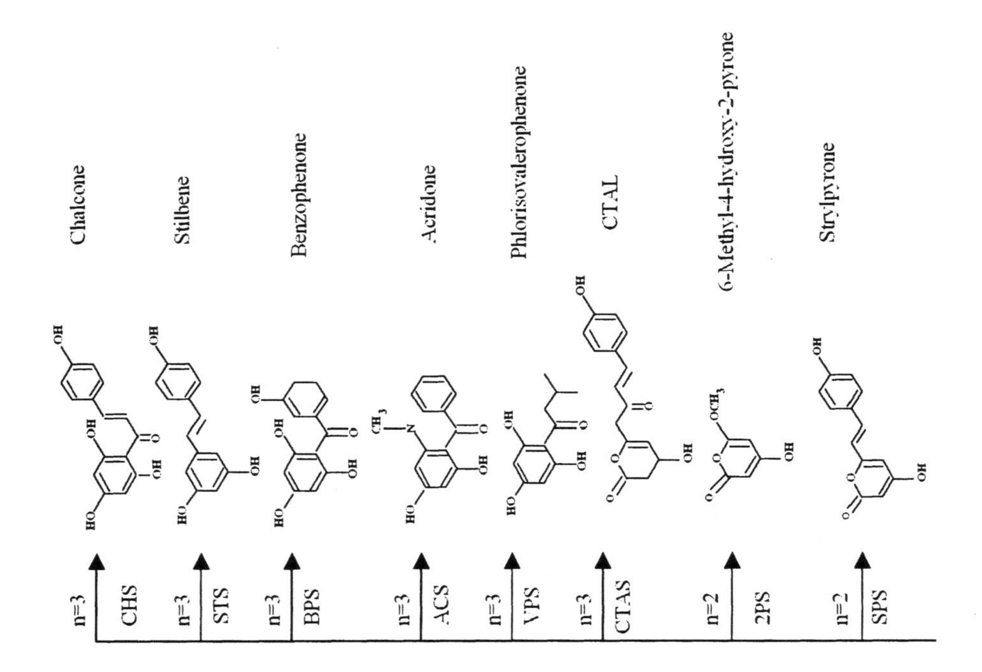

Chalcone
Stilbene
Benzophenone
Acridone
Phlorisovalerophenone
CTAL
6-Methyl-4-hydroxy-2-pyrone
Styrylpyrone
n=3
CHS
n=3
STS
n=3
BPS
n=3
ACS
n=3
VPS
n=3
CTAS
n=2
2PS
n=2
SPS
CoA-S-C-R

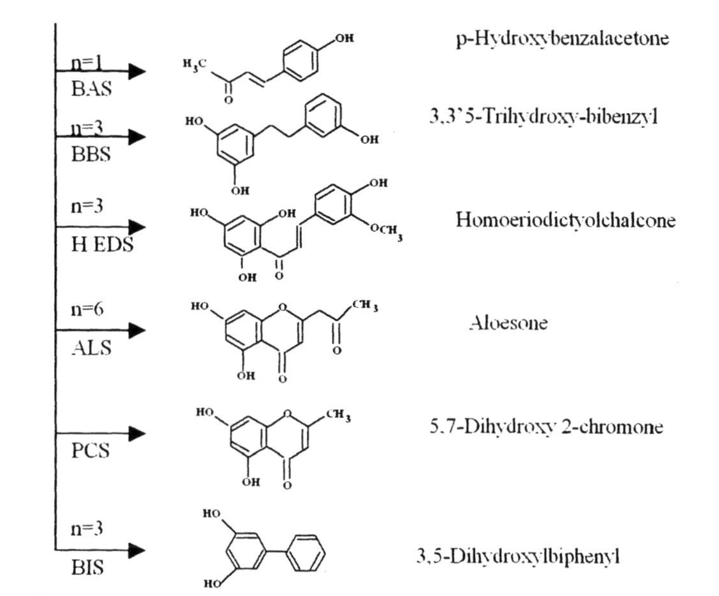

Figure 1. Type III polyketide synthases and their reaction products. CHS: chalcone synthase, STS: stilbene synthase; BPS: benzophenone synthase; ACS: acridone synthase; VPS: valerophenone synthase; CTAS: p-coumaryltriacetic acid synthase; 2PS: 2-pyrone synthase; SPS: styrylpyrone synthase; BAS: benzalacetone synthase; H/EDS: homoeryodyctiol/eriodyctiol chalcone synthase; ALS: Aloesone synthase; PCS: phenylchromone synthase; BIS: biphenyl synthase

Table 1. Features of the five polyketide synthase (PKS) genes and their identity of coding sequences and the deduced amino acid sequences shown in brackets

		PKS1	*PKS2*	*PKS3*	*PKS4*	*PKS5*
Gene features	Intron(bp)	383	384	384	711	577
	Exon1/2(bp)	178/998	178/998	178/998	178/974	178/998
	Deduced protein (aa)	391	391	391	383	391
Identity (%) DNA/AA	*PKS1*	—	99.4(98.2)	99.7(98.7)	85.7(86.2)	91.4(95.9)
	PKS2		---	99.7(99.2)	85.1(84.6)	90.7(94.4)
	PKS3			—	85.3(85.1)	91.0(94.9)
	PKS4				—	89.1(86.9)
	PKS5					—

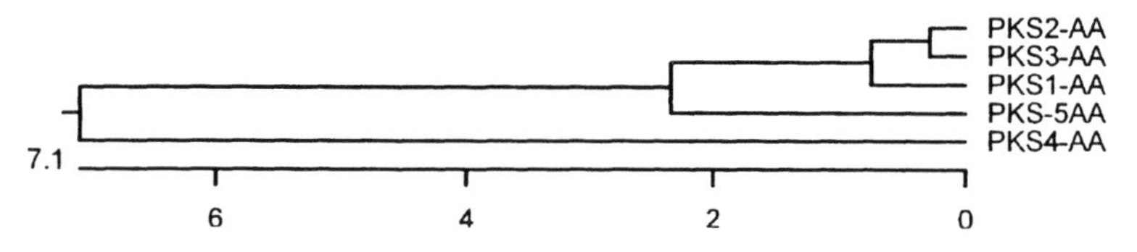

Alignment of the amino acid sequences with other PKS sequences reported in the data banks indicated that the PKS proteins in raspberry are very similar to CHS and CHS-like proteins from other plant species. Sequences with the highest similarity were from *Casuarina glauca* (Laplaze et al, 1999), *Citrus sinensis* (Springob et al., 2000), *Camellia sinensis* (Takeuchi et at., 1994) and *Juglans nigra* x*Juglans regia* cross (Cladot et al., 1997). Nearly all conserved residues in CHS were found in all five proteins, including the four amino acid residues that define the active site (Ferrer et al., 1999). A comparison with a CHS consensus sequence arrived at from twenty CHS sequences with demonstrated enzyme activity suggested that all five raspberry gene products conformed to the consensus with few exceptions. These exceptions were the K49R, M64R, P120L and V188A differences between PKS1 and PKS3. PKS2 showed two additional amino acid differences at R259H and F344L (Zheng et al., 2001). PKS4 and PKS5 showed a larger divergence from the other PKS genes.

Raspberries accumulate a variety of flavonoid and aroma compounds that derive from PKS type III reactions (Borejsza-Wysocki and Hrazdina, 1996, Zheng et al., 2001). CHS produces precursors for flavonol, flavandiol and anthocyanin synthesis. Benzalacetone synthase (BAS) carries out a single condensation step in the synthesis of *p*-hydroxybenzalacetone, the precursor of raspberry ketone, that is responsible for the characteristic aroma of raspberries. To understand the role of the type III PKS' during the ripening of fruits, we cloned and characterized five type III PKS genes/gene products and discuss their properties below.

All type III PKS's show high degree of sequence similarity that makes it presently impossible to assign function based on sequence alone. All PKS' are homodimeric proteins with a subunit size of 40-45 kDa. All carry out decarboxylation and condensation reactions, contain a single intron at the conserved site of Cys 65. Their reaction centers have been reported to contain Cys164, Phe254, His303 and Asn336 (Jez et al., 2001; Ferrer et al., 1999).

To determine the functional relationship between the five raspberry genes and their protein products, their coding regions were inserted into the expression vector pET-9a under the control of the T7 promoter and the proteins were expressed in *E. coli* cells. The crude extracts from the IPTG induced cells containing the individual genes showed the presence of strong bands with an approximate molecular mass of 43 kDa when analyzed by SDS-PAGE. An exception was PKS4 that produced only relatively low amounts of the soluble protein. Extensive efforts to produce a corresponding amount of the soluble protein to the other PKS proteins failed. All five proteins produced by the *E.coli* expression system reacted strongly with a anti-chalcone synthase antibody preparation (Hrazdina et al., 1986) on Western blots. Extracts from non-induced *E.coli* cells containing the empty vector, or expression constructs without induction showed no cross-reacting proteins.

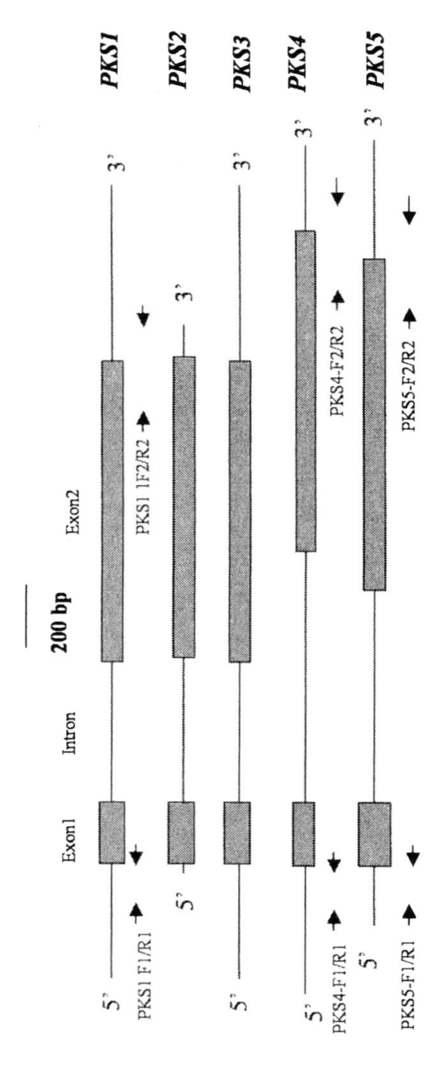

Figure 2. Structure of the five raspberry PKS genes.

To assay the activity of the expressed proteins (Hrazdina et al., 1976) CHS or CHS-like reaction products were extracted from the reaction mixtures, separated on paper chromatograms in the presence of authentic carrier substances, the chromatograms scanned for radioactivity, the radioactive sections of the chromatograms cut out and the amount of radioactivity determined by liquid scintillation spectrometry. Extraction of the reaction mixtures at high pH (8.8) gave naringenin (chemically cyclized from the chalcone) as the sole reaction product from PKS1,PKS3 and PKS5. PKS2 protein containing reaction mixtures showed no incorporation of radioactivity in any of the reaction products, indicating that the gene could be expressed in E. coli, but its protein product was silent. The reaction product of the PKS4 protein was identified as *p*-hydroxybenzalacetone, the precursor of the raspberry aroma compound, *p*-hydroxyphenylbutanone (raspberry ketone) (Borejsza-Wysocki and Hrazdina, 1996).

A number of reaction products were separated on paper chromatograms from the acidic (pH 2.8) extraction of the reaction mixtures, which were identified by LC/MS. The radioactive compound at the front of the chromatogram was malonic acid, liberated by the thioesterase activity from malonyl-CoA (Helariutta et al., 1995). Another radioactive compound was identified in the PKS4 reaction mixture as *p*-hydroxybenzalacetone, the condensation product of *p*-coumarate with a single malonate (Borejsza-Wysocki and Hrazdina, 1996). A major reaction product from PKS3 reaction mixture was identified as *p*-coumaryltriacetic acid lactone (Akiyama et al., 1999). Comparison of the radiochromatograms of the reaction products from the five PKS proteins identified the five raspberry PKS genes as follows: PKS1 and PKS5 are CHS genes, the PKS2 gene codes for a silent protein with no detectable enzyme activity under the assay conditions, PKS3 as a *p*-coumaryltriacetic acid syntase (CTAS), and PKS4 as benzalacetone synthase (BAS).

Structural Architecture of the PKS Genes

We have been intrigued by PKS2, the gene that differs by two bases from PKS1, resulting in R259H and F344L mutations and producing a protein with no detectable enzyme activity. By mutational analysis we have exchanged in the PKS1 gene, CHS, R259 to H and F344 to L, expressed this mutant in *E. coli* and determined the activity of the expressed protein. The R259H and F344L mutations of PKS1, a CHS with demonstrated activity, completely eliminated enzyme activity in the mutated protein. This is rather unique, because both amino acid positions (i.e. 259, 344) are peripheral and are not in the active pocket region of the enzyme (Figure 3). Modeling CHS1 with the POW-RAY

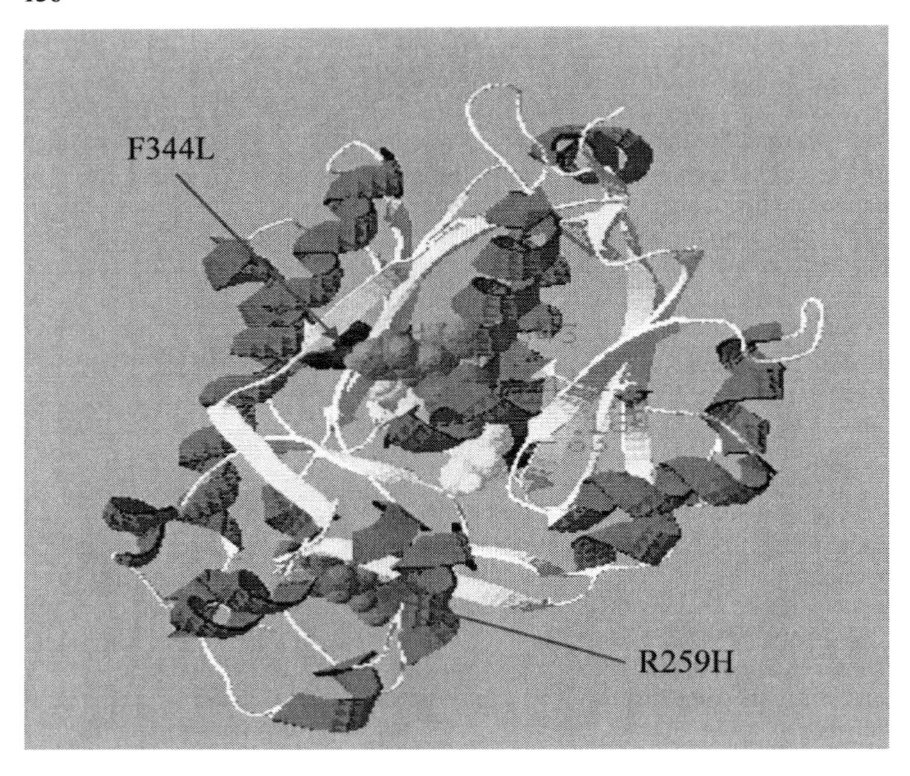

Figure 3: Location of R259H(purple) and F344L(navy) in the PKS2 protein. The F344L mutation is in the spatial proximity of the reaction center, while the R259H mutation is on the β1d strand. Strands β1d and β2d are lining the reaction pocket.

and RASMOL programs clearly indicated that positions 259 and 344 are not within the active center of CHS, but are rather at the periphery of the enzyme. Therefore, we do not expect direct interaction of these amino acids with the substrate loading /condensation mechanism. It is likely that mutation of these amino acids interferes either with the correct folding of the protein, or with the substrate accessibility of the enzyme.

Previously it was expected that only the modification of amino acids in the immediate vicinity of the active pocket has major consequences for the enzyme's activity. Our data with the raspberry PKS2 indicate that modification of amino acids at peripheral locations of the enzyme can also disrupt enzyme activity. A similar conclusion was arrived at with another CHS from *Matthiola incana* that had a single nucleotide mutation in a triplet AGG coding for a conserved arginine into AGT coding for serine (R72S) (Hemleben et al., 2005). The R72S mutation has also completely eliminated activity of the *Matthiola* CHS, apparently by interfering with the correct folding of the protein, or with the channeling of the substrate to the reaction center of the enzyme.

Further analysis of the raspberry PKS genes has shown that the proteins coded by the five genes have the seven conserved cysteines at positions 30,60,84,130,164,190 and 341. Cysteines at these conserved locations are likely to have structural functions, because in addition to the CHS' from which the consensus sequence derives, they were found at identical or similar locations in *Arachis hypogea* STS, in the *Gerbera hybrida* 2-pyrone synthase (2PS), in the *Ruta graveolens* acridone synthase (ACS), the *Rheum palmatum* benzalacetone synthase (BAS) and in the aleosone synthase (ALS) (Abe et al., 2004).

From these seven cysteines, two are located in the active site pocket that has been defined by the Cys164, Phe 215, His 303 and Asn336 tetrad (Jez et al., 1999). Both of these cysteines are paired with phenylalanine, e.g. Phe129-Cys 130 and Cys164-Phe 165. Phe 129-Cys130 is on the ascending ß4 strand of the protein, while Cys164-Phe165 are on the descending loop between the ß5 and α-7 sheet, positioning the two cysteines in proximity of each other (Figure 4). Cys164 has been elucidated as the attachment and decarboxylation site for malonyl-CoA (Jez et al., 1999), however, Cys130's role is not quite clear. It is likely that it may be responsible for the binding of *p*-coumaryl-CoA. In an earlier mutational analysis experiment (Lanz et al., 1991) it was reported that only Cys164 has an essential function in substrate binding and condensation of malonyl-CoA with the starter substrate *p*-coumaryl-CoA. Although mutation in the surrounding amino acid His1612Gln in STS and Gln161His in CHS showed reduced activity both in the absence and presence of cerulenin, the role of the immediate neighboring Phe was not investigated at either positions 139 or 165.

Detailed investigations on the reaction mechanism of plant type III polyketide synthases (Jez et al., 2000) have indicated that Cys164 serves as the single nucleophile for polyketide formation, but is not essential for the decarboxylation reaction of malonyl-CoA. Decarboxylation of malonate took place when Cys164 was mutated to alanine. The presently accepted working hypothesis for the CHS reaction mechanism suggests that *p*-coumaryl-CoA binds to Cys164 as a thioester, transferring *p*-coumarate to this amino acid. Malonyl – CoA then binds and positions the bridging cabon of malonate near the carbonyl of the enzyme bound *p*-coumaryl thioester. Asn336, His303 and Phe215 then orient the thioester carbonyl of malonyl-CoA for decarboxylation. Asn336 and His303 then interact with malonyl-CoA's thioester carbonyl, creating an electron sink that stabilizes the enol tautomere of the acetyl group. The enzyme bound

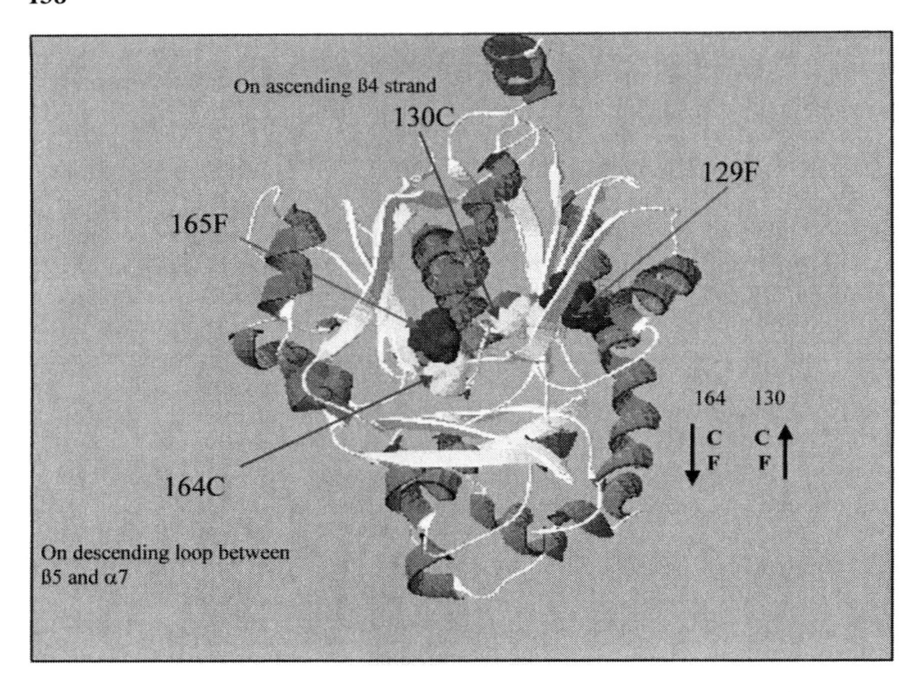

Figure 4. Location of the F129-C130 and C164-F165 pairs within the active pocket site of CHS1.

p-coumaryl thioester then releases the thiolate group of Cys164 and transfers *p*-coumarate to the tautomerized acetyl-CoA thioester. Cys164 than recaptures the elongated *p*-coumarylacetyldiketide, and releases CoA. The so formed *p*-coumarylacetyldiketide then is ready for further elongation, until the *p*-coumaryltriacetate fills the active pocket site and no further elongation is possible. This model is based on the early structure of type II ß-ketoacyl synthases that were believed to have the same four amino acids at similar locations participating in the condensation/chain elongation reactions. Recent crystallographic data which are discussed by Smith et al., (2003) show that the terminal carbon released from malonyl-CoA of the acyl-S-Cys161 is not in the form of CO_2, but as bicarbonate, (see also Witkowski et al., 2002) and that not one, but two histidines (His293, His331) are required for the decarboxylation reaction. According to these studies the active center of type I and type II polyketide synthases is defined by Cys161, His293, His 331, Lys326, Gly394 and Phe395. Mutational analysis of a bacteria type III polyketide synthase has also shown that mutation of Cys160, the catalytic nucleophile, resulted in a protein that was still able to carry out the regioselective construction of an eight carbon dihydroxyphenylacetyl-CoA skeleton with surprising efficiency (Tseng et al., 2004).

Future Prospects

Since the mutational analysis of the cysteines within CHS was carried out at a relatively early time (Lanz et al., 1991) when the available tools were not as refined as those of the present, it seems to be an opportune time to repeat those early experiments on the amino acids defining the catalytic pocket of type III polyketide synthases, including the role of the Phe129-Cys130 and Cys164-Phe165 amino acid pairs. Since the five PKS genes and proteins in raspberries showed considerable differences in their structure and function, we may encounter further surprises during the continuation of our investigation of the PKS genes in *Rubus*.

Acknowledgement

The authors thank Chris Dana, VPI, for rendering the PKS genes.

References

Abe, I.; Utsumi, Y.; Oguro, S.; Noguchi, H. (2004) FEBS Lett. 562: 171-176.

Akiyama, T.; Shibuya M.; Liu H.M.; Ebizuka, Y. (1999) Eur. J. Biochem. 263: 834-839.

Asturias, F.J.; Chadick, J.Z.; Cheung, I.K.; Stark, H.; Witkowski, A.; Joshi, A.K.; Smith, S. (2005) Nature structural & Molecular Biology 12:225-232.

Borejsza-Wysocki, W.; Hrazdina, G. (1996) Plant Physiol. 110: 791-799.

Claudot, A.C.; Ernst, D.; Sandermann, H.; (1997) Planta 203: 275-282.

Ferrer, J.L.; Jez, J.M.; Bowman, M.E.; Dixon, R.A.; Noel, J.P.; (1999) Nature Struct. Biol. 6:775-784.

Helariutta, Y.; Elomaa, P.; Kotilainen, M.; Griesbach, R.J.; Schröder, J.; Teeri T.H. (1995) Plant Mol. Biol. 28:47-60.

Hemleben, V.; Dressel, A.; Epping, B.; Lukacin, R.; Martens, S.; Austin, M.B. (2005) Plant Mol. Biol. 55: 455-465.

Hrazdina, G.; Kreuzaler, F.; Hahlbrock, K.; Grisebach, H. (1976) Arch. Biochem. Biophys. 175: 392-399.

Hrazdina, G.; Lifson, E.; Weeden, N.F. (1986) Arch. Biochem. Biophys. 247: 414-419

Jez, J.M.; Ferrer, J.L.; Bowman, M.E.; Dixon, R.A.; Noel, J.P. (2000) Biochemistry 39: 890-902.

Kreuzaler, F.; Hahlbrock, K. (1972) FEBS Letters 28: 69.

Lanz, T.; Tropf, S.; Marner, F-J.; Schröder, J.; Schröder, G. (1991) J. Biol. Chem. 266: 9971-9976.

Laplaze, L.; Gherbi, H.; Frutz, T.; Pawlowski, K.; Franche, C.; Macheix, J.J.; Auguy, F.; Bogusz, D.; Duhoux, E. (1999) Plant Physiol. 121: 113-122.

Reimold, U.; Kroger, M.; Kreuzaler, F.; Hahlbrock, K. (1983) EMBO J 2 : 1801-1805

Smith, S.; Witkowski, A.; Joshi, A.K. (2003) Progr. Lipid Res. 42: 289-317.

Springob, K.; Lukacin, R.; Ernwein, C.; Groning, I.; Matern, U. (2000) Eur. J. Biochem. 267: 6552-6559.

Takeuchi, A.; Matsumoto, S.; Hayatsu, M. (1994) Plant Cell Physiol. 35: 1011-1018.

Tseng, C.C.; McLoughlin, S.M.; Kelleher, N.L.; Walsh, C.T. (2004) Biochemistry 43: 970-980.

Witkowski, A.; Joshi, A.K.; Smith, S. (2002) Biochemistry 41: 10877-10887.

Zheng, D.; Schröder, G.; Schröder, J.; Hrazdina, G. (2001) Plant Mol. Biol. 46: 1-15.

Chapter 10

Molecular and Biochemical Characterization of Novel Polyketide Synthases Likely to Be Involved in the Biosynthesis of Sorgoleone

Daniel Cook[1], Franck E. Dayan[1], Agnes M. Rimando[1], N. P. Dhammika Nanayakkara[2], Zhiqiang Pan[1], Stephen O. Duke[1], and Scott R. Baerson[1]

[1]National Products Utilization Research Unit, Agricultural Research Service, U.S. Department of Agriculture, P.O. Box 8048, University, MS 38677
[2]National Center for Natural Products Research, School of Pharmacy, University of Mississippi, University, MS 38677

Sorgoleone, an oily exudate secreted from the root hairs of sorghum (*Sorghum bicolor* (L.) Moench), acts as a potent allelochemical. Its phytotoxic properties make the elucidation of the biosynthetic enzymes participating in this pathway desirable. Previous studies suggest that the biosynthetic pathway of sorgoleone involves a polyketide synthase as well as a fatty acid desaturase, an *O*-methyl transferase, and a cytochrome P450 monooxygenase. This polyketide synthase is proposed to use a novel long chain fatty acyl-CoA (C16:3) as a starter unit and catalyzes three iterative condensation reactions using malonyl-CoA to form a transient linear tetraketide that cyclizes to form a pentadecatriene resorcinol. To identify the polyketide synthase gene(s) involved in the biosynthesis of these alkylresorcinols, a root-hair specific EST (expressed sequence tag) collection was mined for potential candidates. A total of nine polyketide synthase-like EST's were identified representing five unique contigs, three of which were preferentially expressed in root hairs. The molecular and biochemical characterization of these three candidate polyketide synthases are presented, two of which represent a novel type of type III plant polyketide synthase.

Natural products represent a diversity of chemical compounds with varied biological activities. Natural products are an important source of novel pharmaceuticals as well as agricultural pesticides *(1,2)*. Natural products are derived from a number of pathways that create basic scaffolds that are further modified by various tailoring enzymes to create the wide diversity of structures that exist in nature. Polyketide synthases are responsible for the synthesis of an array of natural products including antibiotics such as erythromycin in bacteria *(3)* and mycotoxins such as aflatoxin in fungi *(4)*. Furthermore, in plants they are part of the biosynthetic machinery of flavonoids, phytoalexins, and phenolic lipids *(5,6)*.

Phenolic lipids represent an interesting group of natural products with varied bioactivities and uses, which are often derived from a polyketide pathway *(5,6)*. Examples of phenolic lipids are shown in Figure 1 and include urushiol, an allergen from poison ivy, anacardic acid, an anti-feedant, alkylresorcinols from various grasses possessing antifungal activity, and sorgoleone, an allelochemical from sorghum. In addition, phenolic lipids are important for the synthesis of formaldehyde-based polymers used in the automobile industry and are important in some countries in the lacquering process *(6)*.

Sorgoleone (2-hydroxy-5-methoxy-3-[(*Z,Z*)-8′,11′,14′-pentadecatriene]-*p*-benzoquinone (Figure 1), a phenolic lipid of particular interest in plant chemical ecology, is an allelochemical produced by several *Sorghum* species *(7)*. Allelochemicals can be defined as compounds produced by one plant that limits the growth of another plant *(8)*. Allelopathic compounds have been characterized in a number of plants other than sorghum, including spotted knapweed and rice *(8,9)*. Sorgoleone refers to a group of structurally related benzoquinones having a hydroxy and a methoxy substitution at positions 2 and 5, respectively; and having a 15- or 17-carbon chain with 1, 2, or 3 double bonds. These compounds were first isolated from hydrophobic root exudates of *Sorghum bicolor* and are distinguished from one another on the basis of their molecular weights *(10,11,12,13)*. The 2-hydroxy-5-methoxy-3-[(*Z,Z*)-8′,11′,14′-pentadecatriene]-*p*-benzoquinone (Figure 1), in addition to the other sorgoleone congeners, accounts for more than 90% of the root exudates of *Sorghum (11)* that are released into the soil where they repress the growth of other plants present in their surroundings. The biosynthetic pathway of sorgoleone requires four types of enzymes: fatty acid desaturases, polyketide synthases, *O*-methyl transferases, and cytochrome P450 monooxygenases as shown in Figure 2 *(14,15)*. A type III polyketide synthase catalyzes a crucial step in the formation of the pentadecatriene resorcinol intermediate in sorgoleone biosynthesis. This phenolic lipid can be subsequently modified by various tailoring enzymes, such as *O*-methyl transferases, cytochrome P450's, and glycosyltransferases.

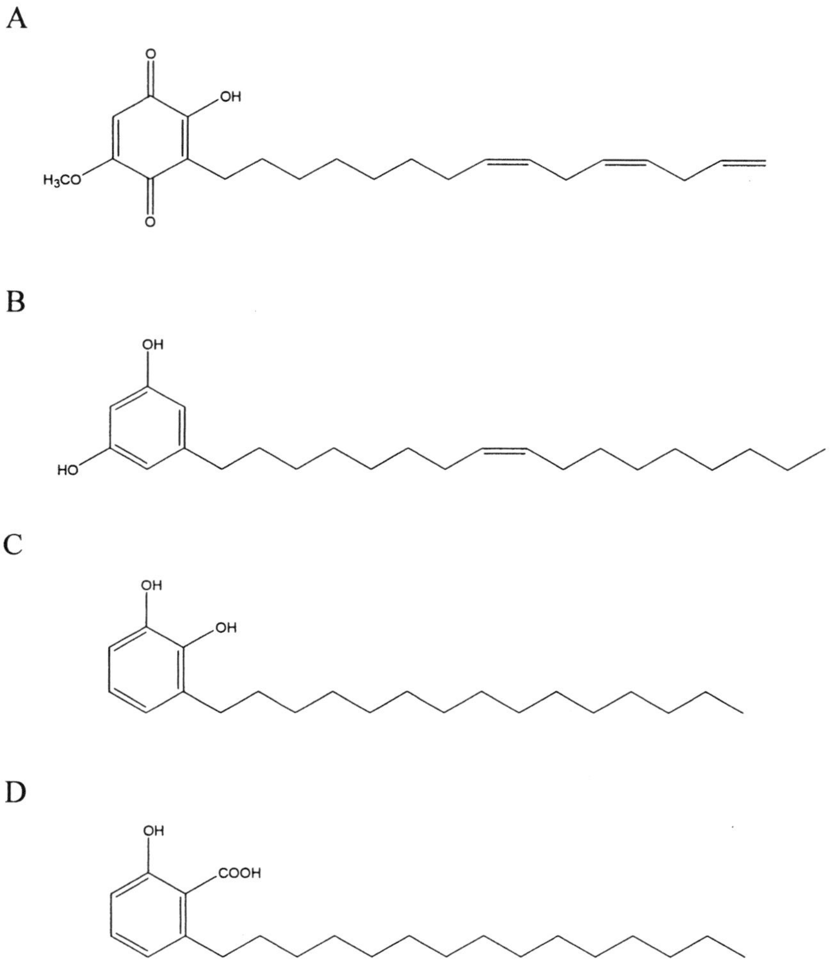

Figure 1. Chemical structures of various phenolic lipids (A) sorgoleone, (B) heptadecenyl resorcinol, (C) urushiol, and (D) anacardic acid.

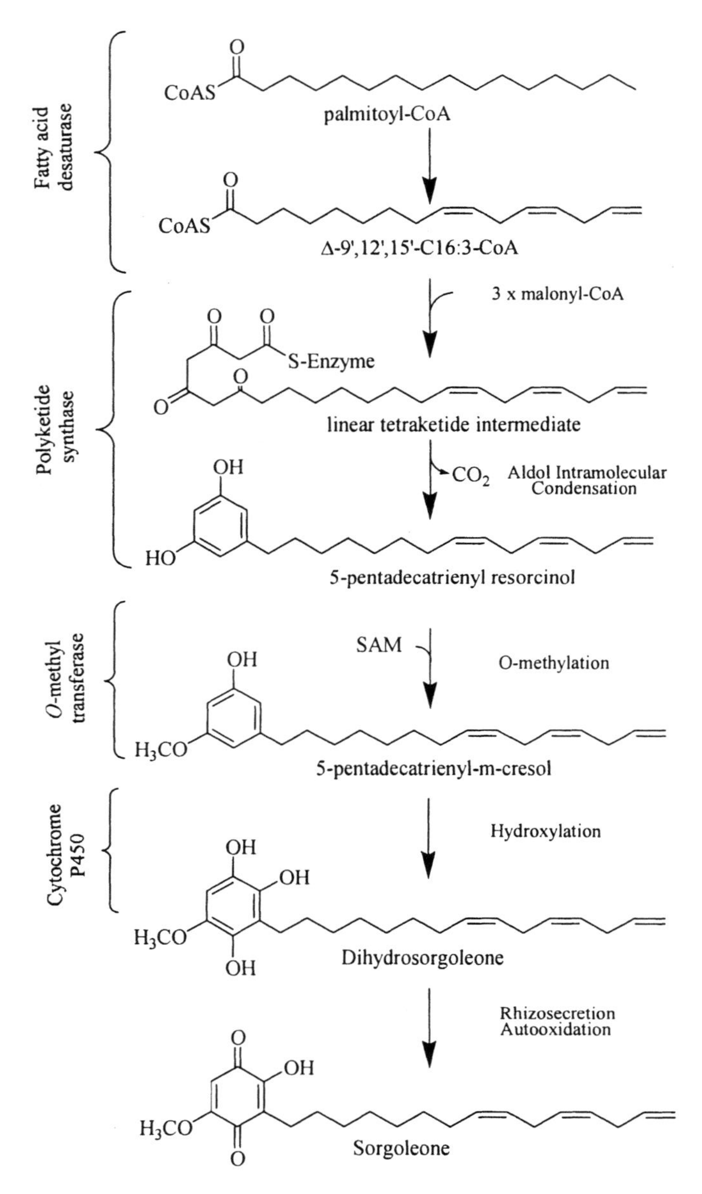

Figure 2. Proposed biosynthetic pathway for sorgoleone. The hydroquinone, produced in vivo, is thought to undergo autooxidation once secreted into the rhizophere to yield the more stable benzoquinone, sorgoleone.

Materials and Methods

cDNA Library Construction and EST Data Analysis

Sorghum bicolor (cv. BTX623) were grown for 5-7 days on a capillary mat system as described previously by Czarnota et al. *(16)*. Root hairs were then isolated using the method of Bucher et al. *(17)*, and total RNAs were isolated using the Trizol reagent (Invitrogen Corporation, Carlsbad, CA) per manufacturer's instructions. Tissue disruption was performed with a hand-held homogenizer at 25,000 rpm. RNA purity was determined spectrophotometrically, and integrity was assessed by agarose gel electrophoresis. Poly-A+ mRNAs were prepared using an Oligotex mRNA Midi Kit (Qiagen, Valencia, CA), then used for the construction of the cDNA library with a Uni-Zap XR cDNA library construction kit (Stratagene, La Jolla, CA). Mass excision of the primary library was performed to generate phagemid clones, which were then randomly chosen for sequencing via the University of Georgia, Laboratory for Genomics and Bioinformatics wet-lab pipeline *(18)*. Raw sequence traces were then filtered for quality control and elimination of contaminating vector sequences via an automated bioinformatics pipeline developed at the University of Georgia *(18)*. Database mining was performed using the Magic Gene Discovery software *(19)*, and by BLASTN and TBLASTN analysis *(20)*.

Real-Time RT-PCR

Real-time PCR was performed by the method of Baerson et al. *(21)* with the following modifications. Total RNAs were prepared using Trizol as described above, then re-purified using an RNeasy Midi Kit (Qiagen, Valencia, CA), including an "on-column" DNase I treatment to remove residual DNA contamination. Real-time PCR was performed in two biological replicates (i.e., two RNA samples from different plants, with three PCR reactions on each RNA sample) for each tissue using an ABI PRISM™ 5700 Sequence Detector (Applied Biosystems, Foster City, CA) with gene-specific primers corresponding to the candidate polyketide synthases and primers specific to 18S rRNA (Forward, 5′- GGCTCGAAGACGATCAGATACC-3′; reverse, 5′-TCGGCATCGTTT ATGGTT- 3′). Gene specific primers for the five candidate polyketide synthases were designed.

Heterologous Expression of Recombinant Polyketide Synthases

Full-length coding sequences were generated using the SMART RACE cDNA Amplification Kit (Clontech Laboratories Inc., Palo Alto, CA) per manufacturer's instructions from RNA prepared from root hairs (described above). PCR products containing complete open reading frames flanked by appropriate restriction sites, were then generated and directly cloned into pET15b (EMD Biosciences, La Jolla, CA), in-frame with a poly-histidine tract and thrombin cleavage site. The resulting *E. coli* expression vectors were transformed into strain BL21/DE3 (EMD Biosciences, La Jolla, CA) for recombinant enzyme studies.

For recombinant protein production, *E. coli* cultures were grown at 37°C to an optical density of 0.6 at 600 nm, then induced with 0.5 mM IPTG and allowed to grow 5 additional hours at 25°C. Cells were harvested by centrifugation at approximately 3000 x g for 20 min at 4°C, washed with cold 0.9% NaCl, then collected by re-centrifugation at 3000 x g. Pellets were resuspended in cold lysis buffer (100 mM potassium phosphate, pH 7.0, 1 M NaCl, 5 mM imidazole, 10% glycerol, 1 μg/mL leupeptin) and extracted using a French Press at a pressure of 1500 p.s.i. Benzonase (25 U/mL) and 1 mM PMSF were added immediately to the lysate. After 15 min incubation at room temperature, the lysate was centrifuged at 15,000 x g for 20 min, and supernatant was loaded onto a Ni-column activated with 2 mL of 0.1 M $NiSO_4$ and washed with 10 mL of distilled water. The Ni-column was previously equilibrated with 10 mL buffer A (100 mM potassium phosphate, pH 7.0, 500 mM NaCl, 5 mM imidazole). The column was washed with 3.5 mL buffer A between each 2 ml of supernatant. Once all of the sample was loaded, the column was washed with 8 mL of buffer A followed with 8 mL of buffer B (100 mM potassium phosphate, pH 7.0, 500 mM NaCl, 50 mM imidazole). Recombinant polyketide synthases were then eluted with 2.5 ml of elution buffer (100 mM potassium phosphate, pH 7.0, 500 mM NaCl, 250 mM imidazole). The recombinant protein-containing fraction (250 mM imidazole) was desalted on a PD-10 column equilibrated with cold desalting buffer (100 mM potassium phosphate, pH 7.0, 10 mM DTT, 10% glycerol). Protein concentrations were determined using a Bio-Rad protein assay kit (Bio-Rad Laboratories, Hercules, CA). Enzyme preparations were stored at –80°C prior to use.

Polyketide Synthase Enzyme Assays

Polyketide synthase enzyme assays contained 100 mM potassium phosphate buffer (pH 7.0), 40 μM malonyl-CoA, 25 μM starter molecule (i.e. palmitoyl-CoA), and 2 μg protein in a 200 μL volume at 30° C for 30 minutes. Reactions were quenched by addition of 10 μL of 20% HCl. The lipid resorcinol products were extracted by phase partitioning with 1 ml of ethyl acetate. The organic phase (upper layer) obtained by centrifugation at ~14,000 x g for 1 minute was transferred to a fresh tube, dried under vacuum, and subsequently analyzed by GC/mass spectrometry as a trimethysilyl derivative. Product formation was quantified using selected ion monitoring at *m/z* 268, a common fragment to all lipid resorcinols. The identity of the product formed was verified by their retention times and mass specta relative to authentic standards for pentadecylresorcinol (Chem Services) and olivetol (Sigma).

Results and Discussion

The proposed pathway for sorgoleone biosynthesis *(15)* as shown in Figure 2 indicates that a minimum of four different classes of enzymes would be required for its biosynthesis starting with a saturated acyl-CoA fatty acid (Figure 2). A targeted approach for obtaining the candidate sequences representing fatty acid desaturases, polyketide synthases, *O*-methyltransferases, and cytochrome P450s from *Sorghum bicolor* was initiated. The allelochemical sorgoleone is thought to be produced exclusively in the root hairs, thus most if not all the biosynthetic enzymes and their corresponding mRNAs are likely to be abundant in this tissue type *(16)*. On this basis, a functional genomics approach was developed to identify and characterize the biosynthetic enzymes of sorgoleone in sorghum root hairs. An expressed sequence tag (EST) analysis was selected as a gene isolation strategy, as this approach is ideally suited for profiling the more abundant transcripts in a specific cell or tissue type *(22)*. A cDNA library was constructed from RNA prepared from root hairs, from which an EST database was generated (see Materials and Methods) for subsequent data mining.

The EST data was mined for candidate polyketide synthases using both the Magic Gene Discovery software *(19)*, and also analyzed by BLAST searches using functionally characterized plant type III polyketide synthase sequences as queries against the EST dataset translated in all possible reading frames. From these analyses, 9 polyketide synthase-like EST's were identified in the root hair EST data set (Table I). Furthermore, candidate EST's were identified for all the other enzymes proposed to be part of the sorgoleone biosynthetic pathway (data not shown).

A secondary screen using real-time PCR was used to identify candidate polyketide synthases whose expression pattern correlated with the accumulation of sorgoleone. Gene-specific primers were designed for the 5 contigs and singletons representing the polyketide synthases (Table I). Real time PCR was

Table I. Summary of polyketide synthase candidate sequences identified in *S. bicolor* root hair EST database

Family	*Clones*	*Contigs*	*Singletons*	*%,Total*	*Root hair-specific*
Polyketide synthase	9	2	3	0.165	3

[a]Three sequences exhibiting root hair-specific expression were identified from a subset of the 5 polyketide synthase-like contigs and singletons.

used to assay expression levels of the candidate genes in the different tissue types from sorghum including mature leaves, immature leaves, panicles, apices, mature stems, roots, and root hairs. Three candidate polyketide synthase genes were identified having their highest expression levels in root hairs (Table I). Full length coding sequences were generated by PCR (see Materials and Methods) and subcloned into *E. coli* expression vectors.

Full-length open reading frames for all three candidates were overexpressed in *E. coli*, and purified using an activated Ni-column (see Materials and Methods). Acyl CoA's varying in length and saturation were tested in enzyme assays with all three recombinant enzymes to determine their substrate specificty (see Materials and Methods) (Figure 3). Furthermore benzyl CoA was tested as a substrate representing a different functional group (Figure 3).

Most type III plant polyketide synthases characterized thus far utilize relatively small starter units, such as coumaryl Co-A *(5)*. Significantly, two of the three polyketide synthase candidates identified from sorghum root hairs, preferentially utilized long chain acyl CoA substrates. Furthermore, $\Delta 9',12',15'$-hexadecatrienoyl CoA, the major *in vivo* substrate for the polyketide synthase involved in sorgoleone biosynthesis, is also used by these same two enzymes. It is notable that these two polyketide synthases do not utilize the smaller substrates, suggesting that these enzymes belong to a new class of polyketide synthases dedicated to the synthesis of phenolic lipids. The third candidate polyketide synthase showed no activity towards any of the substrates tested. The predicted protein sequences of these two active polyketide synthases exhibit a high degree of similarity to a small family of putative polyketide synthases from rice whose function have yet to be established (Table II). Since rice is also known to synthesize phenolic lipids, this small family of polyketide synthases may be responsible for their biosynthesis. In addition, the protein sequence contains the consensus amino acids that define a type III polyketide synthase. Further studies such as over-expression and RNAi with transgenic sorghum plants will be pursued in future efforts, which may provide further evidence that these polyketide synthases catalyze a key step in the sorgoleone biosynthetic pathway.

Summary and Conclusions

The combination of a root hair specific EST approach and expression analysis was an effective strategy for isolating candidate polyketide synthases potentially involved in sorgoleone biosynthesis. As a result of these efforts, two novel type III polyketide synthases have been identified that preferentially use long chain acyl Co-A's and are potentially involved in sorgoleone biosynthesis. These candidate polyketide synthases can form pentadecatriene resorcinol, an intermediate in sorgoleone biosynthesis. Furthermore, these efforts may aid in the identification of other polyketide synthases responsible for the biosynthesis of phenolic lipids in other plant species.

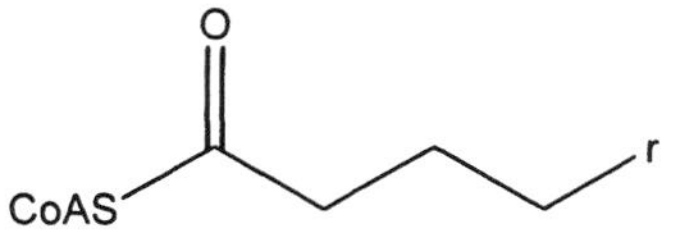

r:d alkyl extension (number of carbon atoms:number of double bonds)=0:0 (butryl), 2:0 (hexanoyl), 4:0 (octanoyl), 6:0 (decanoyl), 8:0 (lauroyl), 10:0 (myristoyl), 12:0 (palmitoyl), 12:1 (palmitoleoyl), 12:3 (hexadecatrienoyl), 14:0 (stearoyl), 14:1 (oleoyl), 14:2 (linoleoyl), 16:0 (arachidoyl), 16:4 (arachidonoyl)

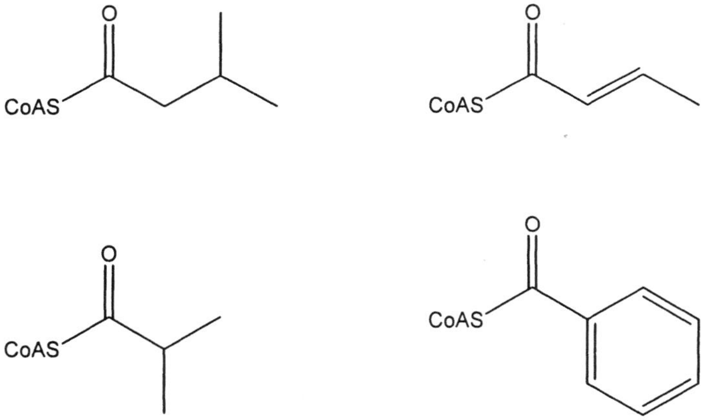

Figure 3. Polyketide synthase substrate specificity. Structures are shown for some of the substrates used for determining specificity of recombinant S. bicolor polyketide synthases.

Table II. Top 10 BLASTP results for clone using the NCBI non-redundant (nr) peptide sequence database

Description	[a]*Score*	[b]*E-value*
putative chalcone synthase [*Oryza sativa* (japonica cultivar-group)]	475	1e-132
putative chalcone synthase [*Oryza sativa* (japonica cultivar-group)]	453	4e-126
putative chalcone synthase [*Oryza sativa* (japonica cultivar-group)]	427	4e-118
putative chalcone synthase [*Oryza sativa* (japonica cultivar-group)]	424	3e-117
putative chalcone synthase [*Oryza sativa* (japonica cultivar-group)]	424	3e-117
putative chalcone synthase [*Oryza sativa* (japonica cultivar-group)]	422	1e-116
hypothetical protein [*Oryza sativa* (japonica cultivar-group)]	401	3e-110
Chalcone synthase [*Lilium hybrid* cv. 'Acapulco']	384	4e-105
putative chalcone synthase [*Oryza sativa* (japonica cultivar-group)]	380	6e-104
Chalcone synthase [*Lilium hybrid* division I]	378	2e-103

[a]The score (S) for an alignment is calculated by summing the scores for each aligned position and the scores for gaps.
[b]The E-value indicates the number of different alignments with scores equivalent to or better than (S) that are expected to occur in a database search by chance. The lower the E-value, the more significant the score.

Literature Cited

1. Dayan, F. E. In *Encyclopedia of Pest Management*; Pimentel, D. Ed; Marcel Dekker, Inc.: New York, NY, 2002; pp 521-525.
2. Duke, S.O.; Rimando, A.M.; Baerson, S.R.; Scheffler, B.E.; Ota E.; Belz, R.G. *J. Pesticide Sci.* **2002**, *27*, 298-306.
3. Rawlings, B.J. *Nat. Prod. Rep.* **2001**, *18*, 231-281.
4. Rawlings, B.J. *Nat. Prod. Rep.* **1999**, *16*, 425-484.
5. Austin, M.B.; Noel J.P. *Nat. Prod. Rep.* **2003**, *1*, 79-110.
6. Kozubek, A.; Tyman, J.H.P. *Chem. Rev.* **1999**, *99*, 1-26.
7. Czarnota, M.A.; Rimando A.M.; Weston L.A. *J. Chem. Ecol.* **2003**, *29*, 2073-2083.

8. Duke, S.O.; Belz, R.G.; Baerson, S.R.; Pan Z.; Cook D.D.; Dayan, F.E. *Out. Pest Manag.* **2005**, *16*, 64-68.
9. Bais, H.P.; Loyola Vargas, VM; Flores, H.E.; Vivanco, J.M. *Plant Physiol.* **2002**, *128*, 1173-1179.
10. Netzly, D.; Riopel, J.L.; Ejeta, G.; Butler, L.G. *Weed Sci.* **1988**, *36*, 441-446.
11. Kagan, I.A.; Rimando, A.M.; Dayan, F.E. *J. Agric. Food Chem.* **2003**, *51*, 7589-7595.
12. Rimando, A.M; Dayan, F.E.; Czarnota, M.A.; Weston, L.A.; Duke, S.O. *J. Nat. Prod.* **1998**, *61*, 927-930.
13. Rimando, A.M.; Dayan, F.E.; Streibig, J.C. *J. Nat. Prod.* **2003**, *66*, 42-45.
14. Fate, G.D.; Lynn, D.G. *J. Am. Chem. Soc.* **1996**, *118*, 11369-11376.
15. Dayan, F.E.; Kagan, I.A.; Rimando, A.M. *J. Biol. Chem.* **2003,** *278*, 28607-28611.
16. Czarnota, M.A.; Paul, R.N.; Weston, L.A.; Duke, S.O. *Internat. J. Plant Sci.* **2003**, *164*, 861-866.
17. Bucher, M.; Schroeer, B.; Willmitzer, L.; Riesmeier, J.W. *Plant Mol. Biol.* **1997**, *35*, 497-508.
18. http://fungen.botany.uga.edu
19. Cordonnier-Pratt, M.-M.; Liang, C.; Wang, H; Kolychev, D.S.; Sun, F.; Freeman. R.; Sullivan. R.; Pratt, L.H. *Comp. Funct. Genom.* **2004**, *5*, 268-275.
20. Altschul, S.F.; Gish, W.; Miller, W.; Myers, E.W.; Lipman, D.J. *J. Mol. Biol.* **1990**, *215*, 403-410.
21. Baerson, S.R.; Sanchez-Moeiras, A.; Pedrol-Bonjoch, N.; Schulz, M.; Kagan, I.A.; Agarwal, A.K.; Reigosa, M.J.; Duke, S.O. *J. Biol. Chem.* **2005**, *280*, 21867-21881.
22. Boguski, M.S.; Lowe, T.M.; Tolstoshev, C.M. *Nat. Genet.* **1993**, *4*, 332-333.

Structural Organization and Reaction Mechanisms Utilized by Polyketide Synthases

Chapter 11

Polyketide Biosynthesis beyond the Type I, II, and III Polyketide Synthase Paradigms: A Progress Report

Ben Shen[1–3,*], Yi-Qiang Cheng[1], Steven D. Christenson[1], Hui Jiang[1], Jianhua Ju[1], Hyung-Jin Kwon[1], Si-Kyu Lim[1], Wen Liu[1], Koichi Nonaka[1], Jeong-Woo Seo[1], Wyatt C. Smith[1], Scott Standage[1], Gong-Li Tang[1], Steven Van Lanen[1], and Jian Zhang[3]

[1]Division of Pharmaceutical Sciences, [2]University of Wisconsin National Cooperative Drug Discovery Group, and [3]Department of Chemistry, University of Wisconsin at Madison, Madison, WI 53705
[*]Corresponding author: email: bshen@pharmacy.wisc.edu, telephone: (608) 263–2673, and fax: (608) 262–5345

The archetypical polyketide synthases (PKSs), known as type I, II, and III PKSs, have accounted for the vast structural diversities embodied by polyketide natural products, but recent progress in polyketide biosynthesis clearly suggests much greater diversity for PKS mechanism and structure. We have previously argued the emergence of novel PKSs and cautioned the oversimplification of polyketide biosynthesis according to the type I, II, and III PKS paradigms on the basis of our studies on the biosynthesis of the enediynes, the macrotetrolides, and leinamycin. We present here a brief progress report on our current effort to mechanistically characterize these novel PKSs.

Introduction

Polyketides are a large family of structurally diverse natural products found in bacteria, fungi, and plants, and many of them are clinically important drugs. They are biosynthesized from acyl CoA precursors by polyketide synthases (PKSs). While the archetypical PKSs, known as type I, II, and III PKSs, have provided the molecular basis to account for the vast structural diversities embodied by the polyketides and the biotechnology platform for combinatorial biosynthesis of "unnatural" natural products in bacteria, recent progress in polyketide biosynthesis clearly suggests that PKSs have much greater diversity in both mechanism and structure. Using results from our studies on the biosynthesis of the enediynes, the macrotetrolides, and leinamycin as examples, we have previously argued the emergence of novel PKSs and cautioned the oversimplification of polyketide biosynthesis according to the type I, II, and III PKS paradigms.[1] Since then, many additional PKSs, deviating from the archetypical PKSs in mechanism, structure, or both, have been reported.[2,3] Here, we present a brief progress report on our current effort to mechanistically characterize these novel PKSs.

NcsB—an iterative type I PKS for aromatic polyketide biosynthesis

We have now fully sequenced and annotated the neocarzinostatin (**1**) biosynthetic gene cluster from *Streptomyces carzinostaticus* ATCC 15944, unveiling two iterative type I PKSs, NcsB and NcsE, for the biosynthesis of the naphthoic acid (**2**) and the enediyne core moieties of **1**, respectively.[4] NcsB catalyzes the biosynthesis of **2** from the acyl CoA substrates, NcsB3 and NcsB1 modify **2** to **3** by sequential hydroxylation and O-methylation, and NcsB2 finally couples **3** to the enediyne core to afford **1** (Figure 1).[4] This proposal has been supported by inactivation of *ncsB1*, *ncsB2*, or *ncsB3* in *S. carzinostaticus* and by expression of *ncsB* in *Streptomyces lividans* TK24.[5] The resulting *ΔncsB1*, *ΔncsB2*, or *ΔncsB3* mutant completely lost its ability to produce **1**, as would be expected for genes that are essential for **1** biosynthesis, and **2** was the major metabolite detected upon expression of *ncsB* in *S. lividans*, consistent with the proposed timing for the regioselective reduction at C-5 and C-9 of the nascent linear polyketide intermediate to occur before the ensuing intramolecular cyclization and aromatization to **2** (Figure 1). We have optimized the heterologous expression of *ncsB*, and NcsB can be overproduced in *E. coli* as a soluble protein, setting the stage to investigate the NcsB-catalyzed synthesis of **2** in vitro.

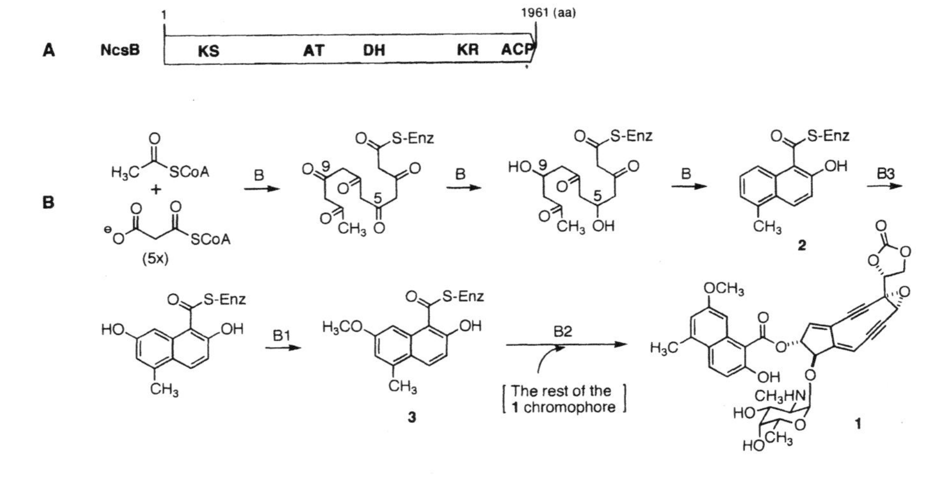

*Figure 1. (A) Domain organization of the NcsB naphthoic acid synthase and (B) proposed pathway for biosynthesis of the naphthoic acid intermediate (**2**) from the acyl CoA substrates by NcsB and its subsequent conversion to **3** by NcsB3 and NcsB1 and incorporation into neocarzinostatin (**1**) by NcsB2. ACP, acyl carrier protein; AT, acyltransferase; DH, dehydratase; KR, ketoreductase; KS, ketoacyl synthase.*

NcsE and SgcE—iterative type I PKSs for enediyne biosynthesis

Since the first report three years ago of the biosynthetic gene clusters for C-1027 (**4**) from *Streptomyces globisporus*[6] and for calicheamicin from *Micromonospora echinospora* ssp, *calichensis*[7] as model systems for 9- and 10-membered enediyne natural product biosynthesis, both of which utilize an iterative type I enediyne PKSs, two additional new enediyne natural products, have been reported,[8,9] four additional enediyne biosynthetic gene clusters have been cloned,[4,10,11] and at least eleven additional enediyne biosynthetic loci have been identified albeit the enediyne structures encoded by the latter are yet to be established.[11] We have now completed the cloning, sequencing, and annotation of the biosynthetic gene clusters for **4** from *S. globisporus*,[6] **1** from *S. carzinostaticus*,[4] and maduropeptin (**5**) from *Actinomadura madurae* and proposed a unified mechanism for enediyne biosynthesis featuring the enediyne PKSs such as NcsE for **1**, SgcE for **4**, and MadE for **5** (Figure 2).[1,10,12] The involvement of the enediyne PKSs in enediyne natural product biosynthesis have been unambiguously established by gene inactivation and complementation experiments in *S. carzinostaticus* for NcsE[4] and in *S. globisporus* for SgcE,[6] respectively, and by cross-complementation experiments in the *S. carzinostaticus ΔncsE* and *S. globisporus ΔsgcE* mutants for MadE.

Successful complementation of the *S. carzinostaticus ΔncsE* or *S. globisporus ΔsgcE* mutants by *ncsE* or *sgcE*, respectively, and cross-complementation of enediyne PKS gene mutants *ΔncsE* and *ΔsgcE* by homologs from other 9-membered enediyne biosynthetic pathways, such as *madE* for **5**, provided an excellent opportunity to map the putative active sites of the enediyne PKS in vivo. The enediyne PKS has been assigned to consist of six domains—ketosynthase (KS), acyltransferase (AT), acyl carrier protein (ACP), ketoreductase (KR), dehydratase (DH), and the terminal domain (TD) that has also been proposed to encode a phosphopantetheinyl transferase (PPTase) activity (Figure 2).[4,6,7,10-14] While the KS, AT, KR, and DH domains are highly homologous to known PKS domains, supporting their sequence-based functional assignments, the ACP and TD domains are unique to enediyne PKSs, whose assigned functions are very speculative. We have systematically mutated selected sets of conserved residues within both the ACP and TD domains of SgcE by site-directed mutagenesis to verify their predicted functions. The resultant mutants were introduced into the *S. globisporus ΔsgcE* mutant strain to examine if they can complement **4** biosynthesis with the native *sgcE* as a positive control. For example, while expression of the native *sgcE* gene restored **4** production to the *ΔsgcE* mutant to the level comparable to that in the *S. globisporus* wid-type strain, **4** production was not detected upon expression of either the *sgcE* (S974A) or *sgcE* (D1829A) mutant, indicating that the Ser974 and D1829 residues are essential for SgcE activity. Since the two mutations, S974A and D1829, are within the predicted ACP and TD domain of SgcE, respectively, which are absolutely conserved among all enediyne PKSs, these experiments provided experimental evidence to support their sequence-based functional assignments, serving as a proof of principle to map the active sites of all domains identified within the enediyne PKS by in vivo experiments.

Complementary to the aforementioned in vivo studies, we have optimized the heterologous expression of *ncsE* and *sgcE* in *E. coli* and overproduced and purified NcsE and SgcE in soluble form to homogeneity. Remarkably, the purified NcsE and SgcE proteins were yellow-colored with a distinct absorption spectrum with several maxima between 325 nm to 475 nm. Since PKSs and PPTases are not known to contain a prosthetic group that absorbs at these wavelengths, we attribute the yellow color to a biosynthetic intermediate, most likely a polyunsaturated linear polyketide that is still covalently tethered to the SgcE PKS via a thioester linkage to the phosphopantetheine group of the ACP domain. This proposal was further strengthened by the production and isolation of the SgcE (S974A) mutant as a negative control. Since ACP cannot be functional unless the conserved Ser residue is phosphopantetheinylated,[13,14] the

S974A mutation therefore renders SgcE incapable of being posttranslationallly modified by the PPTase into the holo-form, thereby abolishing its ability to catalyze any ACP-dependent polyketide synthesis. The purified SgcE (S974A) mutant was indeed colorless, consistent with the hypothesis that the yellow color resulted from an SgcE-bound polyketide intermediate. Experiments are in progress to structurally identify the yellow-colored putative intermediate.

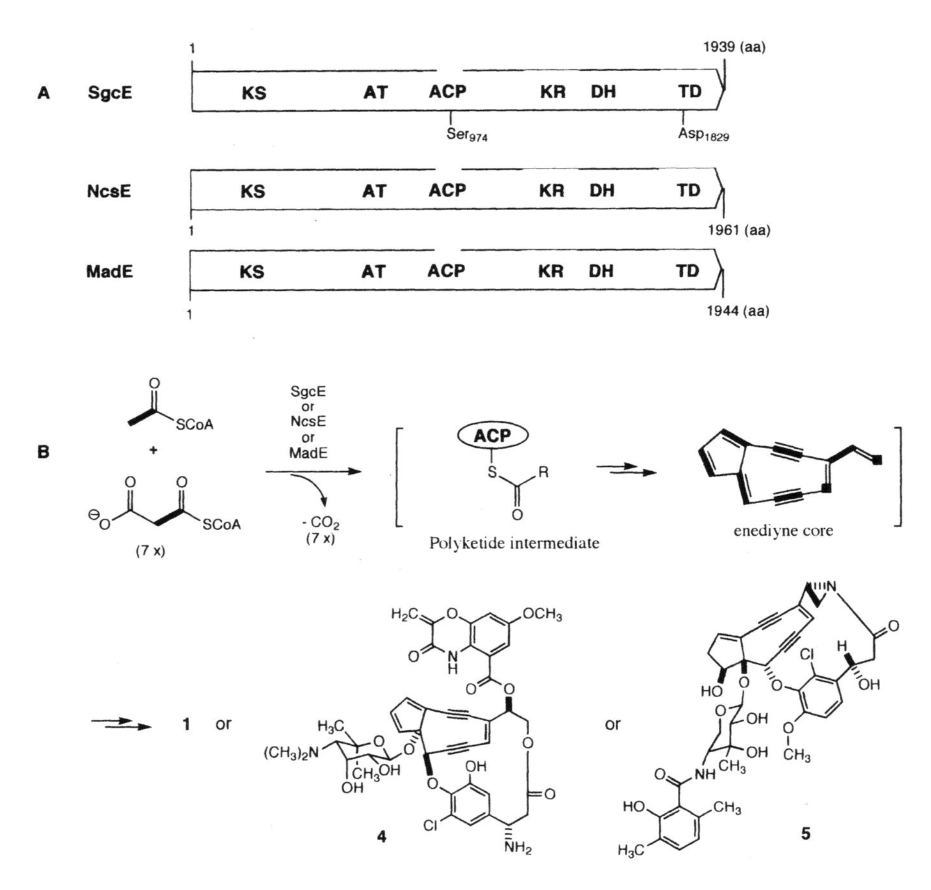

*Figure 2. (A) Domain organization of the MadE, NcsE, and SgcE enediyne PKSs and (B) proposed pathway for biosynthesis of a polyunsaturated intermediate (structure unknown) from the acyl CoA substrates by NcsE, MadE, or SgcE and its subsequent transformations to C-1027 (**4**), maduropeptin (**5**), or neocarzinostatin (**1**) by the rest of their respective biosynthetic machinery. ACP, acyl carrier protein; AT, acyltransferase; DH, dehydratase; KR, ketoreductase; KS, ketoacyl synthase; TD, terminal domain that most likely encodes a phosphopantetheinyl transferase.*

NonJK—Type II PKSs that catalyze C-O bond formation

We have previously demonstrated that NonJKL are sufficient to support the biotransformation of (±)-nonactic acid (**6**) into nonactin (**7**) in *S. lividans* and confirmed that NonL is a CoA ligase that catalyzes the conversion of (±)-**6** into (±)-nonactyl CoA (**8**).[15-18] On the basis of these results we concluded that NonJK represented C-O bond-forming type II PKSs that act directly on (±)-**8** to catalyze their tetramerization and cyclization into **7**, demonstrating the first time that PKS has the intrinsic propensity to evole catalytic activities catalyzing other than C-C bond formation (Figure 3A). We have now co-expressed *nonJK* in *E. coli* and showed that cell-free extract made from *E. coli* overexpressing *nonJK* can efficiently catalyze the synthesis of **7** from (±)-**8**; no activity was detected in cell-free extracts made from *E. coli* carrying the expression vector as a negative control or constructs expressing *nonJ* or *nonK* alone. While these results set the stage to investigate the NonJK PKSs in vitro, current effort to purify the NonJK proteins met with little success. Although soluble NonJK proteins were detected from the active cell-free extract, all attempts to fractionate the NonJK proteins have failed so far to recover any activity.

The discovery of the NonJK C-O bond-forming PKSs inspired us to search for natural products whose biosynthesis may utilize similar chemistry. We reasoned that additional C-O bond-forming PKSs would not only support the generality of this novel chemistry in natural product biosynthesis but also provide an alternative system to circumvent the difficulties associated with the study of the NonJK PKSs in vitro. The paramycins, such as paramyicn 607 (**9**) and paramycin 593 (**10**) produced by *Streptomyces alboniger*, are structural analogs of **7**, and feeding experiments with isotope-labeled precursors unveiled a remarkable biosynthetic similarity between **7** and **9**/**10**.[19] Thus, it is not difficult to imagine that variations of the (±)-**6** pathway could readily afford the synthesis of the penultimate intermediates such as **11**, **12**, or **13** for **9** and **10** biosynthesis. Activation of the former into the corresponding CoA esters, **14**, **15**, **16**, requiring a NonL homolog, and subsequent dimerization and cyclization, invoking the same C-O bond-forming chemistry as the NonJK PKSs, would finally afford **9** and **10** (Figure 3B). Therefore, **9** and **10** would provide an excellent opportunity to further explore and expand the C-O bond-forming PKS chemistry in natural product biosynthesis. We have recently developed an expedient genetic system for *S. alboniger*, and this has set the stage to clone, sequence, and characterize the paramycin biosynthetic gene cluster from this organism, potentially uncovering additional novel C-O bond-forming PKSs.

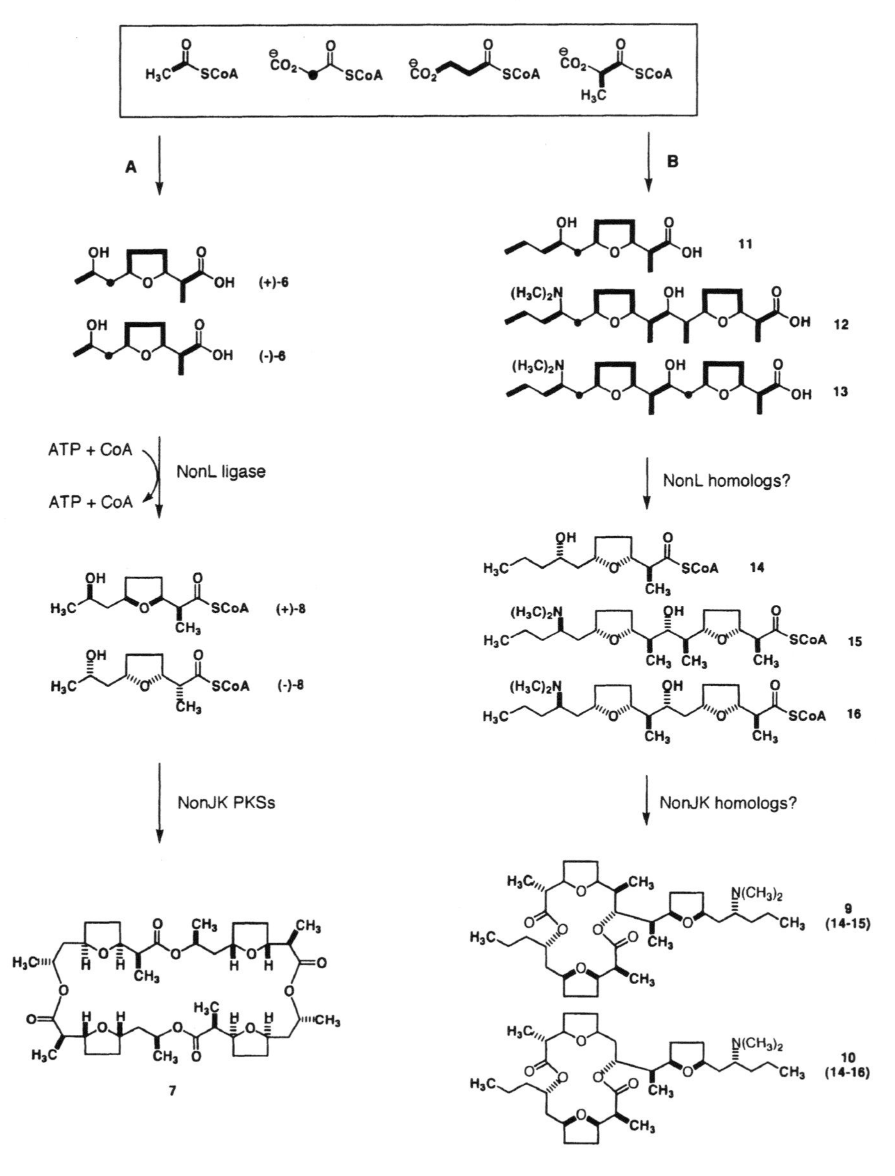

Figure 3. Incorporation patterns from isotope-labeled acyl CoA precursors and proposed unified biosynthetic pathways for (A) nonactin (7) and (B) paramycin 607 (9, coupling between intermediates 14 and 15) and 593 (10, coupling between intermediates 14 and 16) featuring C-O bond-forming PKSs such as NonJK and homologs.

LnmIJ/LnmG—"AT-less" type I PKSs that require a discrete AT enzyme acting iteratively in trans

Since we first demonstrated that LnmIJ are AT-less PKSs that require the LnmG AT enzyme to provide the AT activity iteratively in trans for leinamycin (**17**) biosynthesis in *Streptomyces atroolivaceus*,[20-22] more than a dozen additional gene clusters featuring AT-less PKS have been reported.[23-39] To expand our effort in characterizing AT-less PKSs, we have now completed the cloning, sequencing, and annotation of three additional biosynthetic gene clusters for lactimidomycin (**18**) from *Streptomyces amphibiosporus*, migrastatin (**19**)/iso-migrastatin (**20**)/dorrigocin A (**21**) and B (**22**) from *Streptomyces platensis*,[40,41] and oxazolomycin (**23**) from *Streptomyces albus*,[42] all of which feature AT-less PKSs (Figure 4). The involvement of the cloned AT-less PKSs in the biosynthesis of these metabolites has been confirmed by in vivo gene inactivation and complementation experiments.

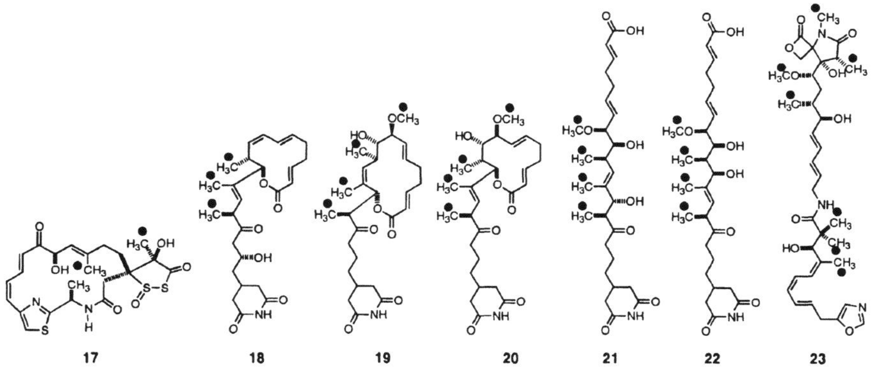

Figure 4. Polyketide natural products whose biosyntheses are governed by AT-less PKSs as exemplified by leinamycin (17), lactimidomycin (18), migrastatin (19), iso-migrastatin (20), dorrigocin A (21), dorrigocin B (22), and oxazolomycin (23). (•), methyl group of S-adenosylmethionine origin.

These additional data not only substantiate the generality of AT-less PKS for polyketide biosynthesis but also enable us to speculate if AT-less PKS clusters could be predicted according to the structures of polyketide natural products. Reduced (also known as complex) polyketide natural products often carry alkyl branches on their carbon backbones. Feeding experiments with isotope-labeled precursors clearly established that the alkyl branches are typically derived from the acyl CoA extender units, such as methylmalonyl CoA, ethylmalonyl CoA, or methoxymalonyl ACP for the resultant methyl, ethyl, or methoxy branches, respectively.[2,43] Noniterative type I PKSs account tisfactorily for their ability to incorporate various extender units regiospecifically into the resultant polyketide products by evolving cognate ATs, with the desired substrate specificity, within the PKS module for each cycle of polyketide chain elongation (Figure 5A). Exceptions to this general observation have been occasionally noticed, in which the alkyl branches, often a methyl group, is derived from S-adenosylmethionine (SAM). This biosynthetic variation has been equally accounted for by noniterative type I PKS modules evolved with an additional methyl transferase (MT) domain. Thus, while a PKS module with malonyl CoA-specific AT and MT domains is functionally equivalent to a PKS module with a methylmalonyl CoA-specific AT domain alone in terms of providing the net extender unit for polyketide chain elongation, a PKS module with methylmalonyl CoA-specific AT and MT domains installs gem-dimethyl groups to the resultant polyketide product, a structural feature that would not be possible to synthesize by a PKS module with a AT domain alone with varying extender unit specificity (Figure 5B).[44-47]

The extender unit specificity for an AT-less PKS, however, is determined by the discrete AT enzyme that acts iteratively in trans and loads the same extender units to every module of the cognate AT-less PKS. Therefore, AT-less PKS has little flexibility in varying extender units, and an AT-less PKS with a malonyl CoA-specific discrete AT will afford a polyketide product with no alkyl branch that could be of extender unit origin. Consequently, a polyketide natural product with its backbone derived from the same extender unit and all its methyl branches originating from SAM is most likely biosynthesized by an AT-less PKS (Figure 5C), as has been exemplified by **17** to **23** via isotope-labeled precursor feeding experiments (Figure 4). These structural features should serve as predictors to search for natural products whose biosyntheses are most likely governed by AT-less PKSs.

While evolution of a MT domain within an AT-less PKS module satisfactorily accounts for the introduction of methyl branches into the resultant polyketides as depicted in Figure 5C, it is difficult to predict a priori how an AT-less PKS with a malonyl CoA-specific discrete AT furnishes an ethyl- or methoxy branch since no such domain, in a mechanistic analogy to MT for a methyl group, to introduce an ethyl or methoxy group into polyketide backbone is known to date. Strikingly, cloning, sequencing, and characterization of the biosynthetic gene cluster for **23** from *S. albus*[42] unveiled an AT-less PKS with a discrete AT enzyme featuring two AT domains. We have proposed that each of the AT domains specifies for malonyl CoA and methoxymalonyl ACP, respectively, to provide these extender units to the oxazolomycin AT-less PKS for **23** biosynthesis (Figure 5D). This finding has set the stage to characterize an AT-less PKS that utilizes a methoxymalonyl ACP as an extender unit. Characterization of an AT-less PKS that can incorporate an extender unit other than malonyl CoA is significant, proving that AT-less PKS is not limited by the choice of extender units. The latter should serve as an inspiration to search for additional AT-less PKSs that incorporate novel or multiple different extender units to further expand polyketide natural product structural diversity.

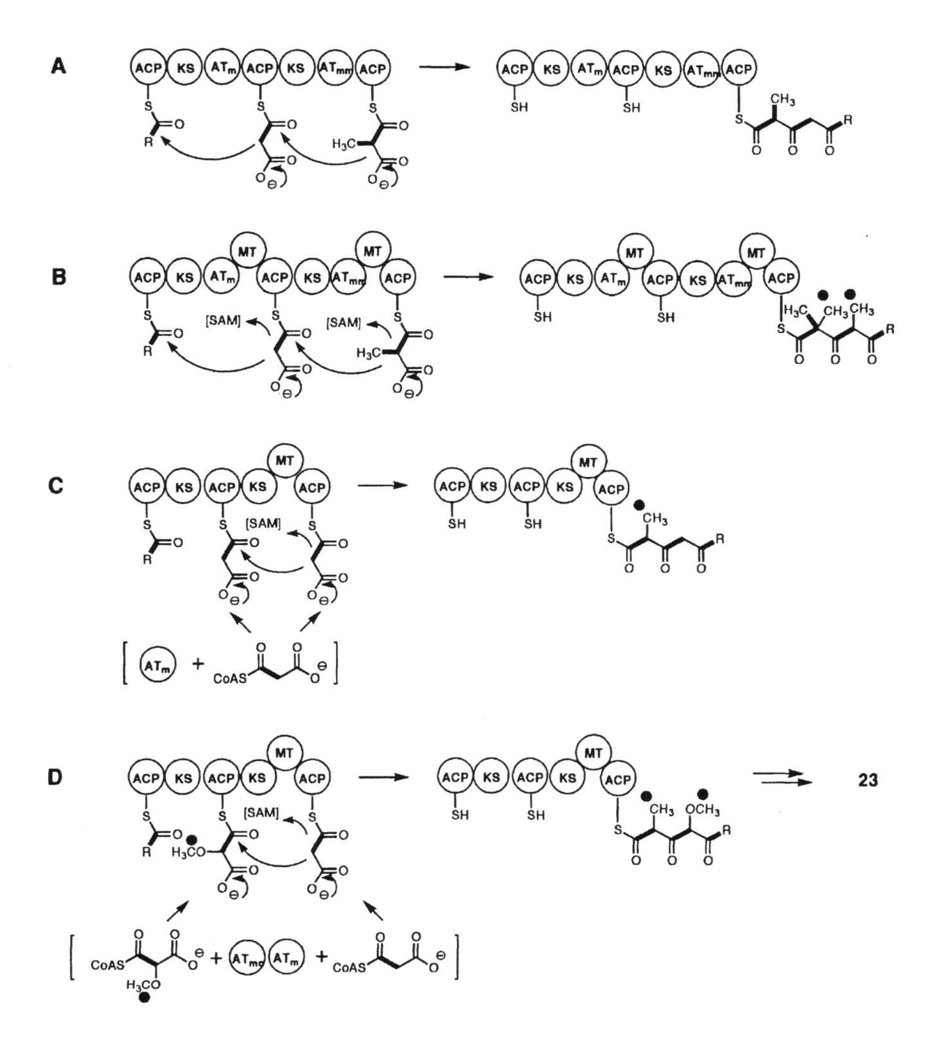

Figure 5. Installation of alkyl branches into polyketides by (A) and (B) noniterative type I PKSs or (C) and (D) AT-less PKSs. (•), methyl group of S-adenosylmethionine origin; ACP, acyl carrier protein; AT_m, malonyl CoA-specific acyltransferase; AT_{mm}, methylmalonyl CoA-specific acyltransferase; AT_{mo}, methoxymalonyl ACP-specific acyltransferase; KS, ketoacyl synthase; MT, methyltransferase; SAM, S-adenosylmethionine.

Concluding remarks

The three types of PKSs described here, the enediyne PKS, the C-O bond-forming PKS, and the AT-less PKS, are only representive examples that reside outside the archetypical PKS paradigms. Continued exploration on the mechanism of polyketide biosynthesis will undoubtly uncover more unusual PKSs. These novel PKSs, in combination with the archetypical ones, will ultimately enhance the toolbox available to facilitate combinatorial biosynthesis and production of "unnatural" natural products. The full realization of the potential embodied by combinatorial biosynthesis of PKSs for natural product structural diversity, however, depends critically on the fundamental characterization of PKS structure, mechanism, and catalysis.

Acknowledgement

Studies on polyketide biosynthesis described from the Shen laboratory were supported in part by the University of California BioSTAR program (Bio99-10045) and Kosan Biosciences, Inc., Hayward, CA, NIH grants CA78747, CA106150, CA113297, an NSF CAREER Award MCB9733938, and an NIH Independent Scientist Award AI51689.

References

1. Shen, B. *Curr. Opinion Chem. Biol.* **2003**, *7*, 285-295.
2. Moss, S. J.; Martin, C. J.; Wilkinson, B. *Nat. Prod. Rep.* **2004**, *21*, 575-593.
3. Wenzel, S. C.; Muller, R. *Curr. Opinion Chem. Biol.* **2005**, *9*, 447-458.
4. Liu, W.; Nonaka, K.; Nie, L.; Zhang, J.; Christenson, S. D.; Bae, J.; Van Lanen, S. G.; Zazopoulos, E.; Farnet, C. M.; Yang, C. F.; Shen, B. *Chem. Biol.* **2005**, *12*, 293-302.
5. Sthapit, B.; Oh, T.-J.; Lamichhane, R.; Liou, K.; Lee, H. C.; Kim, C.-G.; Sohng, J. K. *FEBS Lett.* **2004**, *566*, 201-206.
6. Liu, W.; Christenson, S. D.; Standage, S.; Shen, B. *Science*, **2002**, *297*, 1170-1173.
7. Ahlert, J.; Shepard, E.; Lomovskaya, N.; Zazopoulos, E.; Staffa, A.; Bachmann, B. O.; Huang, K.; Fonstein, L.; Czisny, A.; Whitwam, R. E.; Farnet, C. M.; Thorson, J. S. *Science* **2002**, *297*, 1173-1176.
8. Oku, N.; Matsunaga, S.; Fusetani, N. *J. Am. Chem. Soc.* **2003**, *125*, 2044-2045.
9. Davies, J.; Wang, H.; Taylor, T.; Warabi, K.; Huang, X.-H.; Andersen, R. J. *Org. Lett.* **2005**, *7*, 5233-5236.

10. Liu, W.; Ahlert, J.; Gao, Q.; Wendt-Pienkowski, E.; Shen, B.; Thorson, J. S. *Proc. Natl. Acad. Sci. USA*, **2003**, *100*, 11959-11963.
11. Zazopoulos, E.; Huang, K.; Staffa, A; Liu, W.; Bachmann, B. O.; Nonaka, K.; Ahlert, J.; Thorson, J. T.; Shen, B.; Farnet, C. M. *Nat. Biotechnol.* **2003**, *21*,187-190.
12. Shen, B.; Liu, W.; Nonaka, K. *Curr. Med. Chem.* **2003**, *10*, 2317-2325.
13. Sanchez, C.; Du, L.; Edwards, D.J.; Toney, M. D,; Shen, B. *Chem. Biol.*, **2001**, *8*, 725-738.
14. Walsh, C. T.; Gehring, A. M.; Weinreb, P. H., Quadri, L. E. N.; Flugel, R. S. *Curr. Opinion Chen. Biol.* **1997**, *1*, 309-315.
15. Smith, W. C.; Xiang, L.; Shen, B. *Antimicrob. Agents Chermother.*, **2000**, *44*, 1943-1953.
16. Kwon, H.-J.; Smith, W.C.; Xiang, L.; Shen, B. *J. Am. Chem. Soc.*, **2001**, *123*, 3385-3386.
17. Kwon, H.-J.; Smith, W. C.; Scharon, A. J.; Hwang S. H.; Kurth, M. J.; Shen, B. *Science*, **2002**, *297*, 1327-1330.
18. Shen, B.; Kwon, H.-J. *The Chem. Rec.,* **2002**, *2*, 389-396.
19. Hashimoto, M.; Komatsu, H.; Kozone, I.; Kawaide, H.; Abe, H.; Natsume, M. *Biosci. Biotechnol. Biochem.*, **2005**, *69*, 315-320.
20. Cheng, Y.-Q.; Tang, G.-L.; Shen, B. *J. Bacteriol.*, **2002**, *184*, 7013-7024.
21. Cheng, Y.-Q.; Tang, G.-L.; Shen, B. *Proc. Natl. Acad. Sci. USA*, **2003**, *100*, 3149-3154.
22. Tang, G.-L.; Cheng, Y.-Q.; Shen, B. *Chem. Biol.*, **2004**, *11*, 33-45.
23. Piel, J. *Proc. Natl. Acad. Sci. USA*, **2002**, *99*, 14002-14007.
24. Piel, J.; Hofer, I.; Hui, D. *J. Bacteriol.*, **2004**, *186*, 1280-1286.
25. Hildebrand, M.; Waggoner, L. E.; Liu, H.; Sudek, S.; Allen, S.; Anderson, C.; Sherman, D. H.; Haygood, M. *Chem. Biol.*, **2004**, *11*, 1543-1552.
26. Piel, J.; Hui, D.; Wen, G.; Butzke, D.; Platzer, M.; Fusetani, N.; Matsunaga, S. *Proc. Natl. Acad. Sci. USA*, **2004**, *101*, 16222-16227.
27. Piel, J.; Wen, G.; Platzer, M.; Hui, D. *ChemBioChem* **2004**, *5*, 93-98.
28. Piel, J.; Hui, D.; Fusetani, N.; Matsunaga, S. *Environ. Microbiol.* **2004**, *6*, 921-927.
29. Zhu, G.; LaGeier, M. J.; Steijskal, F.; Millership, J. J.; Cai, X.; Keithly, J. S. *Gene* **2002**, *298*, 79-89.
30. Otsuka, M.; Ichinose, K.; Fujii, I.; Ebizuka, Y. *Antimicrob. Agents Chemother.*, **2004**, *48*, 3468-3476.
31. Mochizuki, S.; Hiratsu, K.; Suwa, M.; Ishii, T.; Sugino, F.; Yamada, K.; Kinashi, H. *Mol. Microbiol.* **2003**, *48*, 1501-1510.
32. Arakawa, K.; Sugino, F.; Kodama, K.; Ishii, T.; Kinashi, H. *Chem. Biol.*, **2005**, *12*, 249-256.
33. El-Sayed, A. K.; Hothersall, J.; Cooper, S. M.; Stephens, E.; Simpson, T. J.; Thomas, C. M. *Chem. Biol.*, **2003**, *10*, 419–430.

34. Kunst, F.; Ogasawara, N.; Moszer, I.; Albertini, A. M.; Alloni, G.; Azevedo, V.; Bertero, M. G.; Bessieres, P.; Bolotin, A.; Borchert, S. et al. *Nature* **1997**, *390*, 249-256.
35. Paitan, Y.; Alon, G.; Orr, E.; Ron, E. Z.; Rosenberg, E. *J. Mol. Biol.*, **1999**, *286*, 465-474.
36. Kopp, M.; Isrschik, H.; Pradella, S.; Müller, R. *ChemBioChem* **2005**, *6*, 1277-1286.
37. Carvalho, R.; Reid, R.; Viswanathan, N.; Gramajo, H.; Julien, B. *Gene* **2005**, *359*, 91-98.
38. Huang, G.; Zhang, L.; Birch, R. *Microbiology* **2001**, *147*, 631-642.
39. Perlova, O.; Gerth, K.; Kaiser, O.; Hans, A.; Muller, R. *J. Biotechnol.* **2005**, in press.
40. Ju, J.; Lim, S.-K.; Jiang, H.; Shen, B. *J. Am. Chem. Soc.* **2005**, *127*, 1622-1623.
41. Ju, J.; Lim, S.-K.; Jiang, H.; Seo, J.-W.; Shen, B. *J. Am. Chem. Soc.* **2005**, *127*, 11930-11931.
42. Zhao, C.; Ju, J.; Christenson, S.D.; Smith, W.C.; Song, D.; Zhou, X.; Shen, B.; Deng, Z. **2005**, submitted.
43. Staunton, J.; Weissman, K. J. *Nat. Prod. Rep.* **2001**, *18*, 380-416.
44. Du, L.; Sanchez, C.; Chen, M.; Edwards, D.J.; Shen, B. *Chem. Biol.*, **2000**, *7*, 623-642.
45. Molnar, I.; Schupp, T.; Ono, M.; Zirkle, R. E.; Milnamow, M.; Nowak-Thompson, B.; Engel, N.; Toupet, C.; Stratmann, A.; Cyr, D. D.; Gorlach, J.; Mayo, J. M.; Hu, A.; Goff, S.; Schmid, J.; Ligon, J. M. *Chem. Biol.*, **2000**, *7*, 97-109.
46. Tang, L.; Shah, S.; Chung, L.; Carney, J.; Katz, L.; Khosla, C.; Julien, B. *Science* **2000**, *287*, 640-642.
47. Julien, B.; Shah, S.; Ziermann, R.; Goldman, R.; Katz, L.; Khosla, C. *Gene* **2000**, *249*, 153-160.

Chapter 12

Structural Enzymology of Aromatic Polyketide Synthase

Tyler Paz Korman, Brian Douglas Ames, and Shiou-Chuan Tsai

Departments of Molecular Biology and Biochemistry and Department of Chemistry, University of California, Irvine, CA 92697–3900

The type II (aromatic) polyketide synthase biosynthesizes many important antibiotic and anticancer polyketides in bacteria and fungi. Similar to the type II fatty acid synthase, the aromatic PKS carries out a series of reactions catalyzed by individual soluble enzymes, each of which are encoded by a discrete gene. The growing polyketide intermediates are transported between the enzymes attached to the acyl carrier protein through a thioester linkage. Using X-ray crystallography and NMR, enzyme structures from the actinorhodin and tetracenomycin PKSs have been solved. The structure of the individual PKS enzymes reveal key molecular features that help explain the observed substrate specificity, enzyme mechanism and protein-protein interactions. These structures are also a valuable resource to guide combinatorial biosynthesis of polyketide natural products.

There are at least three characterized, architecturally different types of PKSs, although with better detection methods, the structural diversity of PKSs continues to increase (1). The focus of this chapter is the aromatic polyketide synthase (also called "Type II PKS") that biosynthesizes aromatic polyketides (Figure. 1), natural products that show an enormously rich and varied range of bioactivities. Examples include antibiotics (such as tetracyclines (2), tetracenomycin (3) and actinorhodin (4)), anticancer agents (such as resistomycin (5), doxorubicin (6) and mithramycin (7)), anti-fungals (such as pradimicin (8)) and anti-HIV therapeutics (such as rubromycin (9) and griseorhodin (10)) (11) (Figure 1). Despite extensive efforts to obtain polyketides synthetically due to their widespread medical applications, many aromatic polyketides have proven to be difficult to obtain via organic synthesis (12-14). By comparison, the biosynthesis of polyketides by polyketide synthase (PKS) offers a more economic and technically simpler protocol. By transforming the PKS gene into a bacterial host system followed by industrial fermentation, kilogram quantities of polyketides are routinely produced overnight (15). As a result, there is great interest in developing biosynthetic systems for the production of medically important polyketides.

The Molecular Logic of Aromatic PKS

The aromatic PKSs are comprised of 5-10 distinct enzymes whose active sites are used iteratively. Past molecular genetic studies in the groups of Hopwood (16), Hutchinson (17), Floss (18), Khosla (19), Robinson (20) and Salas (21) have established the current mechanism for a typical aromatic PKS (Fig. 2, the actinorhodin PKS): (1) Chain initiation: acyl carrier protein (ACP) primed by malonyl-CoA:ACP transacylase (MAT) (the involvement of MAT is an unresolved issue (22)); (2) Iterative chain elongation by the ketosynthase (KS) / chain length factor (CLF) heterodimer; (3) First-ring cyclization, either uncatalyzed or may involve enzyme domains; (4) Chain reduction by ketoreductase (KR); (5) First-ring aromatization by aromatase /cyclase (ARO/CYC); (6) Subsequent cyclization by the same or different ARO/CYCs. During each step, chain transfer between different catalytic domains is mediated by the phosphopentetheinyl group (PPT) covalently attached to a serine of ACP.

The advancement of PKS genetics has led to the development of a set of design rules for the rational manipulation of chain synthesis, reduction of keto groups and early cyclization steps (Figure 3): (1) The aromatic polyketide KR has a high specificity for the C9-carbonyl group (23). The C9-specificity is demonstrated by the product outcome during the biosyntheses of actinorhodin (4), doxorubicin (24), R1128 (25) and enterocin (26,27). In special cases, a highly specific reduction at other positions has also been observed (26,28). The structural basis this highly regio-specific behavior of KR is not well understood

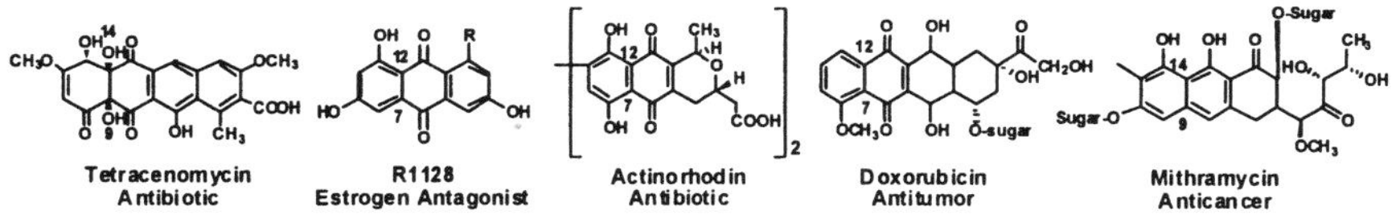

Tetracenomycin
Antibiotic

R1128
Estrogen Antagonist

Actinorhodin
Antibiotic

Doxorubicin
Antitumor

Mithramycin
Anticancer

Figure 1. Representative Aromatic Polyketides that are numbered with the first ring cyclization specificity.

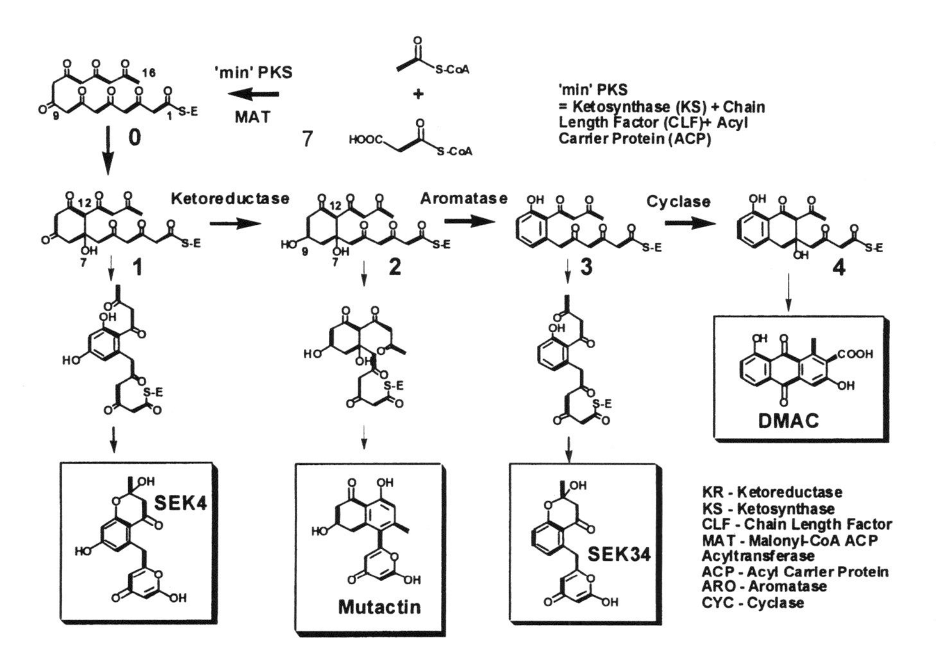

Figure 2. The Biosynthetic Pathway of Actinorhodin PKS

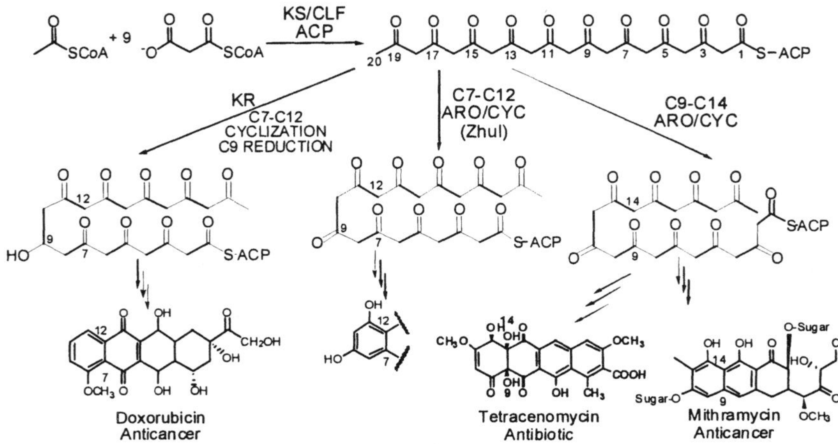

Figure 3. From the same polyketide chain, different cyclization patterns diversify polyketide products

(29). (2) The inclusion of KR often results in products that are cyclized at the C7-C12 positions (28,30), whereas the inclusion of a mono-domain ARO (in the absence of KR) often results in products that are cyclized at the C9-C14 positions (30,31) (Figure 3).

The above molecular rules have been used to rationally design polyketide products. Genetically engineered *Streptomyces* strains have produced new polyketides by expression of combinations of appropriate enzymatic subunits from naturally occurring polyketide synthase (the "mix and match" approach). These experiments have successfully created many novel polyketides (28,31-37) giving rise to > 100 new polyketides (37). The evidence cited above has led researchers in the field to appreciate that the large diversity of naturally occurring polyketides is a result of the controlled variation in chain length, a specific choice of building blocks and by differential chain modifications that are mediated by PKSs. The argument for the success of engineered biosynthsis is that natural selection process may have been very effective in identifying antibacterial polyketides due to the pressure of competition (38), but likely has not resulted in the selection of compounds with different pharmacologic activities. Therefore, the potential for development of new pharmaceutical based on bioengineering of novel polyketides is enormous.

A Summary of Aromatic PKS Structural work

Despite the advances cited above, random domain shuffling often results in complete inactivation of PKSs. For example, Burson and Khosla demonstrated that 90% of the shuffled KS/CLF domains resulted in the loss of activity (39). Further, because of our incomplete understanding of the mechanisms and specificities of individual enzymes, as well as the influence of protein-protein interactions in PKSs, many potential manipulations such as the rational control of cyclization patterns, are currently not available, (40). This lack of understanding about the structure-function relationship between PKS enzymes has recently been highlighted as a problem by Staunton and Weissman in the millennium PKS review (40). An important step toward remedying this limitation is to solve the crystal structures of individual PKS domains and then to use this knowledge to study the structure-function relationship of PKSs. Currently, crystal structures of the *S. coelicolor* MAT (41), the actinorhodin KS/CLF (42), the priming KS of R1128 (43), aclacinomycin-10-hydroxylase (44) and the actinorhodin oxygenase (45) have been solved. Several crystal structures of fatty acid synthase domains have also been solved (46,47), including the fatty acid MAT (48), ACP (49), KSs (50-52), KRs (46,53,54), enoylreductase (55-57) and dehydratases (58). In addition, the NMR structure of two polyketide ACP structures have been solved (59,60). Further, two cyclase structures, tcmI (61) and SnoaL (62), have been reported. These two cyclases catalyze the cyclization of the last ring during the biosyntheses of tetracenomycin and nogalamycin, respectively. They bear no sequence homology to the ARO/CYCs that promote the first and second ring formation (Figure 2). The structure-function relationship of the aromatic PKS domains help elucidate important molecular features such as chain length control, stereo- and regio-selectivity of KR, as well as the cyclization specificity of ARO/CYC. As a result, there remains a need for the crystal structures of polyketide KR and ARO/CYC. In this chapter, we will report our recent crystal structure solution of the actinorhodin KR, tetracenomycin ARO/CYC, as well as structure-based engineering of the actinorhodin KS/CLF.

Structure-Based Bioengineering of KS/CLF

The crystal structure of actinorhodin KS/CLF (which synthesizes a 16C chain) revealed an 18 Å long channel that starts with a conserved cysteine in the KS active site and spans across the KS-CLF dimer interface (Figure 4) (42). Residues lining the sides of the channel from both the KS and the CLF subunits are also well conserved. Sequence alignments of CLF subunits revealed that this

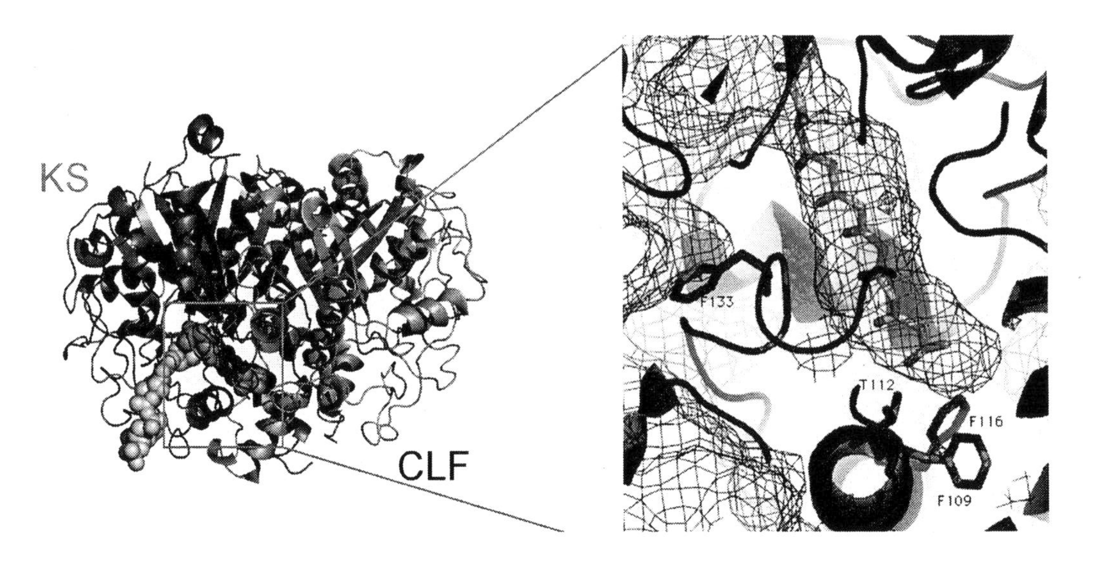

Figure 4. The Crystal structure of the actinorhodin KS/CLF, which catalyzes the formation of polyketide products with a 16-carbon backbone. The substrate binding channel is enlarged in the right panel. F109', T112' and F116' of CLF are located at the bottom of the substrate channel.

region is highly conserved, except for residues 109, 112, 116, and 133 (*act* CLF numbering). The terminal wall of this channel is capped by a helix-turn-loop in the CLF, where F109', T112' and F116' are located. Remarkably, as chain length specificity increases from C_{16} to C_{24}, these four residues are replaced with less bulky amino acids. We hypothesize that these residues determine chain length. Therefore we constructed mutants of the *act* CLF bearing large-to-small changes at these positions using site-directed mutagenesis (Figure 4). The strain containing the wild-type *act* KS/CLF produced the expected 3,8-dihydroxy-1-methylanthraquinone-2-carboxylic acid (DMAC). Replacing the *act* KS/CLF with wild-type tetracenomycin (*tcm*) KS/CLF (a decaketide synthase) yielded RM20b as the major product. Significantly, >65% of the polyketide products of the double mutant F109A/F116A were decaketides. Triple mutations (F109A/T112A/F116A) did not alter the specificity further. The purified mutant actinorhodin KS/CLFs *in vitro* showed that F116A was sufficient to increase the levels of decaketides to 64% of the total polyketides, while the F109A/F116A double mutant synthesized decaketides as >95% of the total polyketides.

To probe the roles of the corresponding residues in a decaketide CLF, small-to-large mutations at G116 and M120 (*tcm* CLF numbering) were introduced into *tcm* CLF. A single G116T mutation in the *tcm* CLF yielded >65% octaketide products with an overall polyketide yield comparable to that of wild-type *tcm* KS/CLF. Thus, manipulating one or two residues located at the helix-turn-loop region is sufficient to convert an octaketide synthase into a decaketide synthase, and vice versa. We have shown through rational mutagenesis that CLF exerts polyketide chain length control by defining the size of the polyketide channel, thereby confirming its role as the chain length factor. Residues 109, 112, and 116 in the *act* CLF serve as gatekeepers that terminate the channel at the KS/CLF interface. Reducing the sizes of these residues lengthens the channel and allows two more chain-extending cycles. Our results also suggest that, under *in vivo* conditions, additional proteins that interact with the KS/CLF further bias the formation of polyketides of a particular chain length, highlighting the complexity of protein-protein interactions among the individual type II PKS catalytic units. Our results should lead to novel strategies for the engineered biosynthesis of hitherto unidentified polyketide scaffolds.

The Regio- and Stereo-Specificity of KR

In order to rationally control the reduction pattern of polyketide biosynthesis, we need to understand the structure-function relationship of KR. Towards this goal, we have solved two co-crystal structures of the actinorhodin KR bound with the cofactor $NADP^+$ or NADPH. The crystal structures provide the following information (63):

1. Helices α4-α7 are important for substrate binding.

Analysis of the crystal structure and native gels indicate that KR exists as a tetramer (Figure 5A) (63,64). It has a highly conserved Rossmann fold with the cofactor NADPH bound near the center between two α–β–α motifs and the polyketide substrate bound in a substrate tunnel that opens from the front to the back side (Figure 5B). The front and back sides of KR are defined as the protein face that contains the NADPH binding pocket or be opposite to the NADPH binding pocket, respectively (Figure 5B is the view of the front side). In comparison to fatty acid KRs, the aromatic polyketide KRs contain an extra loop insertion region near α6-α7 (residues 190 – 210, Figure 5B yellow helices). This insertion region is also the most flexible region of KR. Sequence comparison indicates that this insertion region is a unique, highly conserved feature for nine different aromatic polyketide KRs. The regions between α6-α7 and α4-α5 (Fig. 5B, green helices) define the shape of substrate binding pocket and are therefore important for substrate binding.

2. The proton-relay mechanism anchors the substrate by 3-point docking.

Because the cofactor binding pocket and the proposed active site tetrad (Asn114-Ser144-Tyr157-Lys161, Figure 6A) are highly conserved between actinorhodin KR and existing fatty acid KR structures (46,53,54), we can compare the active site geometry and catalytic mechanism by overlapping the structures of actinorhodin KR and fatty acid KRs. We found that the position of the active site tetrad is completely conserved. This indicates that the catalytic mechanism proposed for fatty acid KRs should also apply to polyketide KRs. The catalytic mechanism of the fatty acid KR has been proposed to proceed through a proton-relay network (46,53). Briefly, the ketone substrate is hydrogen bonded to both Ser144 and Tyr157 that constitute the oxyanion hole (Figure 6). Following hydride transfer from NADPH to the ketone substrate, the alkoxide is stabilized by the oxyanion hole while the tyrosyl proton is transferred to the alkoxide (Figure 6B). An extensive proton relay then takes place to replenish the proton extracted from the tyrosyl-OH, sequentially including the 2-OH of NADPH ribose, lysine-NH and then followed by four water molecules (Figure 6B). If this mechanism is correct, then substrate binding is highly restrained by the intricate hydrogen bonding network. The polyketide substrate is docked into the active site using a three-point docking strategy (Figure 6B). Namely, the C9-carbonyl oxygen of the substrate (in purple, Figure 6A) should be hydrogen-bonded with the side chains of Ser144 and Tyr157, while the C9-carbonyl carbon of the substrate should be within hydride-transfer distance to NADPH (Figure 6B).

3. First ring cyclization must occur before ketoreduction to satisfy the C9 regiospecificity.

An important question for aromatic polyketide biosynthesis is the timing of the first ring cyclization, relative to ketoreduction. We found that the timing

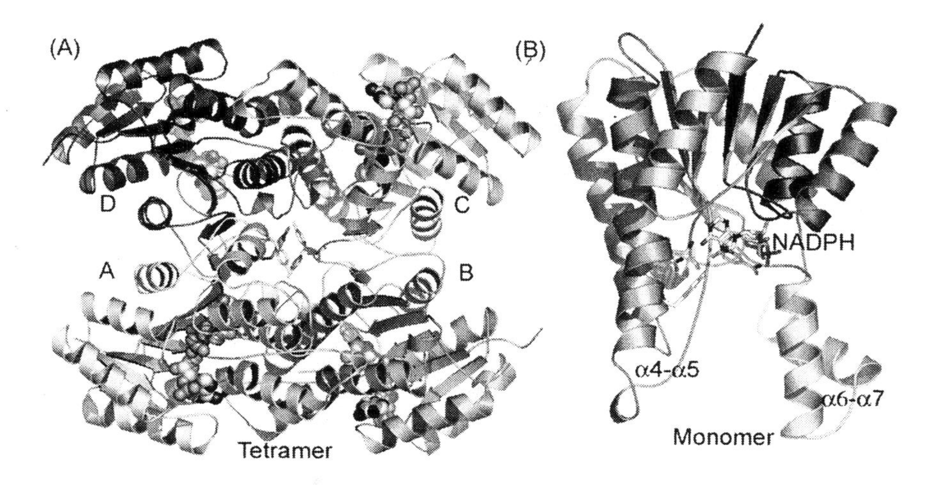

Figure 5. (A) The tetramer model of KR. (B) Enlarged view of the KR monomer

issue is closely related to the C9 regiospecificity. The cyclization of the polyketide chain (producing intermediate **1**, Figure 2) can either occur uncatalyzed, in the active site of KS/CLF, in the interface between KS/CLF, ACP and KR, or in the active site of KR. Substrate docking simulations of the highly flexible linear polyketide chain (intermediate **0**, Figure 2) indicate that it is physically possible to reduce many carbonyl groups of intermediate **0** other than the C9-carbonyl group without evoking protein conformational changes. Due to the flexibility of intermediate **0**, this will result in the loss of the C9 regio-specificity, contradicting results from previous studies that aromatic polyketide KRs are highly specific for C9 reduction. Rather, it is more reasonable to cyclize **0** to **1** prior to the reduction of **1** by KR. Under the dual-constraints imposed by the ring structure of **1** and the three-point docking of the KR active site, the C9-carbonyl group is optimally positioned for ketoreduction when the C7-C12 cyclization takes place (Figure 6B). Similarly, in the case of C5-C10 cyclization, the C7 position (para and meta to the two bulky ring substituents) is selectively reduced. Therefore the cyclization event that leads to **1** is most likely to happen before its reduction by KR.

4. Different binding motifs result in different stereospecificities.

Because of the three-point docking constraint, there are only two possible binding motifs of **1**, either from the front side or the back side of KR. The front side binding motif will lead to the R stereomer, while the back side binding motif will result in the S stereomer (Figure 7). Further, under the dual-constraints

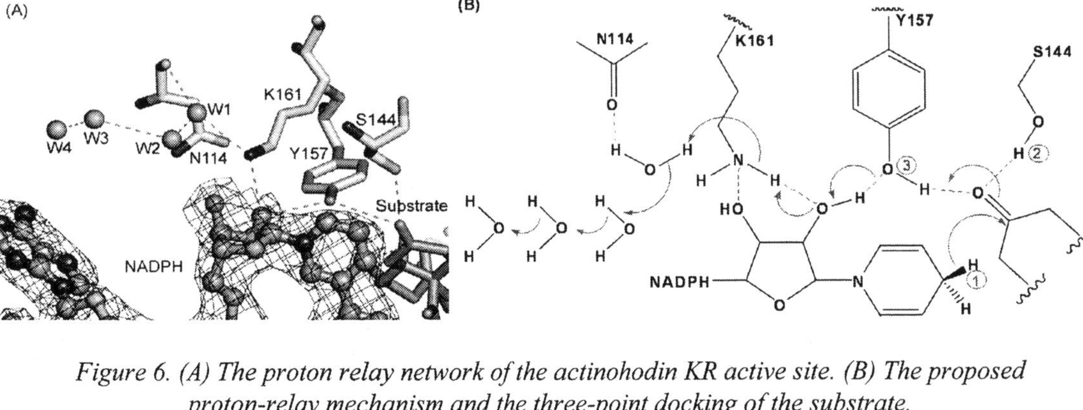

Figure 6. (A) The proton relay network of the actinohodin KR active site. (B) The proposed proton-relay mechanism and the three-point docking of the substrate.

imposed by the ring structure of **1** and the three-point docking of the KR active site, the C9-carbonyl group is optimally positioned for ketoreduction when the C7-C12 cyclization takes place (Figure 2).

The above observations help define the fundamental differences between fatty acid and aromatic polyketide biosynthesis. It is physically possible for fatty acid KR to reduce every carbonyl group of a growing linear polyketide chain (46,53,54). However, for the aromatic polyketide that involves a cyclized polyketide intermediate **1**, it is not energetically favorable for the polyketide KR to reduce carbonyl groups (other than C9) that have an energy penalty imposed by constraints of the active site and substrate geometry. Thus, the first ring cyclization differentiates aromatic PKS from FAS. Finally, sequence comparison indicates that the above features are highly conserved among aromatic polyketide KRs. Therefore mutations of the substrate binding pocket should lead to alternative regio- and stereospecificity for the polyketide substrate. The structures of actinorhodin KR provide an important step toward understanding aromatic polyketide reduction. It also provides the essential base for downstream mutational analyses of its active site and protein surface.

The Cyclization Specificity of ARO/CYC

The aromatic rings are essential for the antibiotic and anticancer activity of polyketides. Understanding the mechanism of aromatic ring formation is crucial for the generation of diverse polyketides with novel or improved therapeutic activity. (27,65). Towards the goal of understanding cyclization specificity, we have crystallized and solved the crystal structure of the tetracenomycin aromatase/cyclase (tcm ARO/CYC) (Figure 8C). Tcm ARO/CYC works directly downstream of tcm KS/CLF to promote the cyclization/aromatization of the first and second rings of tetracenomycin (Figure 1), starting with C9-C14 cyclization (66). Tcm ARO/CYC was cloned from the first 169 residues of the tcmN, a bi-functional enzyme consisting of an N-terminal ARO/CYC domain, and a C-terminal O-methyltransferase domain (66).

The tcm ARO/CYC crystal structure was solved to 1.9 Å using the multiwavelength anomalous diffraction (MAD) method. Our model reveals the following conclusions:

(1) ARO/CYC has a fold that does not resemble any dehydratase fold.

The crystal structure of ARO/CYC consists of seven beta strands, two small alpha helices, and a long C-terminal alpha helix (Figure 8A). This fold resemble the "hot dog in a bun" fold of birch and cherry allergens (67,68). This fold is

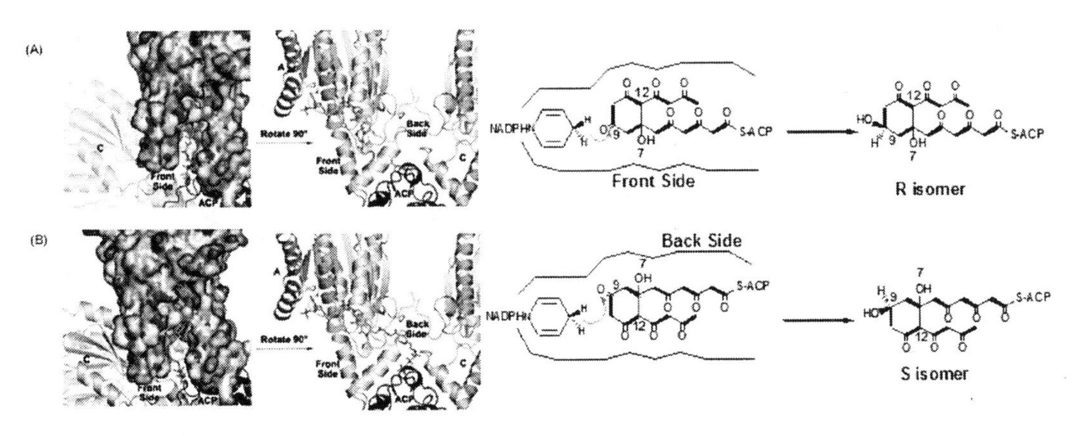

Figure 7. (A) The front side docking results in the R-stereomer during ketoreduction. (B) The back side docking results in the S isomer.

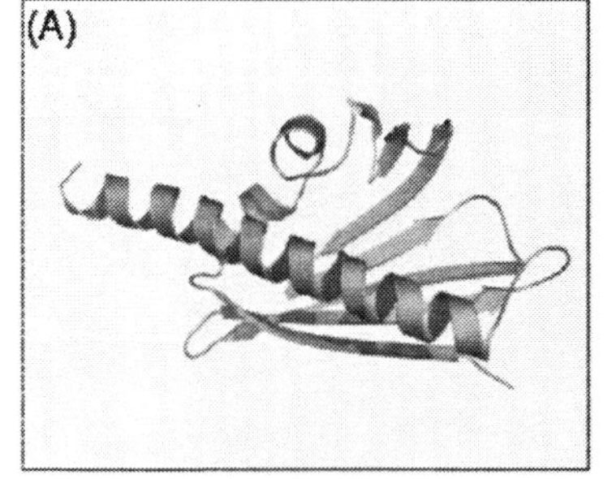

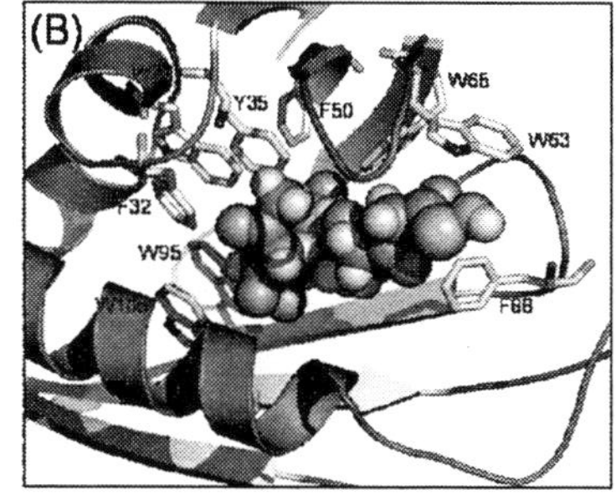

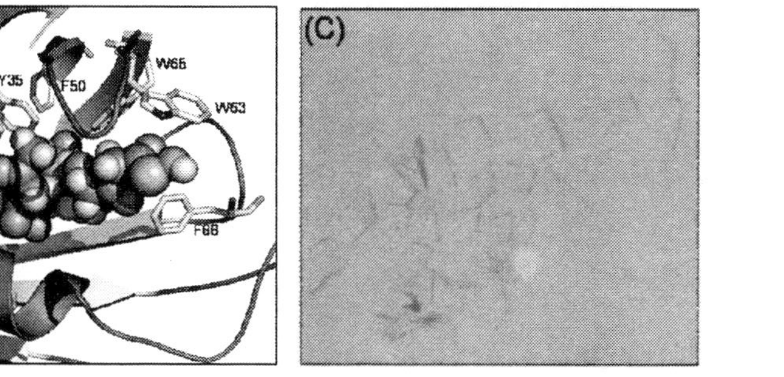

Figure 8. (A) Overall fold of tcm ARO/CYC. (B) Close-up view of binding cavity with a polyketide intermediate docked and hydrophobic aromatic residues labeled. (C) Crystals of ARO/CYC.

similar to that of FabZ and scytalone dehydratase, although tcm ARO/CYC has a uniquely deep substrate cavity (Figure 8B, Figure 9) (58,69). Therefore, the crystal structure of the aromatic polyketide ARO/CYC may represent a unique class of dehydratases or cyclases.

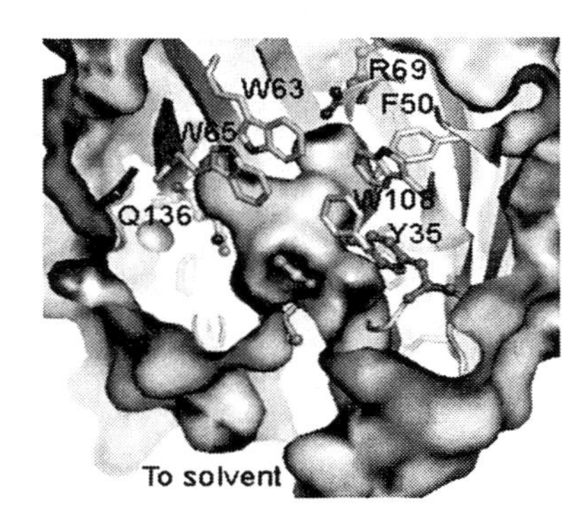

Figure 9. The active site of tcm ARO/CYC

(2) Tcm ARO/CYC has a highly aromatic substrate binding cavity

As shown in Figure 8B, the shape and size of the cavity are defined by many aromatic residues, including Trp28, Trp63, Trp65, Trp95, Trp108, Tyr35, Phe 32, Phe50 and Phe88. The highly aromatic environment will be a strong driving force for aromatic ring formation by stabilizing the transition state of aromatization (Figure 8B, Figure 9).

(3) The active site as a dehydratase

Previously, the crystal structure of a downstream cyclase (tcmI) has been solved. This enzyme cyclizes the fourth ring of tetracenomycin (61). Mutational analysis suggests that tcmI does not actively participate in the chemical reaction. Rather, due to the strong driving force to form the fourth aromatic ring, tcmI serves as a "mold" by steering the auto-cyclization in its substrate-binding cavity (61). TcmI shares no sequence homology with tcm ARO/CYC, and its protein fold is completely different from the tcm ARO/CYC structure. Therefore, tcm ARO/CYC may differ from tcmI by having actual dehydratase function. If this is the case, a general acid or base must be involved. In the active site of tcm ARO/CYC are four semi-conserved hydrophilic residues, Tyr35, Arg69, Gln110, and Asn136, that are identified by substrate docking simulation to serve as the active site base or acid during dehydration . Therefore the dehydration may proceed through a cationic elimination reaction, in which the conserved Arg69 serves as the proton source (Figure 8B, Figure 9).

(4) The molecular basis of ARO/CYC cyclization specificity

Substrate docking also indicates that C9-C14 cyclization is controlled by the size and shape of the binding cavity. Based on sequence alignment between tcm ARO/CYC (a C9-C14 ARO/CYC) and ZhuI (a C7-C12 ARO/CYC), we identified residues that are important to distinguish between these two possible cyclization motifs. For ZhuI, an ARO/CYC that has a C7-C12 cyclization specificity (33), the cavity residues become smaller, such as Tyr28, T35, Asn63, Leu108, Val50 and Val88. The homology model and substrate docking of ZhuI indicates that the polyketide product from C7-C12 cyclization has bulkier substituents than the C9-C14 product, and requires a larger substrate binding cavity. In the future, structure-based mutagenesis will confirm if these cavity residues are indeed important for the cyclization specificity of ARO/CYC.

Concluding Remarks

The structures of KS/CLF, KR and ARO/CYC have provided strong clues to the molecular features that result in the observed chain length, ketoreduction and cyclization specificities during polyketide biosynthesis. Based on structural information, the polyketide chain length has been altered by mutations of the CLF residues at the KS/CLF dimer interface. In the future, it should be possible to mutate residues of KS/CLF, KR and ARO/CYC to change the specificity of ketoreduction and cyclization. Therefore, the crystal structures of PKS domains will serve as the blueprints to guide the combinatorial efforts of polyketide biosynthesis.

Literature Cited

1. Shen, B. *Curr Opin Chem Biol* **2003**, *7*, 285-295.
2. Kim, E.S., Bibb, M.J., Butler, M.J., Hopwood, D.A. & Sherman, D.H. *Gene* **1994**, *141*, 141-142.
3. Motamedi, H. & Hutchinson, C.R. *Proc Natl Acad Sci U S A* **1987**, *84*, 4445-4449.
4. Malpartida, F. & Hopwood, D.A. *Nature* **1984**, *309*, 462-464.
5. Jakobi, K. & Hertweck, C. *J Am Chem Soc* **2004**, *126*, 2298-2299.
6. Otten, S.L., Stutzman-Engwall, K.J. & Hutchinson, C.R. *J Bacteriol* **1990**, *172*, 3427-3434.
7. Blanco, G., Fu, H., Mendez, C., Khosla, C. & Salas, J.A. *Chem Biol* **1996**, *3*, 193-196.
8. Dairi, T., Hamano, Y., Igarashi, Y., Furumai, T. & Oki, T. *Biosci Biotechnol Biochem* **1997**, *61*, 1445-1453.

9. Martin, R., Sterner, O., Alvarez, M.A., de Clercq, E., Bailey, J.E. & Minas, W. *J Antibiot (Tokyo)* **2001**, *54*, 239-249.
10. Li, A. & Piel, J. *Chem Biol* **2002**, *9*, 1017-1026.
11. Khosla, C. & Zawada, R.J. *Trends Biotechnol* **1996**, *14*, 335-341.
12. Woodward, R.B. et al. *Journal of the American Chemical Society* **1981**, *103*, 3210-3213.
13. Woodward, R.B. et al. *Journal of the American Chemical Society* **1981**, *103*, 3213-3215.
14. Woodward, R.B. et al. *Journal of the American Chemical Society* **1981**, *103*, 3215-3217.
15. Pfeifer, B.A., Admiraal, S.J., Gramajo, H., Cane, D.E. & Khosla, C. *Science* **2001**, *291*, 1790-1792.
16. Hopwood, D.A. *Chem Rev* **1997**, *97*, 2465-2498.
17. Hutchinson, C.R. *Chem Rev* **1997**, *97*, 2525-2536.
18. Floss, H.G. *J Ind Microbiol Biotechnol* **2001**, *27*, 183-194.
19. McDaniel, R., Licari, P. & Khosla, C. *Adv Biochem Eng Biotechnol* **2001**, *73*, 31-52.
20. Robinson, J.A. *Philos Trans R Soc Lond B Biol Sci* **1991**, *332*, 107-114.
21. Mendez, C. & Salas, J.A. *Comb Chem High Throughput Screen* **2003**, *6*, 513-526.
22. Bisang, C. et al. *Nature* **1999**, *401*, 502-505.
23. O'Hagan, D. *Nat Prod Rep* **1993**, *10*, 593-624.
24. Meurer, G., Gerlitz, M., Wendt-Pienkowski, E., Vining, L.C., Rohr, J. & Hutchinson, C.R. *Chem Biol* **1997**, *4*, 433-443.
25. Marti, T., Hu, Z., Pohl, N.L., Shah, A.N. & Khosla, C. *J Biol Chem* **2000**, *275*, 33443-33448.
26. Kalaitzis, J.A. & Moore, B.S. *J Nat Prod* **2004**, *67*, 1419-1422.
27. Hertweck, C., Xiang, L., Kalaitzis, J.A., Cheng, Q., Palzer, M. & Moore, B.S. *Chem Biol* **2004**, *11*, 461-468.
28. McDaniel, R., Ebert-Khosla, S., Fu, H., Hopwood, D.A. & Khosla, C. *Proc Natl Acad Sci U S A* **1994**, *91*, 11542-11546.
29. Rix, U., Fischer, C., Remsing, L.L. & Rohr, J. *Nat Prod Rep* **2002**, *19*, 542-580.
30. Kantola, J., Blanco, G., Hautala, A., Kunnari, T., Hakala, J., Mendez, C., Ylihonko, K., Mantsala, P. & Salas, J. *Chem Biol* **1997**, *4*, 751-755.
31. McDaniel, R., Ebert-Khosla, S., Hopwood, D.A. & Khosla, C. *Nature* **1995**, *375*, 549-554.
32. Shen, B. & Hutchinson, C.R. *Proc Natl Acad Sci U S A* **1996**, *93*, 6600-6604.
33. Tang, Y., Lee, T.S. & Khosla, C. *PLoS Biol* **2004**, *2*, E31.
34. Fu, H., McDaniel, R., Hopwood, D.A. & Khosla, C. *Biochemistry* **1994**, *33*, 9321-9326.

35. McDaniel, R., Ebert-Khosla, S., Hopwood, D.A. & Khosla, C. *Science* **1993**, *262*, 1546-1550.
36. Meadows, E.S. & Khosla, C. *Biochemistry* **2001**, *40*, 14855-14861.
37. Carreras, C.W. & Ashley, G.W. *Exs* **2000**, *89*, 89-108.
38. Walsh, C.T. *Science* **2004**, *303*, 1805-1810.
39. Burson, K.K., Huestis, W.H. & Khosla, C. *Abstracts of Papers of the American Chemical Society* **1997**, *213*, 83-Biot.
40. Staunton, J. & Weissman, K.J. *Nat Prod Rep* **2001**, *18*, 380-416.
41. Keatinge-Clay, A.T., Shelat, A.A., Savage, D.F., Tsai, S.C., Miercke, L.J., O'Connell, J.D., 3rd, Khosla, C. & Stroud, R.M. *Structure (Camb)* **2003**, *11*, 147-154.
42. Keatinge-Clay, A.T., Maltby, D.A., Medzihradszky, K.F., Khosla, C. & Stroud, R.M. *Nat Struct Mol Biol* **2004**,
43. Pan, H., Tsai, S., Meadows, E.S., Miercke, L.J., Keatinge-Clay, A.T., O'Connell, J., Khosla, C. & Stroud, R.M. *Structure (Camb)* **2002**, *10*, 1559-1568.
44. Jansson, A., Koskiniemi, H., Erola, A., Wang, J., Mantsala, P., Schneider, G. & Niemi, J. *J Biol Chem* **2004**,
45. Sciara, G., Kendrew, S.G., Miele, A.E., Marsh, N.G., Federici, L., Malatesta, F., Schimperna, G., Savino, C. & Vallone, B. *Embo J* **2003**, *22*, 205-215.
46. Price, A.C., Zhang, Y.M., Rock, C.O. & White, S.W. *Biochemistry* **2001**, *40*, 12772-12781.
47. Huang, W., Jia, J., Edwards, P., Dehesh, K., Schneider, G. & Lindqvist, Y. *Embo J* **1998**, *17*, 1183-1191.
48. Serre, L., Verbree, E.C., Dauter, Z., Stuitje, A.R. & Derewenda, Z.S. *J Biol Chem* **1995**, *270*, 12961-12964.
49. Qiu, X. & Janson, C.A. *Acta Crystallogr D Biol Crystallogr* **2004**, *60*, 1545-1554.
50. Olsen, J.G., Kadziola, A., von Wettstein-Knowles, P., Siggaard-Andersen, M., Lindquist, Y. & Larsen, S. *FEBS Lett* **1999**, *460*, 46-52.
51. Davies, C., Heath, R.J., White, S.W. & Rock, C.O. *Structure Fold Des* **2000**, *8*, 185-195.
52. Price, A.C., Rock, C.O. & White, S.W. *J Bacteriol* **2003**, *185*, 4136-4143.
53. Price, A.C., Zhang, Y.M., Rock, C.O. & White, S.W. *Structure (Camb)* **2004**, *12*, 417-428.
54. Fisher, M., Kroon, J.T., Martindale, W., Stuitje, A.R., Slabas, A.R. & Rafferty, J.B. *Structure Fold Des* **2000**, *8*, 339-347.
55. Rafferty, J.B., Simon, J.W., Baldock, C., Artymiuk, P.J., Baker, P.J., Stuitje, A.R., Slabas, A.R. & Rice, D.W. *Structure* **1995**, *3*, 927-938.
56. Baldock, C., Rafferty, J.B., Stuitje, A.R., Slabas, A.R. & Rice, D.W. *J Mol Biol* **1998**, *284*, 1529-1546.
57. Roujeinikova, A. et al. *J Mol Biol* **1999**, *294*, 527-535.

58. Leesong, M., Henderson, B.S., Gillig, J.R., Schwab, J.M. & Smith, J.L. *Structure* **1996**, *4*, 253-264.
59. Crump, M.P., Crosby, J., Dempsey, C.E., Parkinson, J.A., Murray, M., Hopwood, D.A. & Simpson, T.J. *Biochemistry* **1997**, *36*, 6000-6008.
60. Findlow, S.C., Winsor, C., Simpson, T.J., Crosby, J. & Crump, M.P. *Biochemistry* **2003**, *42*, 8423-8433.
61. Thompson, T.B., Katayama, K., Watanabe, K., Hutchinson, C.R. & Rayment, I. *J Biol Chem* **2004**, *279*, 37956-37963.
62. Sultana, A., Kallio, P., Jansson, A., Wang, J.S., Niemi, J., Mantsala, P. & Schneider, G. *Embo J* **2004**, *23*, 1911-1921.
63. Korman, T.P., Hill, J.A., Vu, T.N. & Tsai, S.C. *Biochemistry* **2004**, *43*, 14529-14538.
64. Hadfield, A.T., Limpkin, C., Teartasin, W., Simpson, T.J., Crosby, J. & Crump, M.P. *Structure (Camb)* **2004**, *12*, 1865-1875.
65. Shen, Y., Yoon, P., Yu, T.W., Floss, H.G., Hopwood, D. & Moore, B.S. *Proc Natl Acad Sci U S A* **1999**, *96*, 3622-3627.
66. Zawada, R.J. & Khosla, C. *J Biol Chem* **1997**, *272*, 16184-16188.
67. Fedorov, A.A., Ball, T., Valenta, R. & Almo, S.C. *Int Arch Allergy Immunol* **1997**, *113*, 109-113.
68. Fedorov, A.A., Ball, T., Mahoney, N.M., Valenta, R. & Almo, S.C. *Structure* **1997**, *5*, 33-45.
69. Wawrzak, Z., Sandalova, T., Steffens, J.J., Basarab, G.S., Lundqvist, T., Lindqvist, Y. & Jordan, D.B. *Proteins* **1999**, *35*, 425-439.

Chapter 13

Mechanisms of Type III Polyketide Synthase Functional Diversity: From 'Steric Modulation' to the 'Reaction Partitioning' Model

Michael B. Austin and Joseph P. Noel*

Howard Hughes Medical Institute, The Jack H. Skirball Center for Chemical Biology and Proteomics, The Salk Institute for Biological Studies, La Jolla, CA 92037

Subtle active site changes often result in the redirection of type III polyketide synthase (PKS) reaction pathway intermediates toward different product fates. Complementary insights from two of our recently published structural and mechanistic studies prompt us to revise the existing 'steric modulation' mechanistic model for type III PKS functional divergence and resultant product specificity. Besides allowing for the active control of the reactivity of polyketide intermediates through mechanisms other then pure steric shape-dependent factors, our new 'reaction partitioning' model of functional divergence also recognizes the mechanistic importance of intramolecular chemical features of an extended polyketide that control the intrinsic reactivities of these highly reactive enzyme generated intermediates. This new model better explains how subtle active site changes can alter catalytic steps downstream of alternative reaction pathway branch points. Although formulated based upon work on type III PKS product specificity, the concept of 'reaction partitioning' can be used to describe control mechanisms in many other PKS biosynthetic systems that do not employ complete reduction of the growing polyketide intermediates.

'Steric Modulation' and Type III PKS Functional Diversity

Natural products include a diverse collection of small molecules, which confer upon their hosts a multitude of attractive, defensive, communicative, or other biological advantages. An organism's suite of natural products often determines both intra- and inter-species interactions, thus contributing to both the identity and individuality of a species. Remarkably, most natural products are built up from a handful of simple building blocks, usually derived from one or more primary metabolic pathways. Likewise, the majority of functionally diverse enzymes participating in secondary or specialized metabolic pathways belong to just a few enzyme superfamilies, grouped by similarity of tertiary protein structure and/or mechanism of catalysis. For example, the reaction at the core of both fatty acid and polyketide biosynthesis is an iterative cysteine-mediated intermolecular Claisen condensation of acyl thioester substrates to form β-keto-esters in a process catalyzed by evolutionarily conserved condensing enzymes sharing the thiolase- or αβαβα-fold*(1)*. In fatty acid biosynthesis, subsequent reduction of each successive β-keto position by associated pathway enzymes produces a relatively inert hydrocarbon tail. In contrast, preservation of all (or some) polar β-keto carbonyls during polyketide biosynthesis results in considerably more reactive polyketide intermediates that can undergo a number of additional intramolecular condensation reactions to produce a diverse collection of hydroxylated and aromatic cyclized natural products exploited by nature for a multitude of purposes.

Our experimental focus over the past few years has been the elucidation of mechanistic principles behind the impressive evolutionary functional divergence of enzymes belonging to the structurally simple chalcone/stilbene synthase (or type III) superfamily of polyketide synthases (PKSs), which differ in their preferences for starter molecules, the number of acetyl additions they catalyze, the nature of the extender moiety and their mechanism of terminating polyketide chain extension (See Figure 1)*(1, 2, 3)*. Using bioinformatic, crystallographic, mutagenic, and protein engineering techniques, we have sought to isolate and characterized the key variables that determine enzymatic product specificity, with special attention to alternative mechanisms and specificities of intramolecular polyketide cyclization. Parallel insights from two such recently-published projects *(4, 5)* have reshaped our understanding of type III PKS catalysis, and further demonstrated the mechanistic relevance of the intrinsic reactivities of various polyketide intermediates.

The first PKS crystal structure, that of alfalfa chalcone synthase (CHS) in 1999 *(6)*, had previously revealed the type III PKS internal active site cavity and conserved Cys/His/Asn catalytic triad responsible for starter substrate loading and iterative polyketide extension *(6)*. CHS provides the first committed chemical intermediate in flavonoid biosynthesis by catalyzing the sequential decarboxylative addition of three acetate units from malonyl-CoA to a *p*-

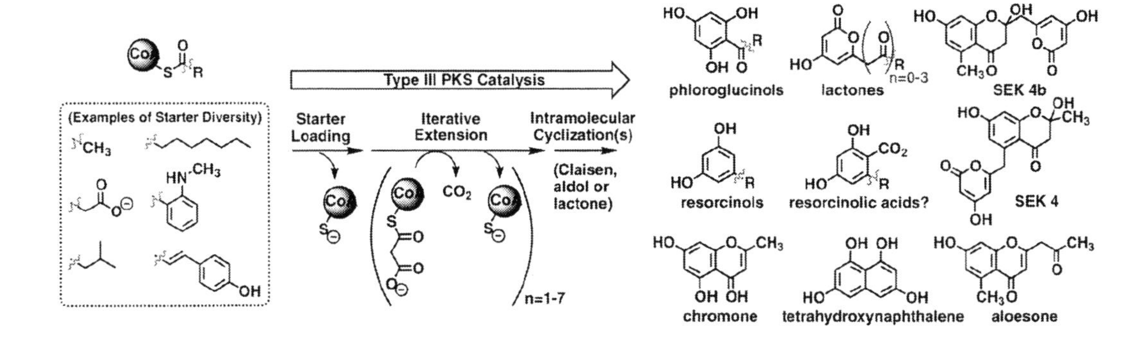

Figure 1. Diversity of Type III PKS catalytic specificity for substrate, number of polyketide extension steps, and ring formation. Functional divergence of CHS-like enzymes involves a wide range of acyl-CoA starter molecules and several distinct intramolecular cyclization modes.

coumaroyl-CoA starter molecule derived from phenylalanine via the general phenylpropanoid pathway. CHS then catalyzes chalcone formation via the intramolecular cyclization and aromatization of the resulting linear phenylpropanoid tetraketide *(6)*. Interestingly, the CHS active site cavity revealed no other "reactive" residues likely to catalyze the final intramolecular cyclization of CHS's tetraketide intermediate. Furthermore, sequence comparisons showed that other type III PKS enzyme sequences vary in the identity and steric bulk of residues aligning with CHS active site residues. The obvious interpretation was that the divergent functionality of CHS-like enzymes depends, to a first approximation, solely upon side chain volume and polarity across the dozen or so residues lining their active site cavities. This "passive" model of enzyme-directed polyketide biosynthesis invokes steric and polar influences to confer a unique topology upon functionally divergent active site cavities, in order to filter out starter molecules based upon size and shape and then enforce specific cyclization-conducive folded conformations of their respective linear polyketide intermediates. The subsequently published 2-PS crystal structure supported this initial model by confirming that the increase in steric bulk of just a few active site residues resulted in a major reduction of active site cavity volume *(7)*, and mutagenic substitutions at these same positions further confirmed the importance of the available space and shape of the active site cavity in determining specificity for starter molecules, number of polyketide extension cycles, and selection between Claisen and lactone modes of intramolecular cyclization *(7, 8)*.

Development of 'Reaction Partitioning' Model

In order to further refine the existing 'steric modulation' model of PKS catalysis, and also to determine whether this explanatory model was sufficient to explain all type III PKS functional diversity, we undertook a homology-guided bioinformatic evaluation of known CHS-like protein sequences and existing functional data. Our detailed application of the existing model to each known type III PKS enzyme family produced a number of new and useful predictions regarding active site cavity shape and size, as well as specific amino acid substitutions responsible for the distinct substrate and product specificities of most of these enzymes. This analysis was subsequently published in review form *(1)*. More importantly, this preliminary examination diagnosed two type III PKS enzyme families, the stilbene synthases (STSs) and tetrahydroxynaphthalene synthases (THNSs), whose substrate and product specificities seemed inconsistent with the enzymatic principles embodied in the steric modulation model. Our subsequent structural elucidation of STS and THNS enzymes revealed unanticipated displacements in each of their protein backbones, relative to CHS and 2-PS *(4, 5)*. In each case, our further mutagenic analyses of their reactions also unexpectedly implicated a chemical (rather than steric or polar) role for residues lining their active site cavities *(4, 5)*. Each of these studies generated specific mechanistic hypotheses for further testing, and insights gleaned from both projects facilitated our formulation of a revised and more comprehensive "reaction partitioning" catalytic model to explain the evolution and expansion of type III PKS metabolic diversity.

In the reaction sequence of CHS, *(6)* there are more than 20 distinct events that must occur in a stepwise fashion between binding of the CoA-linked starter and formation of the aromatic chalcone product. While the occurrence of all or nearly all of these processes within a single active site cavity renders mutagenic characterization of isolated steps nearly impossible, it is common practice to think of each of these steps as independent catalytic events, as one would separately consider the consecutive action of distinct enzymes in a classical linear biosynthetic pathway. While this reductionist approach is necessary for the scientific analysis of these complex multi-functional enzymes, it can also foster an overly myopic focus upon the enzyme's catalytic role during each step, at the expense of an appreciation for the inherent reactivity of unstable chemical intermediates in transit as PKS-mediated steps iteratively alter this intrinsic reactivity. In simple linear biosynthetic pathways carried out by a succession of discrete enzymes, the biochemical or genetic inactivation of one enzyme typically leads to accumulation of the upstream enzyme's product. In contrast, mutagenic or natural variation of residues in most type III PKS active site cavities (other than the conserved catalytic triad necessary for iterative polyketide extension) in multi-functional CHS-like enzymes often results in the redirection of polyketide cyclization fate, rather than accumulation of linear intermediates. The most direct implication of this observation is that multiple modes of cyclization are available to type III PKS active sites, apparently without the need for direct chemical intervention of active site residues as general acids or bases. Consistent with all data available at the time, the original steric modulation model further explained this lack of reactive residues as an indication that type III PKS cyclization specificity is achieved via the strict enforcement, by variation in the shape and polar features of the active site cavity, of specific cyclization-conducive conformations *(6)*.

Our subsequent comparative structural and mechanistic analyses of the STS and THNS active site cavities and reactions produced results that failed to adhere to this initial model on several counts *(4, 5)*. Neither of these functionally divergent type III PKS active site topologies seemed restrictive enough to enforce a single cyclization-conducive polyketide conformation. In the case of THNS, this mystery was confounded by the biosynthesis of products of quite different size within the same or quite similar active site cavities *(5)* (See Figure 2). A final indication of the need to revise the existing model was obviated by the apparent mechanistic involvement of chemically reactive residues other than the catalytic triad in both the STS and THNS physiological reactions *(4, 5)*. All

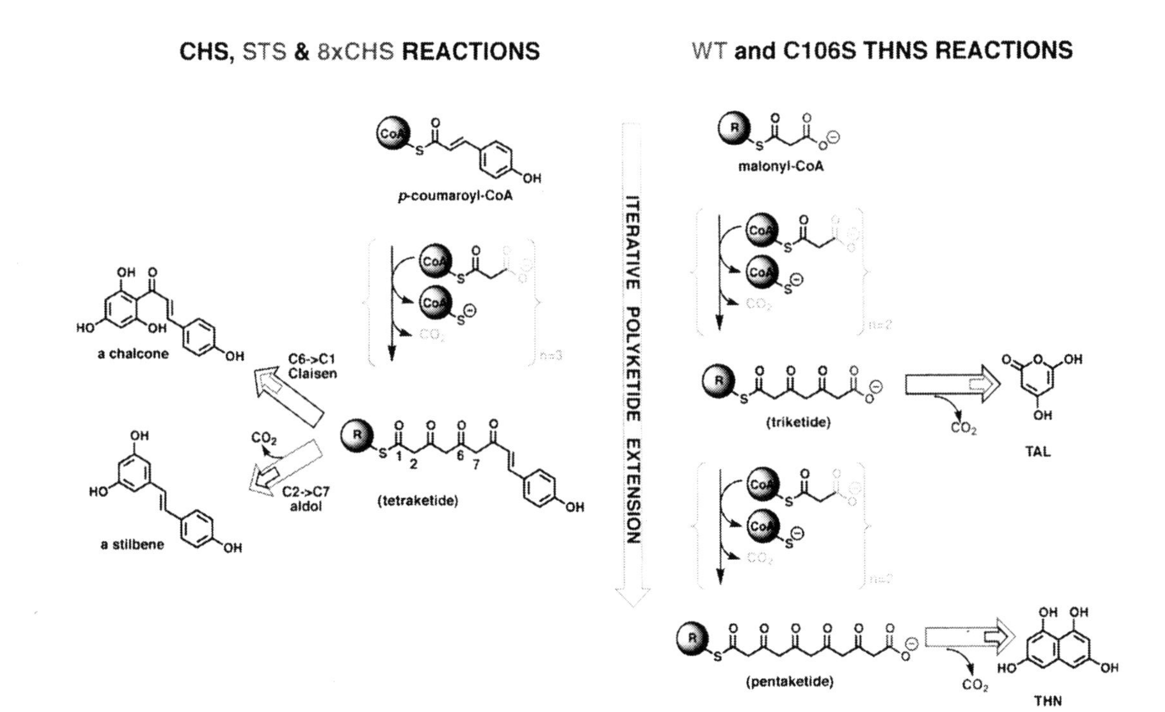

Figure 2. Reaction partitioning by wild type and mutant enzymes discussed in the text. Color of reaction arrows matches enzyme name color, and large and small arrows of the same color represent major and minor products of a given enzyme. R=CoA or catalytic cysteine of the associated PKS.

of these observations, however, are consistent with our revised 'reaction partitioning' general model of type III PKS catalysis and functional divergence, which recognizes the significant role of the intrinsic intramolecular reactivity of transient polyketide intermediates in determining type III PKS cyclization specificity.

This revised model rests with the assumption that iterative polyketide chain extension constituted the fundamental gain of catalytic function behind the subsequent evolutionary explosion of type III PKS functional diversity *(1)*. While this may seem an obvious statement, it redirects our attention toward the intrinsic chemical reactivity of the resulting polyketide intermediates to undergo a number of alternative and competing intramolecular cyclizations *(9)*. The more than 20 distinct steps leading to chalcone formation constitutes one of several branches of alternative reaction pathways available to *p*-coumaroyl-primed polyketide extension by CHS *(1)*. The detection of minor lactone and pyrone byproducts formed in vitro during chalcone biosynthesis served early on as mechanistic evidence for the existence of alternative reaction paths available to CHS-derived reaction intermediates. While formation of these off-pathway heterocyclic lactones and pyrones, especially when resulting from impairment of the expected extension or cyclization fates of reactive intermediates, was always recognized as evidence of linear polyketide intermediate reactivity, this inherent instability was largely dismissed, other than as a possible evolutionary vestige of an earlier and less catalytically complex type III PKS activity *(1)*.

The mechanistic importance of intrinsic polyketide reactivity to the product specificities of modern CHS-like enzymes was not apparent until our unexpectedly smooth structure-guided mutagenic conversion (8xCHS) of alfalfa CHS to a functional STS, without significant modification of active site topology *(4)* (See Figure 2). This simultaneous loss of CHS-like C6->C1 (Claisen) cyclization function and gain of STS-like C2->C7 (aldol) cyclization function, with no expected intervening 'derailment' (i.e. lactone) cyclization products was unexpected, as was the apparent ease with which these functionally-converted CHS mutants achieved the other unique steps of the STS reaction pathway (hydrolytic thioester cleavage and decarboxylative elimination of C1). These results implied that a single mechanistic event (emergence of the necessary STS thioesterase activity) is also sufficient to fully convert the reaction pathway of CHS to that of STS. More specifically, the critical 'aldol switch' mechanistic determinant that distinguishes CHS and STS enzymatic catalysis appeared to be confined to alternative enzymatic partitioning of an identical intermediate between two different competing reactions *(4)*. The consistency of these new results with previously observed differences in the spontaneous cyclization specificity of esterified versus free acid polyketides in solution *(9)*, for the first time clearly indicated the important contribution of intrinsic polyketide reactivity in determining type III PKS cyclization specificity.

The mechanistic relevance of this reaction partitioning concept was again illustrated by our structural and mutagenic analysis of the iterative THNS reaction, where an isosteric mutation of cysteine to serine in the active site cavity resulted in the dramatic reallocation of malonyl-primed THNS intermediates away from the physiologically expected pentaketide-derived fused-ring naphthalene product, instead forming a triketide-derived lactone (TAL) *(5)*. The similarity of a thiol theoretical pKa (but not an isosteric alcohol pKa) to that of a polyketide methylene carbon implicates some critical and relatively early chemical role for this cysteine residue in the THN-pathway reaction, either during or prior to formation of the tetraketide intermediate (i.e. the point of THN and TAL pathway divergence) *(5)*. This result is again inconsistent with the steric modulation model. The dramatically different wild-type THN and C106S TAL reaction pathways occurring within presumably isosteric active site cavities again suggests that a single reaction partitioning enzymatic event controls the flux between these two competing pathways *(5)*. While several alternative mechanistic roles for this residue remain plausible, a continued structural and mechanistic focus upon the point of THN and TAL pathway divergence will significantly simplify future experimentation.

From these specific cases, we can construct a general "reaction partitioning" model for type III PKS cyclization specificity that seems likely to apply to all non-reductive polyketide biosynthetic steps. In any iterative type III PKS multi-step reaction sequence, each successive polyketide intermediate has an increasing number of potentially competing reactions available to it, only one of which is on-pathway to any specific final product (See Figure 3).

This inherent potential of linear polyketides to undergo alternative cyclization reactions forms the basis for our concept of enzymatic 'reaction partitioning' as an important mechanistic focal point for the study of type III PKS product specificity. Thus, it is the complex interplay between multifunctional type III PKS active sites and their reactive polyketide intermediates that ultimately determines product specificity. The actual ratio of alternative products resulting from iterative polyketide extension from a given starter depends upon 'reaction partitioning' between the competing reaction pathways at just a few key diverging mechanistic branch points. Our structural and mutagenic analyses of the reaction pathways catalyzed by STS and THNS clearly indicate that an important evolutionary mechanism for type III PKS product divergence involves modulation of the degree of enzymatic control exerted over various reactive intermediates*(4, 5)*. In other words, the type III PKS active site can actively prevent, passively allow or actively promote each of several potential competing reactions, and a subtle change in the degree of enzymatic control at a crucial mechanistic branch point can result in the dramatic redirection of intermediates toward a new fate. In most cases, our query regarding divergent product specificities should not be 'How does the enzyme catalyze its intramolecular cyclization step', but rather 'How does the enzyme favor a given reaction pathway from among the several alternatives available to its polyketide intermediates'?

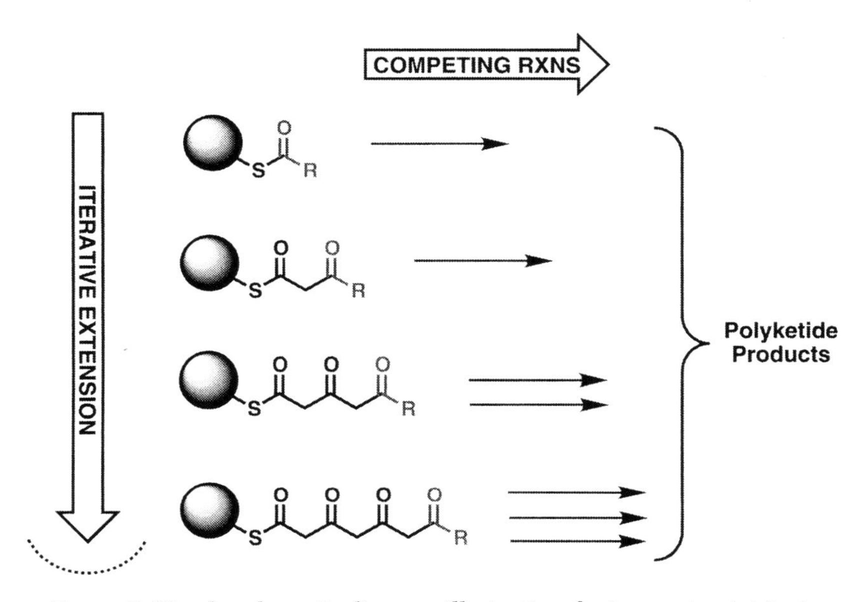

Figure 3. Simple schematic diagram illustrating the increasing intrinsic susceptability of each subsequent polyketide intermediate toward alternative competing reactions. Our "reaction partitioning" model of diversity states that evolutionary divergence of PKS product specificity results from alternative partitioning of intermediates at key mechanistic branchpoints.

Although the focus of our discussion here has been on the mechanistic implications of more recent data inconsistent with the original steric modulation hypothesis, our previous bioinformatic analysis of type III PKS functional diversity suggested that much of the product specificity of CHS-like enzymes is nonetheless indeed controlled via steric and polar modulation of active site shape and volume *(1)*. The initial mechanistic role played by a given active site topology is clearly to favor a particularly size and shape of starter molecule, although *in vitro* promiscuity toward a range of CoA-activated starters confirms that *in vivo* substrate availability remains an important determinant of physiological starter usage. Generally speaking, after loading of an acceptable and available starter, iterative polyketide extension continues until some alternative reaction becomes more favorable. In addition to effects upon starter specificity, steric restriction of active site volume to prevent tetraketide formation also appears to be the preferred method for selecting triketide lactonization (the only favorable intramolecular cyclization available to most triketides) *(7)*.

It would appear from most *in vitro* assays with mutant or incorrectly primed enzymes that triketide or tetraketide lactonization via attack on the C1 thioester by the C5 carbonyl oxygen is the easiest product-releasing cyclization for type III PKS enzymes to achieve *(1)*. Tetraketide intermediates possess an additional and favorable off-loading cyclization reaction possibility, namely the C6->C1 Claisen condensation typified by CHS. While *in vitro* experiments reveal that this cyclization route is easily diverted to lactonization by mutagenesis or acyl starter substitutions, there appears to be a notable selective advantage associated with production of the more physiologically stable phloroglucinol ring, as evidenced by the preponderance of plant type III PKS enzymes whose physiological reactions feature both unusual starter specificity and CHS-like C6->C1 intramolecular cyclization *(1)*. Such enzymes appear to have utilized steric modulation of active site topology to adapt their preferred CHS-like cyclizations to tolerate acyl starter substitution.

As previously discussed, a third alternative for tetraketide cyclization, the C2->C7 aldol condensation of STS enzymes, is triggered by the emergence in CHS enzymes of a thioesterase-like hydrogen bonding network *(4)*. It remains to be seen whether engineered incorporation of this 'aldol switch' into other phloriglucinol-forming type III PKS active sites will allow enzymatic production of resorcinols from a range of starters. And finally, given the range of intrinsic cyclization choices that terminate polyketide extension at or before the tetraketide stage, it is perhaps not surprising that very few type III PKS enzymes manage to form longer polyketides. The THNS active site topology suggests that maintenance of a deep but narrow active site cavity is likely an important factor in achieving the unusual combination of small starter selection and several polyketide extension steps *(5)*. Steric prevention of cyclization-productive conformations is one obvious means to control reactivity, but in the case of THNS an additional active site cysteine residue also appears to play a critical chemical role in preventing the otherwise facile biosynthesis of malonyl-primed triketide lactones *(5)*.

Conclusions and Future Directions

This revised model of PKS mechanistic regulation provides an improved framework for studying the functional divergence of evolutionarily related PKS enzymes. Our model posits that alternative product specificities from a given acyl-CoA starter must initially arise due to the selection of some mutation or set of mutations that specifically alters enzymatic reaction partitioning of key intermediates into other competing reaction pathways. Whether this relative kinetic effect is due to an increase or decrease of enzymatic control over the intrinsic reactivity of key intermediates may not be readily apparent from comparing the overall kinetics of evolutionarily refined CHS-like enzymes, as the critical branch point reactions may not be rate-limiting, and other subsequent mutations may also alter the kinetics of other non-branchpoint steps in the overall polyketide biosynthetic reaction. In light of our reaction partitioning model, it is not surprising that efforts to trap or crystallographically observe polyketide intermediates of either wild-type or mutant type III PKS enzymes have thus far been unsuccessful, which has contributed to our inability to unambiguously establish the mechanistic effects of specificity-altering mutations, as evidenced in both the STS and THNS projects *(4, 5)*. While our improved theoretical understanding of the structural and mechanistic principles underpinning the evolution of type III PKS functional divergence will facilitate future mechanistic analyses, a few complementary experimental approaches seem most likely to lead to further theoretical and practical advances. These areas of focus include the targeted design and synthesis of specific intermediate and transition state analogs, the development of methods to characterize the kinetic and structural parameters of individual type III PKS reaction pathway steps, on-going and future bioengineering experiments to control and modulate type III PKS substrate and product catalytic specificity, and finally the continued discovery and characterization of functionally novel type III PKS enzymes.

The symmetric nature of conformationally flexible polyketide intermediates often means there are both multiple plausible mechanistic pathways to a given product and multiple plausible mechanistic interpretations of mutagenic data. While this complication has frustrated efforts to obtain snapshots of intimate interactions of CHS-like enzymes with their reactive intermediates, our recent theoretical advances, including generation of novel mechanistic hypotheses and improved appreciation for intrinsic polyketide reactivity, should facilitate the future design of better reaction and transition state synthetic mimics for this purpose. This latter point is an important one, as very few mechanistically useful type III PKS intermediate or transition state analogs are presently available. Besides such mechanistic snapshots, the future development of stopped-flow or NMR-based methods for kinetic analysis of individual steps in various type III PKS reactions is also likely to facilitate a better kinetic understanding of enzymatic reaction partitioning.

While our own protein engineering efforts have thus far been confined to elucidation of mechanistic principles, a number of other and more applied engineering projects have focused upon the practical benefits of harnessing this enzymatic machinery. Transgenic expression of STS has conferred biosynthesis of the natural antifungal agent resveratrol in several agricultural plant species lacking this useful pathway *(10)*. Altered flower color, nitrogen fixation, increased digestability of forage crops, and increased production of health-promoting natural products in food crops *(11)* are other goals of past and present flavonoid pathway engineering efforts (reviewed in *(12, 13)*). There has also been recent interest in reconstituting and engineering plant biosynthetic pathways in microbes *(14, 15, 16)*. The combination of flavonoid pathway enzymes from different plants *(17)*, as well as the combinatorial use of enzymes for hydroxylation, methylation, prenylation, glycosylation and acylation *(11)*, promises to yield even greater chemical diversity. While these experiments, as well as extensive domain-swapping in various type I and II PKSs *(18)*, demonstrate the utility of mixing and matching naturally evolved enzymes to achieve new pathways, our increasing understanding of individual enzymes will allow a more sophisticated approach. The ability to engineer substrate and product specificities (natural and non-natural) into mutant enzymes or catalytic domains will increasingly remove our dependence upon nature to provide desirable biocatalytic activities, as well as modified biosynthetic products.

In addition to the promise of these and other bioengineering projects, it is clear that there is still much mechanistic information to learn from the wealth of functional diversity created by natural evolution. The frequent discovery and characterization of new CHS-like enzymes possessing novel substrate or product specificities continues to expand the known range of natural type III PKS catalytic diversity. The still-rapid pace of discovery is illustrated by the presence of several newly discovered type III PKS cyclization modes in our introductory overview of type III PKS cyclization diversity (Figure 1), including heptaketide- and octaketide-derived natural products *(2, 3)* surpassing the type III polyketide chain length record previously held by THNS *(19)*. From a broader perspective, a recent crystal structure of 3-hydroxy-3-methylglutaryl-CoA (HMG-CoA) synthase unexpectedly revealed this enzyme of primary metabolism to constitute a third structural branch *(20)* (in addition to KAS III and type III PKS enzymes) of the Cys-His-Asn catalytic triad-utilizing subgroup of thiolase-fold enzymes *(1)*. Structural and mechanistic characterization of enzymes such as these will undoubtedly reveal further clues regarding the range of enzymatic strategies available to the type III PKS active site, as well as the evolutionary basis for the current functional diversity of these architecturally simple but mechanistically complex enzymes.

Acknowledgments

Work described in this manuscript was supported by the National Institutes of Health under Grant No. AI052443 and by the National Science Foundation under Grant No. MCB-0236027. Joseph P. Noel is an investigator of the Howard Hughes Medical Institute.

References

1. Austin, M. B.; Noel, J. P. *Nat Prod Rep* 2003, Vol. 20, 79-110.
2. Abe, I.; Utsumi, Y.; Oguro, S.; Noguchi, H. *FEBS Lett* 2004, Vol. 562, 171-176.

3. Abe, I.; Utsumi, Y.; Oguro, S.; Morita, H.; Sano, Y.; Noguchi, H. *J Am Chem Soc* 2005, Vol. 127, 1362-1363.
4. Austin, M. B.; Bowman, M. E.; Ferrer, J.; Schroder, J.; Noel, J. P. *Chem Biol* 2004, Vol. 11, 1179-1194.
5. Austin, M. B.; Izumikawa, M.; Bowman, M. E.; Udwary, D. W.; Ferrer, J. L.; Moore, B. S.; Noel, J. P. *J Biol Chem* 2004, Vol. 279, 45162-45174.
6. Ferrer, J. L.; Jez, J. M.; Bowman, M. E.; Dixon, R. A.; Noel, J. P. *Nat Struct Biol* 1999, Vol. 6, 775-784.
7. Jez, J. M.; Austin, M. B.; Ferrer, J.; Bowman, M. E.; Schröder, J.; Noel, J. P. *Chem Biol* 2000, Vol. 7, 919-930.
8. Jez, J. M.; Bowman, M. E.; Noel, J. P. *Biochemistry* 2001, Vol. 40, 14829-14838.
9. Harris, T. M.; Harris, C. M. *Pure Appl. Chem.* 1986, Vol. 58, 283-294.
10. Hipskind, J. D.; Paiva, N. L. *Mol Plant Microbe Interact* 2000, Vol. 13, 551-562.
11. Muir, S. R.; Collins, G. J.; Robinson, S.; Hughes, S.; Bovy, A.; Ric De Vos, C. H.; Van Tunen, A. J.; Verhoeyen, M. E. *Nat Biotechnol* 2001, Vol. 19, 470-474.
12. Winkel-Shirley, B. *Plant Physiol* 2001, Vol. 126, 485-493.
13. Dixon, R. A.; Lamb, C. J.; Masoud, S.; Sewalt, V. J.; Paiva, N. L. *Gene* 1996, Vol. 179, 61-71.
14. Watts, K. T.; Lee, P. C.; Schmidt-Dannert, C. *Chembiochem* 2004, Vol. 5, 500-507.
15. Yan, Y.; Chemler, J.; Huang, L.; Martens, S.; Koffas, M. A. *Appl Environ Microbiol* 2005, Vol. 71, 3617-3623.
16. Jiang, H.; Wood, K. V.; Morgan, J. A. *Appl Environ Microbiol* 2005, Vol. 71, 2962-2969.
17. Dong, X.; Braun, E. L.; Grotewold, E. *Plant Physiol* 2001, Vol. 127, 46-57.
18. Khosla, C.; Gokhale, R. S.; Jacobsen, J. R.; Cane, D. E. *Annu Rev Biochem* 1999, Vol. 68, 219-253.
19. Funa, N.; Ohnishi, Y.; Ebizuka, Y.; Horinouchi, S. *J Biol Chem* 2002, Vol. 277, 4628-4635.
20. Campobasso, N.; Patel, M.; Wilding, I. E.; Kallender, H.; Rosenberg, M.; Gwynn, M. N. *J Biol Chem* 2004, Vol. 279, 44883-44888.

Biotechnological Advances in Polyketide Biosynthesis

Chapter 14

Novel Polyketides from Genetic Engineering (... and Lessons We Have Learned from Making Them)

Leonard Katz, Jonathan Kennedy, Sarah C. Mutka, John R. Carney, Karen S. MacMillan, and Sumati Murli

Kosan Biosciences, Inc., 3832 Bay Center Place, Haywood, CA 94545

Knowledge of the domain organization of type I modular polyketide synthases (PKS) usually allows accurate prediction of the structure of the corresponding polyketide. Genetic engineering of such PKSs can be employed to produce novel compounds of predicted structure. Genetic engineering of the epothilone PKS has given unexpected results that have revealed subtleties of the biosynthesis not previously understood. In one instance, an unprecedented mechanism was uncovered.

Epothilone is a mixed peptide-polyketide produced from the myxobacterium *Sorangium cellulosum* (*1*) that inhibits proliferation of eukaryotic cells through the stabilization of microtubules (*2*). Epothilone D, as well as a number of derivatives of various epothilone congeners (Fig. 1), are currently in clinical trials for the treatment of various cancers. The compounds are made from a peptide synthetase (PS)-polyketide synthase (PKS) whose genes have been cloned, sequenced (*3, 4*), and expressed in various heterologous hosts (*4-7*). Organization of the genes, their corresponding proteins and constituent modules and domains, are shown in Fig. 1. As in the case of many type I modular polyketide synthases, the epothilone (epo) PKS has been engineered to produce novel analogs. Some of the analogs thus produced were what would be predicted from current understanding of polyketide biosynthetic pathways. Other changes to the PKS resulted in the production of novel compounds not predicted by standard polyketide biosynthesis models. Analysis of these compounds, discussed in this chapter, revealed aspects of the biosynthesis of epothilone not previously understood.

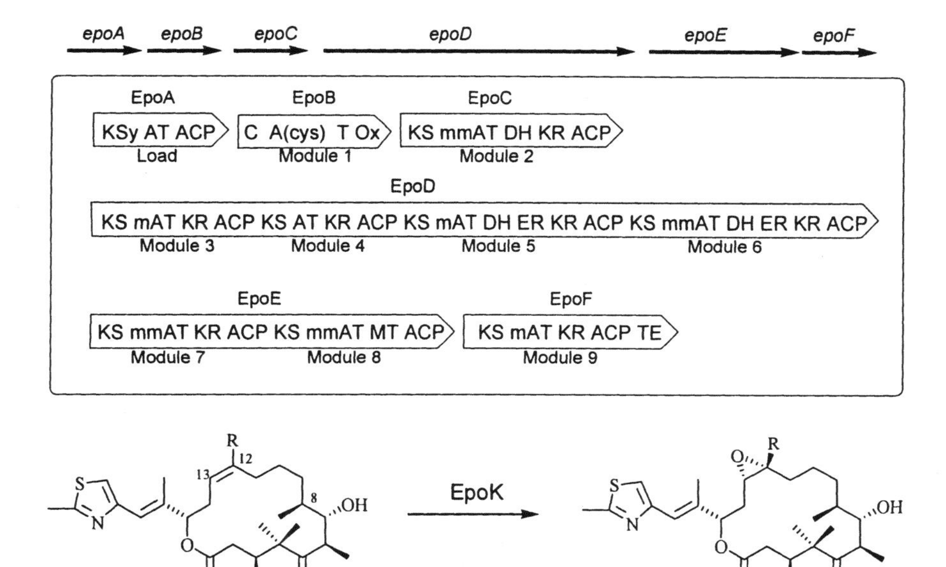

Figure 1. Domain organization of the epothilone PKS and structures of epothilone congeners. The top line shows the gene organization and gene names. The boxed segment below shows the organization of the domains and modules corresponding to each gene. Domain abbreviations: ACP, acyl carrier protein; AT, acyltransferase; mAT and mmAT, AT domain specifying malonate and methylmalonate, respectively; DH, dehydratase; ER, enoylreductase; MT, C-methyltransferase; KR, β-ketoreductase; KS, β-ketoacyl ACP synthase; KSy, KS domain containing tyrosine residue in the active site; TE, thioesterase.

Biosynthesis of Epothilone

Epothilone is produced from the progressive condensation of acyl thioesters employing the PKS shown schematically in Fig. 1. Synthesis starts with the loading of malonyl CoA on the AT domain of EpoA, followed by its transfer to the ACP domain and decarboxylation through the action of the KSy domain, leaving acetyl-ACP. EpoB is a peptide synthetase specific for cysteine. Condensation between the acetyl moiety on EpoA and cysteine on EpoB produces the acetyl-cysteinyl depsipeptide, and subsequent internal cyclization and oxidation produce the methylthiazole group found in the finished compounds. Condensation between the methylthiazole and the malonyl-ACP on module 2 (EpoC) produces a triketide. The β-carbonyl of the triketide is subsequently reduced to an OH group through the function of the KR domain in EpoC. This decarboxylative (Claisen) condensation resembles the condensations that take place during fatty acid biosynthesis. Subsequent progressive condensations catalyzed by modules 3 through 9 result in production of the decaketide which undergoes internal lactonization to generate epothilone C or epothilone D. Module 8 contains a methyltransferase domain that adds a second methyl group to C-2 of the acyl chain formed after the 8th condensation step to produce the geminal dimethyl found in the finished compounds. Epothilones A and B are produced from epothilones C and D through the action of the epoxidase EpoK, produced from a gene that is adjacent to the PKS-encoding genes.

The stepwise biosynthesis of epothilone follows a predictable route through the epo PS-PKS, but two aspects of the synthesis are not explained by the domain organization. It is clear that both epothilones C and D, which differ in the composition of the side chain at C-12, are produced from the same PKS, although only a single compound is produced from each round of biosynthesis. The AT domain in module 4 (epo AT4), therefore, must be able to utilize as a substrate either malonyl CoA (to produce epothilone C) or methylmalonyl CoA (to produce epothilone D). In general, AT domains in bacterial type I modular PKSs are highly specific for a single substrate, and epo AT4 represents the only known example of an AT domain with a lack of malonate-methylmalonate selectivity. By sequence comparisons, epo AT4 resembles the malonate-specifying AT domains (*8*), and there is no clear basis for the apparent lack of selectivity.

Secondly, both epothilones C and D contain a 12,13-*cis* double bond, which does not appear to be determined by the corresponding PKS domains in module 4 since it does not contain the required DH domain. With rare exception, the degree of reduction of a polyketide is determined exclusively by the presence (or absence) of functional reductive domains in the corresponding module. Two exceptions are found in the macrolide family of antibiotics. The actinomycete *Streptomyces venezuelae* produces pikromycin as well as the reduced congener

10,11-dihydropikromycin (Fig. 2). The pik PKS contains both a KR and DH domain in module 2 that would appear to direct the introduction of a *trans* double bond during the synthesis corresponding to the 10,11 positions in the completed molecule (*9*). This was proven through the finding that only the pikromycin aglycone, narbonolide, containing the 10,11-*trans* double bond, was produced when the pik PKS was expressed in the heterologous host *Streptomyces lividans* (*10*). 10,11-Dihydronarbonolide was not seen. Thus, the 10,11-dihydropikromycin observed in the fermentation broth of *S. venezuelae* resulted from the adventitious reduction of the 10,11-double bond by an enzyme present in the host not associated with the gene cluster for pikromycin biosynthesis. The second example is chalcomycin (chm), a 16-membered diglycosidic macrolide that contains a 2,3-*trans* double bond (Fig. 2). Sequencing of the chm PKS indicated that module 7, responsible for the portion of the structure of chalcomycin corresponding to positions 2 and 3, contained only KS, AT, and ACP domains, and lacked the required KR and DH domains to direct the introduction of the 2,3-double bond (*11*). Heterologous expression of the chm PKS in the tylosin-producing host *Streptomyces fradiae,* from which the tylosin PKS genes had been removed, resulted in the production of 5-*O*-mycaminosylchalcolactone (Fig. 2). This is the predicted macrolactone of the chm PKS, associated at C5 with the deoxyaminosugar mycaminose normally found in tylosin (*11*). This result indicated that the 2,3-*trans* double bond present in chalcomycin is introduced after completion of synthesis of the macrolactone. In both the pikromycin and chalcomycin examples, therefore, the changes in degree of reduction were the result of post-polyketide processing events.

The opposite is true, however, for the introduction of the 12,13-*cis* double bond in epothilone. Expression of the epo PKS in the heterologous hosts *Myxococcus xanthus* and, recently, *Escherichia coli* resulted, in both cases, in the production of epothilones C and D (containing the 12,13-*cis* double bond) (*5, 7*), suggesting that the epo PKS itself directs the introduction of the double bond during the synthesis by a module other than module 4. Further evidence to support this hypothesis will be presented below.

The Non-Selectivity of epo AT4 is Context Dependent

To see whether the unusual lack of substrate selectivity of the AT domain of module 4 could be mimicked in other modules of the epo PKS, AT4 was used to replace the native AT domain in module 3 in an *M. xanthus* host that contained the epothilone biosynthesis gene cluster. The exchange of the DNA encoding AT3 for the DNA segment encoding AT4 was accomplished employing a two-step recombination process (Fig. 3). The resulting strain was fermented and the epothilone products were isolated and analyzed. As found in the parent strain

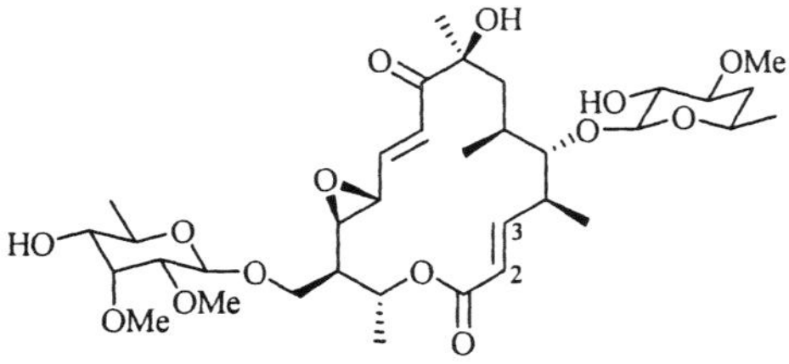

Chalcomycin

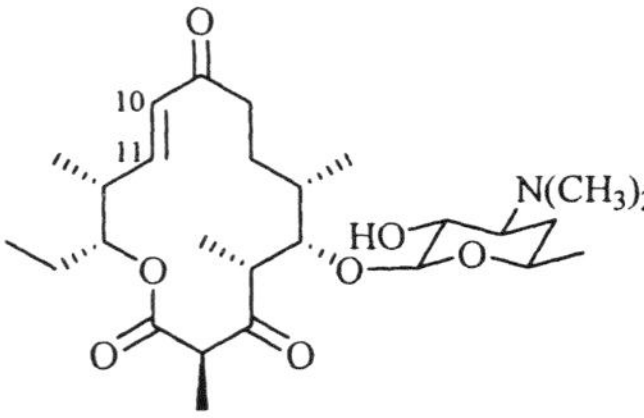

Pikromycin

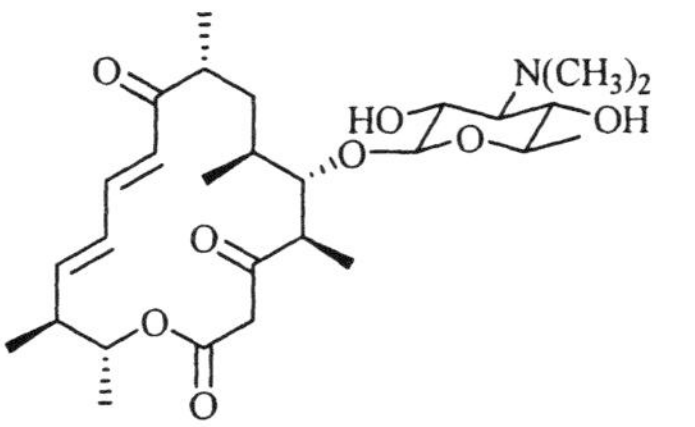

5-*O*-Mycaminosylchalcolactone

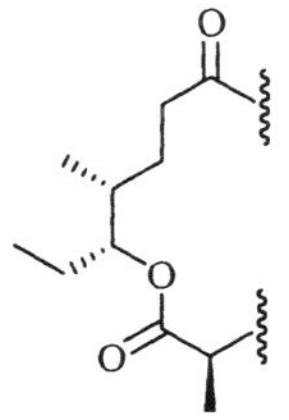

10,11-dihydropikromycin

Figure 2. Structures of macrolides.

with the wt epo PKS, two-thirds of the molecules contained H at C-12; one-third contained a methyl side chain, as previously observed and expected for the absence of substrate specificity of the AT4 domain (Fig. 3). In contrast, however, all the molecules contained H at C-14, indicating functional utilization of malonyl CoA alone as the substrate by the AT domain in module 3. None of the molecules contained the C-14 methyl side chain. One explanation of these findings is that the architecture of the AT4 domain placed in module 3 was altered so that it could not utilize methylmalonyl CoA as a substrate, although it was clearly functional since it was able to incorporate malonyl CoA. It is also possible that the AT domain in module 3 could incorporate methylmalonyl CoA into the growing polyketide, but that the nascent chain so formed could not be used as a substrate for the subsequent condensation conducted by module 4. The KS domain of module 4, therefore, would have acted as a gatekeeper, not permitting passage of the nascent chain containing an α-methyl side chain. Gatekeeping roles for KS domains in modular PKSs are well-established; for example, the KS domains of the erythromycin and pikromycin PKSs are highly selective for the relative orientation of the α-methyl-β-OH side chains of their substrates (*12-14*).

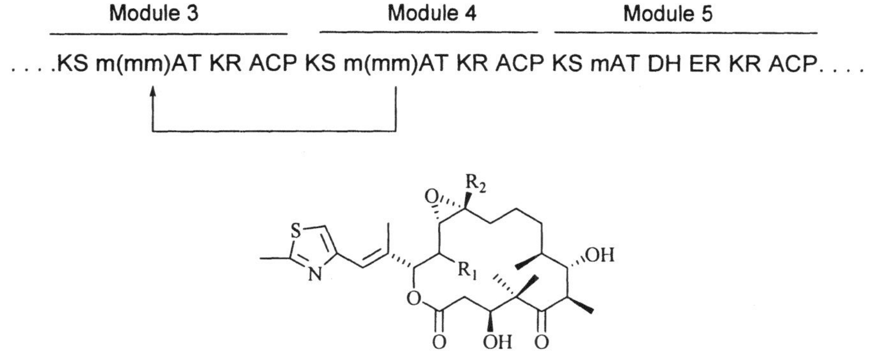

Module 3 Module 4 Module 5

. . . .KS m(mm)AT KR ACP KS m(mm)AT KR ACP KS mAT DH ER KR ACP. . . .

Proportion of Molecules

R_1		R_2	
H	CH_3	H	CH_3
1.0	0	0.67	0.33

Figure 3. Replacement of AT3 with AT4 does not produce 14-methylepothilones. The domains of modules 3 – 5 are shown. The arrow indicates that the AT3 domain was replaced with the AT4 domain. The structure of the compounds produced and their relative proportions are shown in the table.

The DH Domain in Module 5 Catalyzes Two Successive Dehydrations

As described above, although module 4 determines the structure at positions 12 and 13 of epothilone, the 12,13-*cis* double bond in epothilones C and D, the products of the PKS, must be introduced by a function outside of this module. In a recent paper (*15*), we reported that replacement of the complete set of reduction domains in module 5 of the epothilone PKS resulted in the production of an epothilone derivative that had lost the 12,13-*cis* double bond and carried an OH group at C-13, as predicted from the domain organization of module 4 (Fig. 4). This structure led us to propose that the DH domain of module 5 acted on the nascent chain produced by the upstream module to insert a *cis* double bond. The presence of the 10,11-*trans* double bond in the resulting compound was puzzling in that it would normally originate from the action of the DH domain of module 5, which was absent from the modified PKS. We

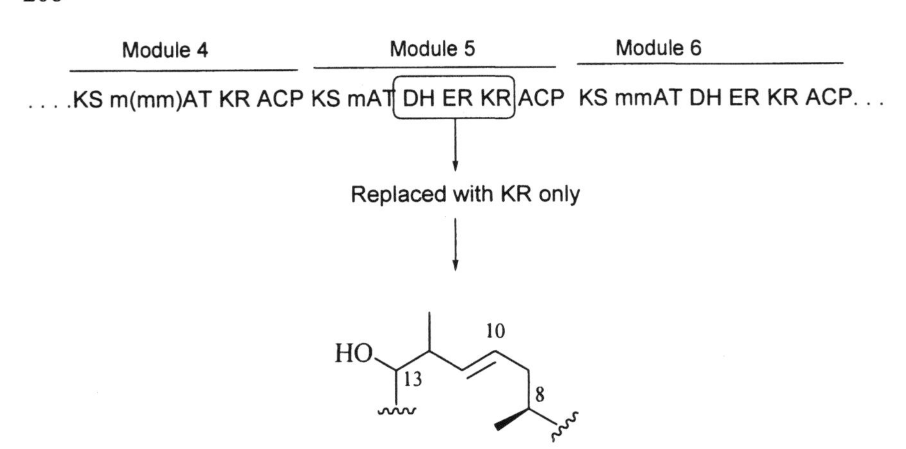

Figure 4. Replacement of the DH/ER/KR domains of module 4 leads to loss of the 12,13-cis double bond.

speculated that the DH domain of module 6 might be responsible for introducing the 10,11-double bond into the 13-hydroxy analog in this unusual instance. During the regular events of epothilone biosynthesis, however, the DH5 domain would normally introduce the 10,11-double bond. DH5 would act twice, therefore, first to introduce a *cis* double bond during or after the fourth condensation cycle, and then to introduce a *trans* double bond during the fifth condensation cycle.

To test this hypothesis, the following experiment was performed. The segment of DNA designated *epo* mod 3-4-5-TE (shown in Fig. 5) was constructed by cloning an *E. coli*-optimized synthetic ORF consisting of modules 3, 4, and 5 of the *epoD* gene. The ORF was fused to a synthetic segment corresponding to the TE domain of the erythromycin PKS in the high copy plasmid pKOS392-97 (*7*), thereby placing it under the control of the *ara*BAD promoter. The resulting plasmid, pKOS455-042, was introduced into *E. coli* K207-3 (F^- *ompT* $hsdS_B$ $(r_B\text{-}m_B\text{-})$ *gal dcm* (DE3) *ΔprpRBCD*::P_{T7}-*sfp*-P_{T7}-*prpE ygfG*::P_{T7}-*accA1*-P_{T7}-*pccB* $panD^{S25A}$), a modified version of BL21 (DE3) that is capable of generating methylmalonyl-CoA when fed with propionate through overexpression of a propionyl-CoA ligase (*prpE*) and propionyl-CoA carboxylase (*accA1, pccB*). The strain also contains a chromosomal copy of *sfp,* a 4-phosphopantetheine transferase gene necessary for phosphopantetheinylation of PKSs in the host (*16*). L-arabinose-induced expression of this construct in *E. coli* at 22 °C resulted in the production of a 569 kDa protein, the size predicted from the sequence. Co-expression of the epo segment with the genes encoding the chaperone proteins GroES/EL, DnaKJ, and GrpE, (on plasmid pG-KJE8) resulted in the appearance of least one-fourth of the total amount of epo 569 kDa protein in the soluble fraction (data not shown).

The thiazoacyl moiety of the *N*-acetylcysteamine (NAC) thioester, compound **A** (Fig. 5), represents the proposed structure of the polyketide chain that is utilized by module 3 for the third condensation in the biosyntheses of epothilones C and D. The thiazoacyl moiety would be expected to serve as the starting material for three successive condensations (and attendant reductions) in the *E. coli* host expressing *epo* mod 3-4-5-TE. The TE domain would remove the acyl chain from the PKS at the end of the third cycle allowing re-utilization of the protein and resulting in the accumulation of product. Compounds **B** and **C** would be expected as products after compound **A** was fed to the *E. coli* host expressing *epo* mod 3-4-5-TE (Fig. 5) if two conditions were met: (1) the AT domain of module 4 utilized either malonyl CoA or methylmalonyl CoA as a substrate for the first condensation, and (2) the DH domain of module 5 introduced the *cis* double bond during (or after) the second condensation and the *trans* double bond during the third condensation, which would subsequently be reduced to the dihydro form by the action of ER5.

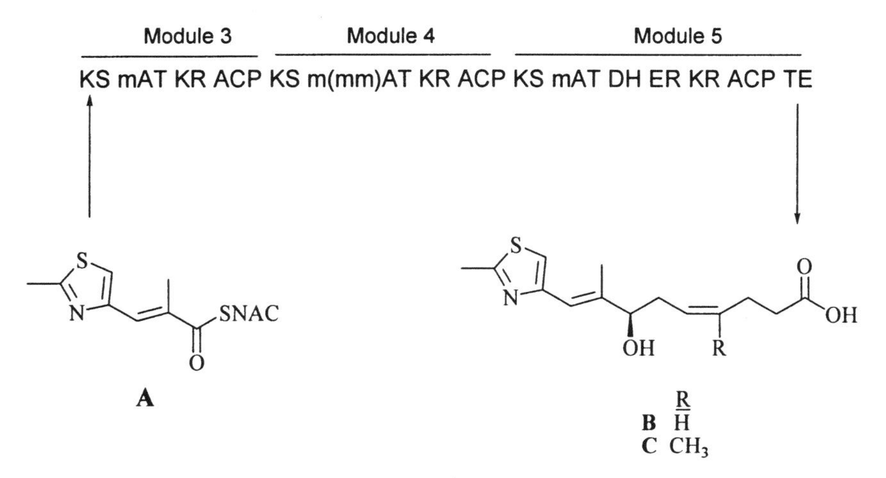

Figure 5. Proposed scheme for the biosynthesis of epothilone intermediates from modules 3-5 in E. coli. The segment containing modules 3, 4, 5 and the TE domain were constructed in an expression vector and expressed as described in the text.

Compound **B** (Fig. 5) was synthesized for use as a standard following the scheme shown in Fig. 6. Compound **A** (Fig. 5) was prepared using aldehyde **2** (Fig. 6) as starting material in a Horner-Emmons homologation reaction followed by *trans*-thioesterification with *N*-acetylcysteamine. Compound **A** was fed to the *E. coli* host growing under conditions that allowed production of soluble Epo Mod3-4-5-TE protein, employing published methodology of precursor-directed polyketide production in *E. coli* (*17*). Concentrated extracts of cultures were analyzed by LC/MS/MS, using multiple reaction monitoring of the *m/z* 282 ($[M+H]^+$)/264 ($[M+H-H_2O]^+$) parent-daughter pair for detection of compound **B**. Chromatograms of the standard and an extract are shown in Figure 7 (A and B, respectively). This finding indicates that **B** was produced from the SNAC-thioester precursor by the activities contained in Epo mod3-4-5-TE. A compound with LC/MS/MS data consistent with compound **C** was also detected (data not shown). Without a reference standard, however, it is not possible to predict the retention time of compound **C**, thus it cannot be determined whether compound **C** was produced.

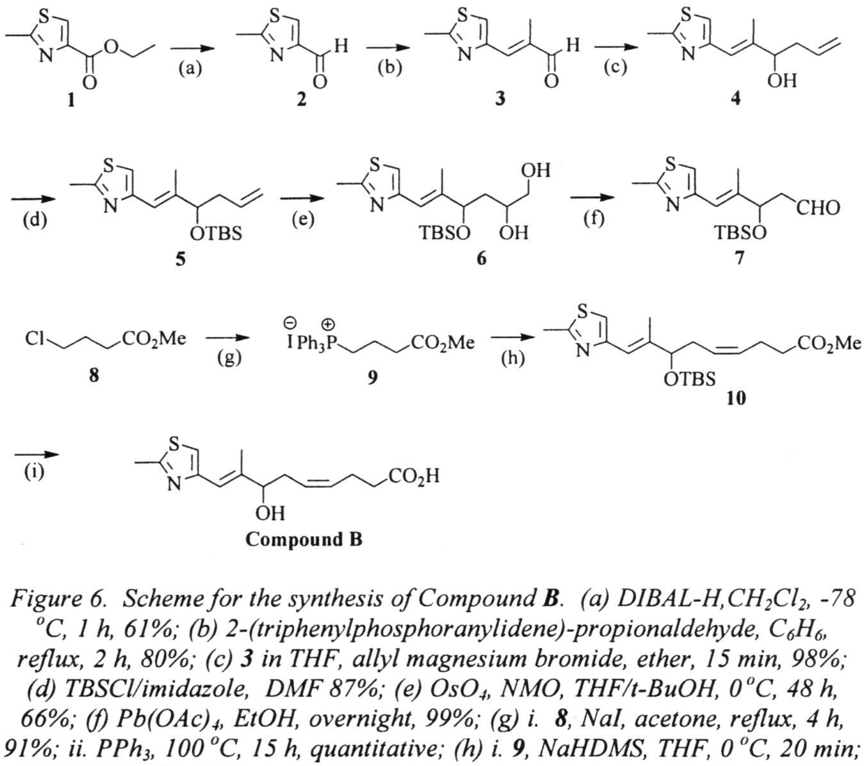

*Figure 6. Scheme for the synthesis of Compound **B**. (a) DIBAL-H, CH_2Cl_2, -78 °C, 1 h, 61%; (b) 2-(triphenylphosphoranylidene)-propionaldehyde, C_6H_6, reflux, 2 h, 80%; (c) **3** in THF, allyl magnesium bromide, ether, 15 min, 98%; (d) TBSCl/imidazole, DMF 87%; (e) OsO_4, NMO, THF/t-BuOH, 0 °C, 48 h, 66%; (f) $Pb(OAc)_4$, EtOH, overnight, 99%; (g) i. **8**, NaI, acetone, reflux, 4 h, 91%; ii. PPh_3, 100 °C, 15 h, quantitative; (h) i. **9**, NaHDMS, THF, 0 °C, 20 min; ii. **7**, THF, 0 °C, 36% as a 9:1 mixture of cis:trans isomers; (i) i. **10**, LiOH, 2-propanol:H_2O (3:1), 99%; ii. HF-pyridine, THF, 0 °C, 93%.*

These findings provide evidence that the DH domain of module 5 is responsible for the introduction of two successive double bonds in epothilones C and D: the *cis*-12,13 and the *trans*-10,11, the latter of which, in the normal course of biosynthesis, is reduced by the action of ER5. 10,11-Dehydroepothilones C and D (containing both the 12,13-*cis* and 10,11-*trans* double bonds) are produced in strains in which the ER5 domain has been inactivated (*18*).

The iterative activity of a domain in bacterial type I PKS is highly unusual but not completely unprecedented. One module each in the aureothin and stigmatellin PKSs is used iteratively to produce the corresponding polyketide (*19, 20*). In addition, under certain circumstances, module 4 of the erythromycin PKS can be used iteratively to produce unusual octaketides (*21*). In each of these cases, collectively known as "stuttering", synthesis proceeds through the module twice, employing the condensation and reduction functions in each passage. In the case of epothilone biosynthesis, only the DH domain of module 5 is used twice. The ER domain of module 5 does not reduce the double bond produced after the fourth condensation. Furthermore, there is evidence to support the notion that the AT and KR domains of module 5 do not participate in the fourth condensation during epothilone biosynthesis. Replacement of the AT4 domain with an AT domain of single substrate specificity results in the production of only one of the two epothilones produced by the wild type PKS (unpublished results). This finding indicates that AT4, rather than AT5, is used in the fourth condensation. Knockout of the KR4 domain results in the production of the corresponding 13-oxo derivative, indicating that the KR domain of module 5 cannot substitute for an inactive KR4 domain (*22*). This result is consistent with the interpretation that KR4 is used in the fourth condensation cycle and that KR5 is used in the fifth. Furthermore, from sequence comparisons of many KR domains, it was predicted that the reduced carbonyls of KR4 and KR5 would have opposite stereochemistries (*23, 24*). Dehydration mediated by DH5, therefore, would be expected to yield one *cis* and one *trans* double bond.

The precise mechanism employed by the DH5 domain to introduce the *cis* double bond after the fourth condensation is not yet known. Three models to describe the events are presented here. The "intermodular" model has the DH5 domain acting directly on the 2,3 atoms of nascent thiazoacyl chain tethered to ACP4 to produce the *cis* double bond (Fig. 8). The DH5 domain, therefore, would have to orient itself to interact with ACP4. After the dehydration event, the chain would be passed to the KS domain of module 5 for the fifth condensation event and subsequent β-carbonyl processing, including the DH5-mediated dehydration, followed by ER5-mediated enoylreduction. Direct interaction of a β-carbonyl processing domain of one module with the ACP domain of the upstream module has not been described previously and would represent a truly unique process in polyketide biosynthesis.

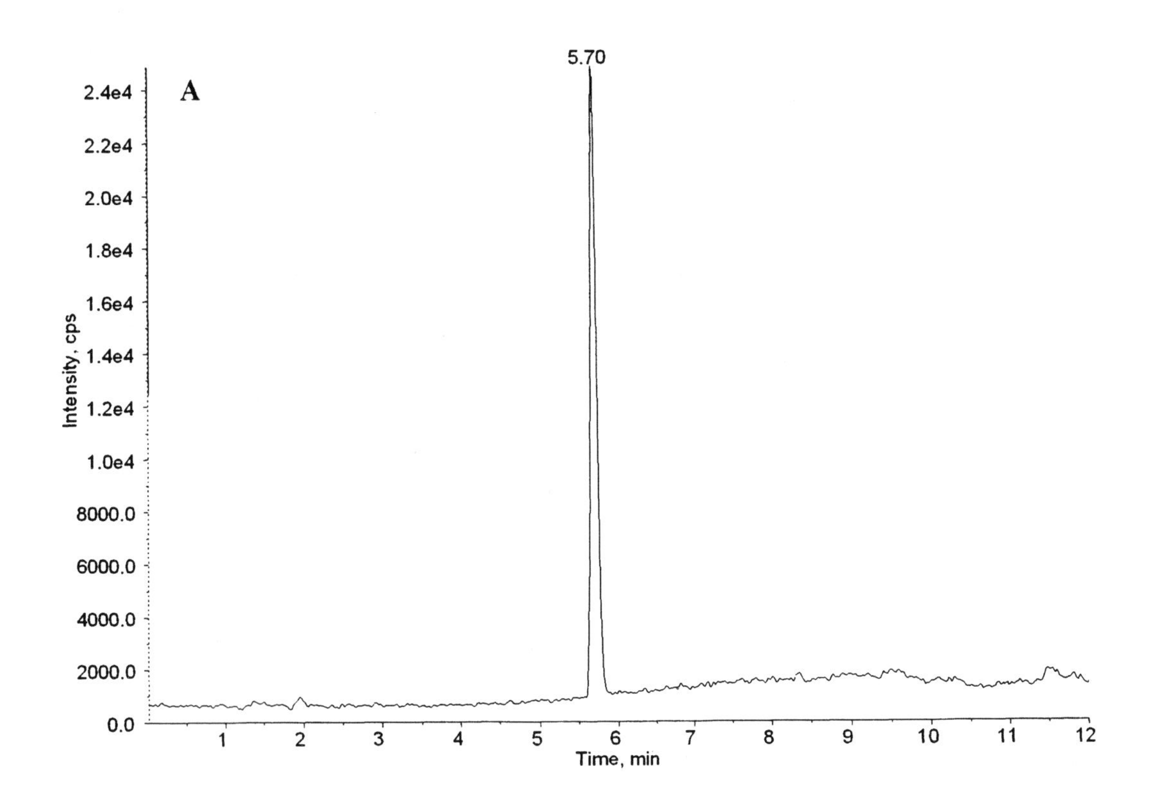
5.70
A
Intensity, cps
2.4e4
2.2e4
2.0e4
1.8e4
1.6e4
1.4e4
1.2e4
1.0e4
8000.0
6000.0
4000.0
2000.0
0.0
1
2
3
4
5
6
7
8
9
10
11
12
Time, min

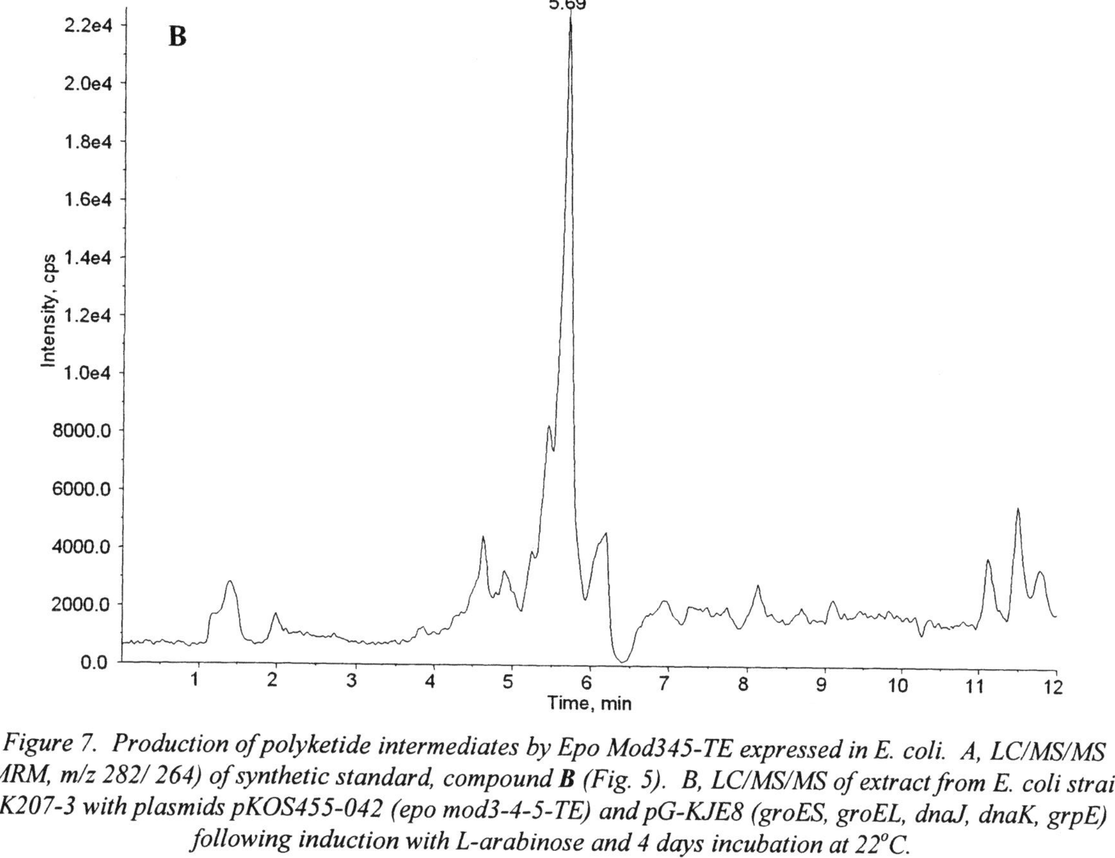

Figure 7. Production of polyketide intermediates by Epo Mod345-TE expressed in E. coli. A, LC/MS/MS (MRM, m/z 282/ 264) of synthetic standard, compound ***B*** *(Fig. 5). B, LC/MS/MS of extract from E. coli strain K207-3 with plasmids pKOS455-042 (epo mod3-4-5-TE) and pG-KJE8 (groES, groEL, dnaJ, dnaK, grpE) following induction with L-arabinose and 4 days incubation at 22°C.*

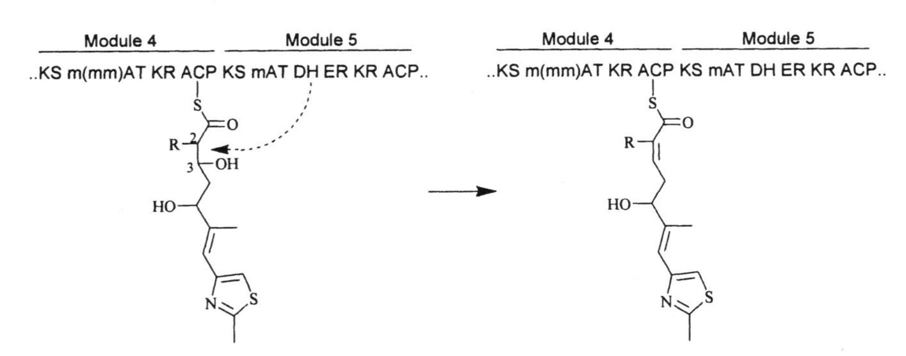

Figure 8. "Intermodular" model of DH5-mediated dehydration to make a cis double bond. The DH5 domain acts on the nascent chain tethered to the ACP4 domain to create a cis double bond. The chain is then transferred to the KS5 domain for the next condensation cycle.

The second and third models each propose that the *cis* double bond is introduced with the nascent acyl chain tethered to the ACP5 domain. The models differ as to the structure of the substrates employed for the dehydration events. In both models, the acyl chain containing the reduced β-carbonyl on ACP4, is passed to the fifth module without undergoing 2,3-dehydration. In the second model (Fig. 9), the fifth condensation takes place extending the chain by two carbons. β-Ketoreduction mediated by KR5 produces the acyl chain tethered to ACP5 containing hydroxyls at positions 3 and 5. The DH5 domain then catalyzes, in either order, the 2,3 dehydration to produce the *trans* double bond, and the 4,5 dehydration to produce the *cis* double bond. The 2,3-*trans* double bond is subsequently reduced by the ER5 domain. Consecutive DH domain-mediated dehydrations of a common acyl chain, including the 4,5 atoms during polyketide synthesis, is unprecedented.

The third, and most complicated model, involves events described previously in the biosynthesis of other complex polyketides. In this model, the polyketide chain containing the 2-OH group and tethered to ACP 4 [**11**], is moved to the ACP domain of module 5 without undergoing the condensation event. This process has been termed "skipping" and is thought to take place in the biosynthesis of the polyketide mupirocin (*25*) as well as in the production of some engineered hybrid polyketides (*26, 27*). Once tethered to the ACP5 domain, the DH5 domain catalyzes the 2,3-dehydration of **11** to yield the *cis* double bond seen in **12**. The ER5 domain, which normally reduces a 2,3-*trans* double bond, cannot act on **12**, hence the *cis* double bond is preserved throughout the remainder of the synthesis. The next step involves the back transfer of **12** from ACP5 to the KS5 domain. This process is termed

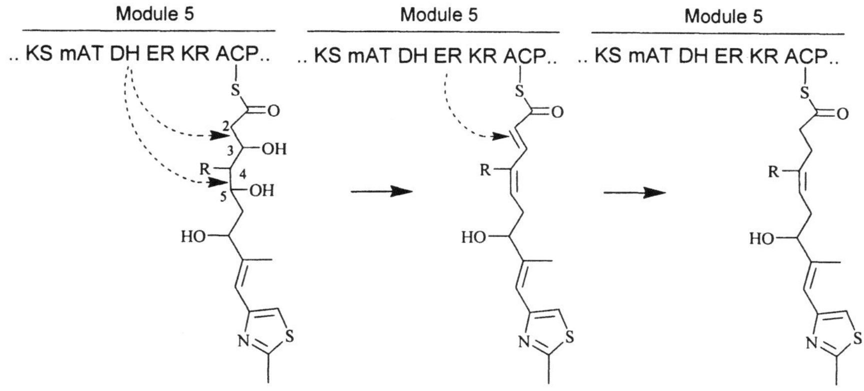

Module 5 Module 5 Module 5

.. KS mAT DH ER KR ACP.. .. KS mAT DH ER KR ACP.. .. KS mAT DH ER KR ACP..

Figure 9. Model of iterative DH5-mediated dehydration with the substrate tethered to ACP5. The order of dehydrations is not specified. The 2,3-trans double bond is reduced by the ER5 domain.

"stuttering" and has been described above. Once on the KS5 domain, the ACP5 domain can be loaded with the malonate extender unit and condensation can take place to produce the extended molecule tethered to ACP5. The full cycle of reduction then takes place, β-ketoreduction (to produce **13**), dehydration mediated by DH5 to introduce the 2,3-*trans* double bond, and enoyl reduction to reduce it, generating **14**. What causes the stuttering, ACP5 to KS5 transfer, rather than the usual ACP5 to KS6 transfer is not known. It is possible that the KS6 domain acts a gatekeeper and does not permit thiotransfer of **12.** As described above, gatekeeper roles have been uncovered for KS domains.

A gatekeeping role for KS5 may also exist in the intermodular model. KS5 might not accept the 3-OH thiazoacyl chain as a substrate, thereby requiring the dehydration step to occur before passage of the nascent chain to KS5. On the other hand, the dehydration event to yield the 12,13-*cis* double bond can be bypassed entirely; the 13-oxo derivatives of epothilones C and D were produced in a strain in which the KR domain of module 4 was inactivated (*22*).

Conclusions

The work reviewed here describes two examples where genetic engineering of the epo PKS resulted in the generation of unpredicted products and led to the uncovering of aspects of the biosynthesis not previously understood. Disruption of the DH5 domain revealed that it is involved in two successive dehydration

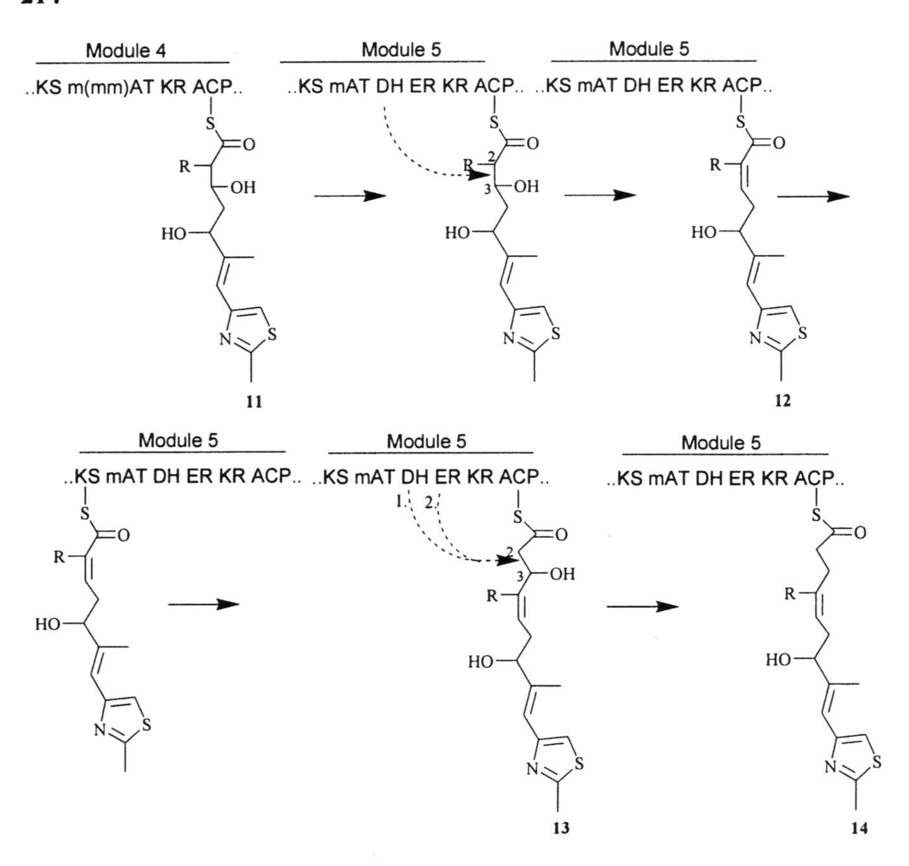

Figure 10. The "Skipping-Stuttering" model for iterative action of the DH5 domain.

steps, potentially in two condensation cycles. Understanding the details of these events is beyond the scope of genetic analysis and will likely require biochemical approaches to distinguish among the models proposed herein. Similarly, the molecular basis for the absence of substrate selectivity of the AT4 domain during the fourth condensation, and the appearance of malonate-specificity when the domain is placed in a different context, will require biochemical and, perhaps structural analysis for clarification. Nonetheless, preliminary information derived from genetic engineering approaches have provided the framework for future studies to more completely understand the nature of epothilone biosynthesis.

Acknowledgements

We thank Gary Liu for providing the thiazoacyl SNAC and Chau Tran for performing LC/MS analysis.

References

1. Gerth, K., Bedorf, N., Höfle, G., Irschik, H., and Reichenbach, H. (1996) *J Antibiot (Tokyo) 49*, 560-563.
2. Bollag, D. M., McQueney, P. A., Zhu, J., Hensens, O., Koupal, L., Liesch, J., Goetz, M., Lazarides, E., and Woods, C. M. (1995) *Cancer Res 55*, 2325-2333.
3. Molnar, I., Aparicio, J. F., Haydock, S. F., Khaw, L. E., Schwecke, T., Konig, A., Staunton, J., and Leadlay, P. F. (1996) *Gene 169*, 1-7.
4. Tang, L., Shah, S., Chung, L., Carney, J., Katz, L., Khosla, C., and Julien, B. (2000) *Science 287*, 640-642.
5. Julien, B., and Shah, S. (2002) *Antimicrob Agents Chemother 46*, 2772-2778.
6. Boddy, C. N., Hotta, K., Tse, M. L., Watts, R. E., and Khosla, C. (2004) *J Am Chem Soc 126*, 7436-7437.
7. Mutka, S. C., Carney, J. R., Liu, Y., and Kennedy, J. (2006) *Biochemistry (in press)*.
8. Haydock, S. F., Aparicio, J. F., Molnar, I., Schwecke, T., Khaw, L. E., Konig, A., Marsden, A. F., Galloway, I. S., Staunton, J., and Leadlay, P. F. (1995) *FEBS Lett 374*, 246-248.
9. Xue, Y., Zhao, L., Liu, H. W., and Sherman, D. H. (1998) *Proc Natl Acad Sci U S A 95*, 12111-12116.
10. Tang, L., Fu, H., Betlach, M. C., and McDaniel, R. (1999) *Chem Biol 6*, 553-558.
11. Ward, S. L., Hu, Z., Schirmer, A., Reid, R., Revill, W. P., Reeves, C. D., Petrakovsky, O. V., Dong, S. D., and Katz, L. (2004) *Antimicrob Agents Chemother 48,* 4703-4712.
12. Chuck, J. A., McPherson, M., Huang, H., Jacobsen, J. R., Khosla, C., and Cane, D. E. (1997) *Chem Biol 4*, 757-766.
13. Weissman, K. J., Bycroft, M., Cutter, A. L., Hanefeld, U., Frost, E. J., Timoney, M. C., Harris, R., Handa, S., Roddis, M., Staunton, J., and Leadlay, P. F. (1998) *Chem Biol 5*, 743-754.
14. Aldrich, C. C., Beck, B. J., Fecik, R. A., and Sherman, D. H. (2005) *J Am Chem Soc 127*, 8441-8452.
15. Tang, L., Ward, S., Chung, L., Carney, J. R., Li, Y., Reid, R., and Katz, L. (2004) *J Am Chem Soc 126*, 46-47.
16. Murli, S., Kennedy, J., Dayem, L. C., Carney, J. R., and Kealey, J. T. (2003) *J Ind Microbiol Biotechnol 30*, 500-509.

17. Murli, S., MacMillan, K. S., Hu, Z., Ashley, G. W., Dong, S. D., Kealey, J. T., Reeves, C. D., and Kennedy, J. (2005) *Appl Environ Microbiol 71*, 4503-4509.
18. Arslanian, R. L., Tang, L., Blough, S., Ma, W., Qiu, R. G., Katz, L., and Carney, J. R. (2002) *J Nat Prod 65*, 1061-1064.
19. He, J., and Hertweck, C. (2003) *Chem Biol 10*, 1225-1232.
20. Gaitatzis, N., Silakowski, B., Kunze, B., Nordsiek, G., Blocker, H., Hofle, G., and Muller, R. (2002) *J Biol Chem 277*, 13082-13090.
21. Wilkinson, B., Foster, G., Rudd, B. A., Taylor, N. L., Blackaby, A. P., Sidebottom, P. J., Cooper, D. J., Dawson, M. J., Buss, A. D., Gaisser, S., Bohm, I. U., Rowe, C. J., Cortes, J., Leadlay, P. F., and Staunton, J. (2000) *Chem Biol 7*, 111-117.
22. Tang, L., Chung, L., Carney, J. R., Starks, C. M., Licari, P., and Katz, L. (2005) *J Antibiot (Tokyo) 58*, 178-184.
23. Reid, R., Piagentini, M., Rodriguez, E., Ashley, G., Viswanathan, N., Carney, J., Santi, D. V., Hutchinson, C. R., and McDaniel, R. (2003) *Biochemistry 42*, 72-79.
24. Caffrey, P. (2003) *Chembiochem 4*, 654-657.
25. El-Sayed, A. K., Hothersall, J., Cooper, S. M., Stephens, E., Simpson, T. J., and Thomas, C. M. (2003) *Chem Biol 10*, 419-430.
26. Rowe, C. J., Bohm, I. U., Thomas, I. P., Wilkinson, B., Rudd, B. A., Foster, G., Blackaby, A. P., Sidebottom, P. J., Roddis, Y., Buss, A. D., Staunton, J., and Leadlay, P. F. (2001) *Chem Biol 8*, 475-485.
27. Thomas, I., Martin, C. J., Wilkinson, C. J., Staunton, J., and Leadlay, P. F. (2002) *Chem Biol 9*, 781-787.

Chapter 15

Biosynthesis of the Antifungal Polyketide Antibiotic Soraphen A in *Sorangium cellulosum* and *Streptomyces lividans*

István Molnár[1,2], Ross Zirkle[1,3], and James M. Ligon[1,4]

[1]Syngenta Biotechnology, Inc., Research Triangle Park, NC 27709
[2]Current address: Southwest Center for Natural Product Research and Commercialization, University of Arizona, Tucson, AZ 85706
[3]Current address: Martek Biosciences, 4909 Nautilus Court North, Boulder, CO 80301
[4]Current address: BASF Agricultural Products, Research Triangle Park, NC 27709

The derivatization of soraphen by combinatorial biosynthesis has been a longstanding goal of the Natural Product Genetics group at Ciba Geigy – Novartis – Syngenta. We have cloned and characterized the soraphen A biosynthetic gene cluster of *Sorangium cellulosum* So ce26. To facilitate the genetic manipulation of So ce26, a central regulator of gliding motility was identified and its encoding gene disrupted to create a non-swarming strain. Despite this improvement, engineering of soraphen biosynthesis remained elusive in So ce26. Thus, the genes for the soraphen polyketide synthase, the post-polyketide tailoring steps, and methoxymalonate biosynthesis were introduced into *Streptomyces lividans*. Production of soraphen A in this genetically tractable strain also required provision of benzoyl-CoA by feeding benzoate or cinnamate.

The 18-membered macrolide polyketide soraphen A is produced by the myxobacterium *Sorangium cellulosum* So ce26 (*1*). Soraphen A shows a strong fungicidal activity against a wide range of species, notably against plant pathogenic fungi. It has a unique mode of action in inhibiting the biotin carboxylase domain of eukaryotic acetyl-coenzyme A carboxylases (ACCs) at nanomolar concentrations, thereby inhibiting fatty acid biosynthesis (*2*). Soraphen A was under development in the nineties at the Ciba Geigy – Novartis Agribusiness – Syngenta succession line of companies as an agricultural fungicide. In spite of an extensive chemical derivatization program, a weak but persistent teratogenic activity associated with all soraphen congeners and active semisynthetic derivatives eventually led to the discontinuation of development activities.

Soraphen A, like several other previously "discarded" natural product leads [see the eventual success of daptomycin, (*3*)], however, might enjoy a revival not only as a "last line of defense" antifungal in immunocompromised patients with systemic fungal infections, but also in the unexpected fields of obesity and diabetes. Recent research suggests that malonyl-CoA, the product of ACCs, is a key metabolite in the regulation of energy homeostasis, with ACC2- (the mitochondrial isoform of ACCs) knockout mice maintaining low body weight with normal insulin- and blood glucose levels even on high fat / high carbohydrate diets (*4, 5*). The availability of the crystal structure of soraphen A bound to the yeast ACC biotin carboxylase domain (*6*) should facilitate the design of new soraphen derivatives and analogs with reduced toxicity but unchanged ACC inhibitory activity.

The Natural Product Genetics group at the Research Triangle Park (NC) research facility of the Ciba Geigy– Novartis Agribusiness – Syngenta line of companies had a longstanding interest in soraphen A biosynthesis (*7-10*). We have proposed that the identification of the soraphen biosynthetic gene cluster will open the way towards the combinatorial biosynthetic manipulation (*11, 12*) of the soraphen structure, and might yield soraphen derivatives that are not easily accessible by chemical derivatization. Although the cancellation of the soraphen development program diminished the company's interest in soraphen research, we were allowed to pursue a limited soraphen-related research program until 2003.

The biosynthetic gene cluster for soraphen A

We have identified a clone from a cosmid library of *So. cellulosum* So ce26 that contains genes involved in soraphen A biosynthesis (*10*). Gene knockouts conducted with several restriction fragments from this cosmid led to the abolition of soraphen A production in So ce26. One of these fragments was sequenced, and found to encode two incomplete modules from a Type I polyketide synthase (PKS). The coverage of the soraphen A locus was extended by isolating overlapping cosmid clones, and a 67.5 kb segment from this cosmid contig was eventually sequenced (*9*). Sequence analysis revealed two large Type I PKS genes collectively encoding eight modules: the gene sequence of this soraphen synthase exhibited colinearity with the biosynthetic sequence of soraphen assembly (Figure 1). The predicted substrate specificities of AT domains (*13, 14*), and the presence of active ketone processing activities in each module were in a good correspondence with the chemical structure of soraphen.

The predicted product of the PKS, the putative soraphen X, is expected to be further processed by tailoring enzymes also encoded in the soraphen biosynthetic gene cluster (Figure 1). Thus SorR that shows homology to Type I PKS ketoreductase domains might convert the C3 hydroxyl of soraphen X to a ketone that collapses to the hemiketal after attack of the hydroxyl at C7. The double bond at C9-C10 might be introduced by Orf4 that shows no homology to other sequence databank entries, but whose gene forms an operon with, and is nested between SorB and SorM, both involved in soraphen biosynthesis. The last tailoring step, the formation of the C11 methoxy moiety, is predicted to be catalyzed by the O-methyltransferase SorM (Figure 1).

Apart from malonyl-CoA and methylmalonyl-CoA, the soraphen PKS apparently uses an unusual extender unit derived from glycerate or glycolate (*15*) to produce the methoxy side groups at C4 and C12. The biosynthesis of this extender unit, expected to be methoxymalonate, is encoded by a subcluster of genes upstream of the soraphen PKS that shows significant similarities to other methoxymalonate biosynthetic subclusters in the ansamytocin and the FK520 biosynthetic gene clusters (*16, 17*). Uniquely, the putative soraphen methoxymalonate biosynthetic subcluster contains a gene, *sorC*, whose multifunctional protein product houses AT, ACP and O-methyltransferase (MT) domains. Thus, the AT domain of SorC is expected to load the unknown glycolytic intermediate onto the ACP domain of SorC or the free-standing ACP protein SorF (Figure 1). The SorD and then the SorE dehydrogenases could oxidize the glyceryl-ACP to hydroxymalonyl-ACP that would be converted to methoxymalonyl-ACP by the MT domain of SorC. The methoxymalonyl-ACP of SorC or SorF could then interact directly with the ATs in modules 3 and 7 to transfer the methoxymalonate moiety to the cognate ACP domains of the soraphen PKS.

The soraphen PKS uses benzoyl-CoA, derived from phenylalanine (*18*), as the starter unit for polyketide synthesis, but the soraphen cluster contains no genes for the biosynthesis of this precursor. The first module of the soraphen PKS contains two ACP and two AT domains (Figure 1) in a "starter module" arrangement proposed to perform the functions of both a loading and the first extension module, with AT1a acting to load the benzoyl starter unit to ACP1a (*19*). The KS within this starter module would then catalyze the first condensation event between the benzoyl starter and the malonate unit loaded onto ACP1b by AT1b.

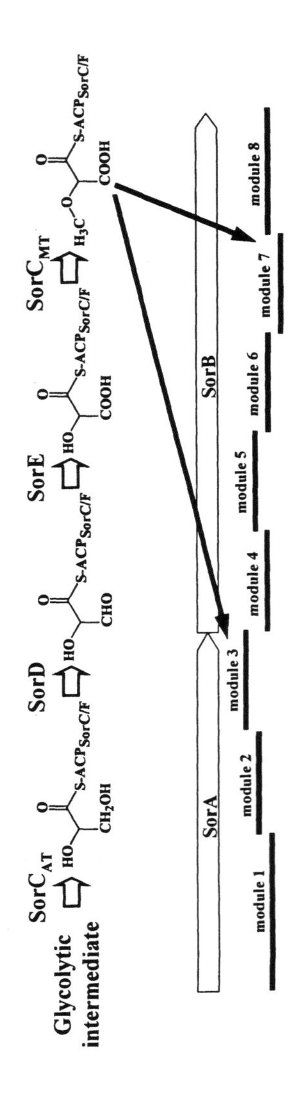

Glycolytic intermediate
SorC$_{AT}$
HO
CH$_2$OH
O
S-ACP$_{SorC/F}$
SorD
HO
CHO
O
S-ACP$_{SorC/F}$
SorE
HO
COOH
O
S-ACP$_{SorC/F}$
SorC$_{MT}$
H$_3$C
O
COOH
O
S-ACP$_{SorC/F}$
SorA
SorB
module 1
module 2
module 3
module 4
module 5
module 6
module 7
module 8

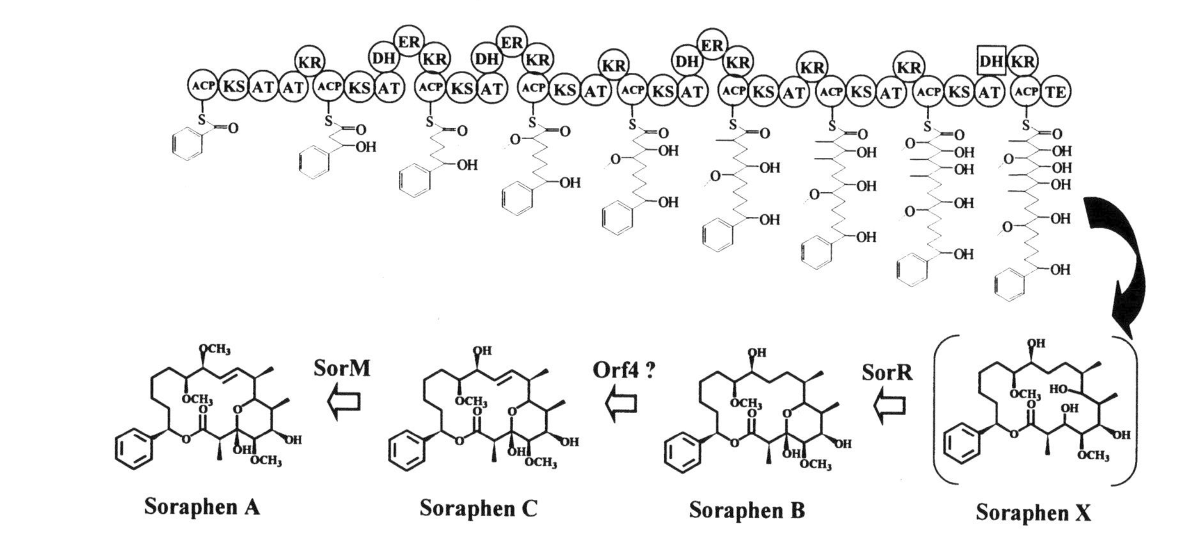

Figure 1. Proposed pathway for soraphen A biosynthesis. ACP, acyl carrier protein domain; AT, acyl transferase; DH, dehydratase; ER, enoyl reductase; KR, ketoacyl reductase; KS, ketoacyl synthase; TE, thioesterase. The inactive DH in module 8 is shown as a square. Adapted with permission from reference (9). Copyright 2002 Elsevier Science B.V.

Swarming motility as an obstacle for *Sorangium* genetic manipulation

While the identification and analysis of the soraphen biosynthetic gene cluster set the stage for the combinatorial genetic manipulation of the soraphen biosynthetic machinery, the peculiarities of *Sorangium* biology and genetics presented substantial barriers towards our progress. The Myxobacteria, and especially *Sorangia*, have been identified as prolific sources of small molecule natural products with promising biological activities (*20*). Until recently, however, there has been little progress in the genetic manipulation, microbiological strain improvement or fermentation process development of *Sorangia*. There are no plasmid or viral vectors useful for these bacteria; very few antibiotic resistance markers are applicable (*21, 22*); and genetic transformation has only been achieved by low-frequency intergeneric conjugation followed by homologous integration of the incoming DNA into the genome (*21*). Gene replacement events are also exceedingly rare in many strains. *Sorangia* grow slowly, with their generation times exceeding 16 hrs (*23*), making their industrial scale fermentation contamination-prone and the volumetric productivity low.

Sorangium cells display coordinated gliding motility on solid surfaces, leading to thin, spreading and merging growth on agar media termed as "swarms". *Sorangium* cultures do not grow readily if plated at low cell densities (*23*). Plated above the threshold density, swarming leads to cross-contamination of the initial microcolonies: consequently, distinct macroscopic colonies derived from single cells do not form. While serial plating of cells from the swarm edge, or selecting for introduced antibiotic resistance markers can be used for strain purification (*9, 22, 24*), the isolation and propagation of strains with discrete genetic changes is both problematic and tedious.

In *Myxococcus xanthus*, the protein product of the *mglA* gene has been identified as a central regulator of swarming motility (*25, 26*): although *mglA* mutants still glide on the microscopic scale, no net macroscopic movement is achieved due to an exceedingly high frequency of cell reversals. Although gliding is achieved by completely different mechanisms in different bacteria, and several different gliding motility systems might operate simultaneously in a single organism (*27*), we reasoned that a homolog of the *M. xanthus* MglA could play a similar central regulatory role in gliding motility in the related myxobacterium *So. cellulosum* So ce26. Thus, we have cloned a gene from So ce26 whose putative product shows 82.5% identity to the *M. xanthus* MglA, and to other putative MglA-like proteins from different bacteria (*8*). The So ce26 MglA, similar to its *M. xanthus* counterpart, also shows homologies to the Ras family of eukaryotic GTPases.

Insertional inactivation of the *mglA* gene on the *So. cellulosum* So ce26 chromosome by homologous recombination led to a mutant that displayed a non-swarming phenotype on agar media (*8*). While the wild type swarms were characterized upon microscopic observation by undulating swarm edges with flares of cells venturing outwards and visible slime trails left behind by the taxiing cells, the *mglA* mutant displayed well-defined colony edges and a complete lack of slime trails or flares, indicating an essential role of MglA in swarming motility in So ce26. While the *mglA* mutant still has to be plated above the threshold cell density, the lack of swarming motility allowed this mutant to readily form separated, compact, single colonies, allowing the easy isolation of strains derived from single cells.

While the *mglA* mutant displayed a longer initial lag phase in liquid media, it showed a similar growth rate in the exponential phase to the wild type strain and reached the same final density in the stationary phase. Importantly, the yields of soraphen A were statistically indistinguishable between the wild type and the mutant strains in standard 12-day flask fermentations (*8*), raising the hope that *mglA* mutant *So. cellulosum* strains might find utility wherever the isolation of distinct genetic events is important, including industrial strain and fermentation process development as well as strain genetic engineering.

A heterologous host for soraphen A production

The heterologous expression of whole biosynthetic gene clusters has emerged recently as an increasingly feasible alternative to strain optimization and fermentation development for every native producer strain, some of which are not amenable to genetic modifications (as the soraphen producer *So. cellulosum* So ce26), or those that have not (environmental libraries) or cannot (sponges, symbionts, "uncultivable" strains) be grown in the laboratory. Heterologous production of complex type I polyketides like soraphen A is still challenging though on account of the exceedingly large size of the PKS genes, the requirement for post-translational modification of PKS enzymes by phosphopantheteinylation, the need for the coenzyme A-activated polyketide building blocks, and the complexity of the post-polyketide tailoring steps. Despite these difficulties, a number of complex type I polyketides have been successfully produced in *S. coelicolor* and its close relative *S. lividans*, or even in *E. coli* (*28*), with usually low initial fermentation titers (in the range of 0.2-50 mg l^{-1}) that could be later improved by classical microbiological or metabolic engineering methods (*29, 30*).

In spite of the improved characteristics of the non-swarming *mglA* mutant *So. cellulosum* So ce26 strain, the lack of genetic tools including multiple selection markers or replicating plasmids, the prohibitively low frequency of gene replacements, and the slow growth of the strain frustrated our efforts to replace or alter the soraphen biosynthetic genes within the native producer and thus to create novel soraphens *via* biological derivatization. To construct a soraphen producer strain more amenable to genetic manipulation, we decided to transfer the soraphen biosynthetic gene cluster to *S. lividans*.

The downstream border of the soraphen cluster was determined to lay downstream of the soraphen C methyltransferase gene *sorM* (Figure 2A): gene disruptions in the *orf5-6* region had no effect on soraphen production (*9, 10*). Similarly, gene disruptions were used to locate the upstream border of the cluster between *orf2* that was shown to play no role in soraphen biosynthesis, and *sorC*, whose multifunctional protein product is proposed to take part in the biosynthesis of the soraphen extender unit methoxymalonyl-CoA (*7, 9*).

The soraphen A biosynthetic gene cluster was reconstructed in *S. lividans* (Figure 2B) using compatible integrative and replicating expression plasmids containing expression cassettes with operons of soraphen biosynthetic genes under the control of the thiostrepton-inducible PtipA promoter (*32*). The soraphen PKS gene *sorA* was integrated into the host chromosome using the ΦC31 phage integrase (*33*) while the *sorB-orf4-sorM* operon, including the rest of the PKS and two putative tailoring genes were integrated into the host chromosome using the transposition functions of IS117 (*34*), to create *S. lividans* SorAB. The *sorR* tailoring gene, and the *sorCDFE* methoxymalonate biosynthetic operon were fused in an artificial operon, and introduced into *S. lividans* SorAB on a replicative expression vector to create *S. lividans* SorABRCDFE. To supply benzoyl-CoA for soraphen biosynthesis, the *badA* gene encoding the benzoyl-CoA ligase of *Rhodopseudomonas palustris* (*31*) was also fused to the *sorRCDFE* operon to create *S. lividans* SorABRCDFE+BL.

The *S. lividans* strains SorAB, SorABRCDFE, and SorABRCDFE+BL were fermented in the presence of thiostrepton to induce the expression of the soraphen biosynthetic genes, but no production of soraphen A, or any other soraphen congener was found by LC-MS analyses (Table I). As there is no evidence that *S. lividans* has the capacity to biosynthesize the soraphen starter unit benzoyl-CoA as part of its normal metabolism (*35*), we tried to ferment these strains while feeding benzoate: SorABRCDFE+BL, but not the other two strains, produced soraphen A as shown by antifungal bioassay and LC-MS analysis (Figure 2C and D, Table I). In conclusion, both an external source of benzoate and the introduction of a suitable benzoyl-CoA ligase [the BadA enzyme of *R. palustris* (*31*)] were necessary for soraphen A production in *S. lividans*.

Benzoyl-CoA supply routes for soraphen A production in *S. lividans*

Several distinct biosynthetic routes for benzoyl-CoA have been described in different organisms (Figure 3). In the soraphen producer *So. cellulosum* So

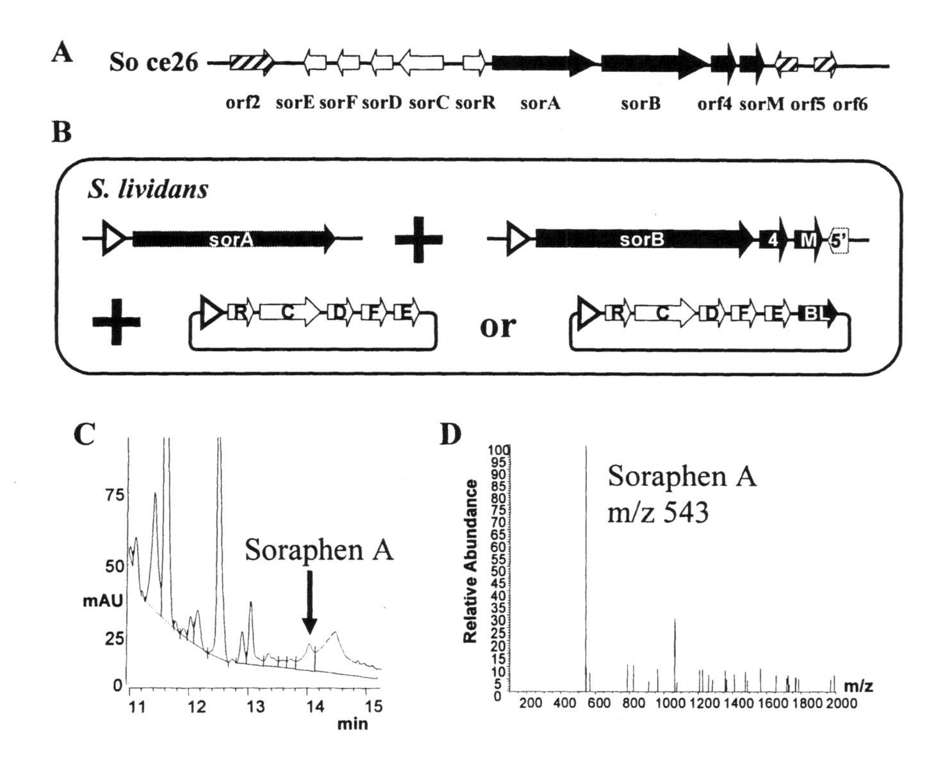

Figure 2. Heterologous expression of the soraphen A biosynthetic gene cluster and production of soraphen A in S. lividans. The arrows are not drawn to scale. Arrows with stripes represent ORFs not proposed to take part in soraphen biosynthesis. (A) The soraphen biosynthetic locus in So. cellulosum So ce26. (B) Expression cassettes in S. lividans, reconstructing the soraphen cluster in the heterologous host. Triangles denote the PtipA promoter. BL, badA benzoate-coenzyme A ligase gene (31). (C) HPLC trace of an extract from a benzoate-supplemented fermentation of S. lividans SorABRCDFE+BL (soraphen A retention time 14.0 min). (D) LC-MS trace of the soraphen A peak in the same extract as in (C) (soraphen A M+Na ion, m/z 543). Adapted with permission from reference (7). Copyright 2004 Society of General Microbiology.

Table I. Soraphen A production by *S. lividans* strains

S. lividans strain	*Soraphen A production upon feeding*						
	NF	*B*	*Phe*	*Cin*	*PP*	*PA*	*BA*
SorAB	—	—	—	—	—	—	—
SorABRCDFE	—	—	—	+	—	—	—
SorABRCDFE+BL	—	+	—	+	—	—	—

NOTE: NF, no feeding; B, benzoate; Phe, phenylalanine; Cin, cinnamate; PP, phenylpyruvate; PA, phenylacetate; BA, benzaldehyde. Soraphen A production was detected by LC-MS analysis.

SOURCE: Adapted with permission from reference (*7*). Copyright 2004 Society of General Microbiology.

ce26, precursor feeding studies showed incorporation of label into the soraphen phenyl moiety from phenylalanine and cinnamate, but not from benzoate (*18, 36*). Phenylalanine might be converted to cinnamate by phenylalanine ammonia lyase (PAL; route 2 on Figure 3). Although this enzyme is rare in bacteria in general, PALs have been identified in several myxobacterial strains (*36*). No PAL-related sequences could be detected, however, in the *S. coelicolor* genome sequence (*37*), thus it was not surprising that soraphen production could not be supported by feeding phenylalanine to the closely related *S. lividans* strains used in this study [Table I, (*7*)]. Feeding *trans*-cinnamate, however, led to the production of soraphen A in fermentations with *S. lividans* SorABRCDFE and SorABRCDFE+BL (Table I), indicating that the cinnamate/4-coumarate CoA ligase of the *S. lividans* host (*38*) is able to commit cinnamate to β-oxidation, yielding benzoyl-CoA for soraphen biosynthesis (Route 3 on Figure 3). Since the last step of β-oxidation, the thiolase-catalyzed conversion of β-ketophenylpropionyl-CoA provides directly the CoA-activated form of benzoate, the presence of the BadA CoA ligase was not necessary for soraphen A production (Table I).

In *Thaurea aromatica*, phenylalanine was found to be converted to benzoyl-CoA by an anaerobic α-oxidation route *via* phenylpyruvate and phenylacetate [route 4 on Figure 3, (*39*)]. Further, a non-oxidative, retro-aldol route to benzoyl-CoA *via* benzaldehyde was also described in plants (route 5 on Figure 3). Feeding these intermediates to our engineered *S. lividans* strains did not lead to the production of soraphen A (Table I), indicating that neither of these two metabolic routes are operational in this host under the tested conditions.

Thus, the benzoyl-CoA starter unit could be provided for soraphen A biosynthesis in *S. lividans* by two biosynthetic routes. The first route relied on benzoate feeding and the expression of a benzoyl-CoA ligase in *S. lividans*. The

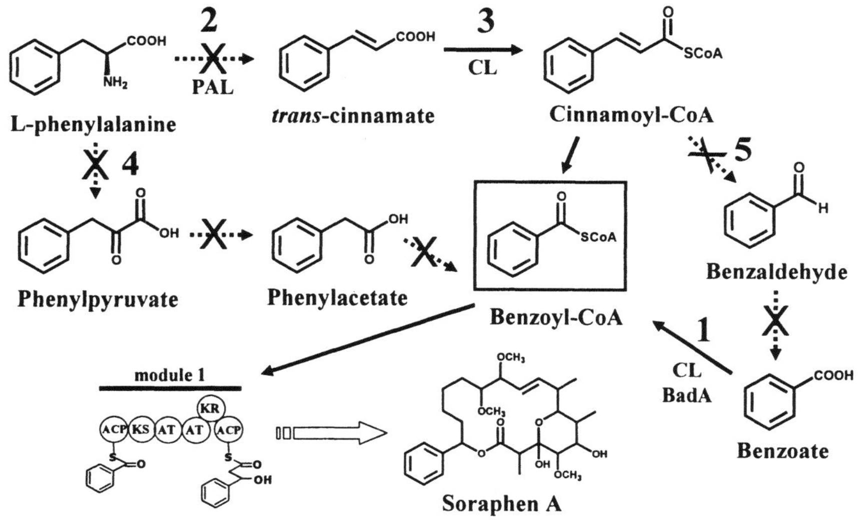

Figure 3. Biosynthetic routes toward benzoyl-CoA used for soraphen A biosynthesis in S. lividans. PAL, phenylalanine ammonia lyase; CL, coenzyme A ligase. Crossed-out arrows symbolize metabolic routes not found to supply benzoyl-CoA for soraphen biosynthesis in the host. See text for detailed explanation. Adapted with permission from reference (7). Copyright 2004 Society of General Microbiology.

second route utilized the intrinsic cinnamate-CoA ligase and β-oxidation enzymes of the host upon feeding *trans*-cinnamate. Introduction of a PAL into *S. lividans* SorABRCDFE might link soraphen A biosynthesis to the existing phenylalanine pool in the heterologous host, and thus might eliminate the need to feed benzoate or cinnamate, both of which are moderately toxic to *S. lividans*.

Conclusions

In a project that spanned about 10 years until the dissolution of the Natural Product Genetics group at Syngenta Biotechnology in 2003, we pursued the elusive goal of manipulating the soraphen A biosynthetic machinery in order to produce soraphen derivatives with reduced toxicity but unchanged acetyl-CoA carboxylase inhibitory activity. We have cloned and sequenced the biosynthetic gene cluster for soraphen A from *So. cellulosum* So ce26, and deduced the biosynthetic sequence for soraphen A assembly from malonate, methylmalonate, benzoate and a glycolytic intermediate that is converted first to methoxymalonate. We have tried to overcome the intransigence of the soraphen producer *Sorangium* strain towards genetic and microbial manipulations by creating a non-swarming, colony-forming mutant by identifying the *mglA* gene as a central regulator of motility, and disrupting this gene on the So ce26 chromosome. Realizing that the barriers of the genetic manipulation of soraphen biosynthesis in the native producer *Sorangium* strain are still formidable, we have reconstructed the biosynthesis of soraphen in the genetics-friendly actinomycete *S. lividans* by transferring the identified soraphen biosynthetic genes into this host and rendering all these genes under the control of the inducible streptomycete promoter PtipA. We had to augment the biosynthetic capabilities of the heterologous host with both a methoxymalonyl-CoA and a benzoyl-CoA biosynthetic route that relied on an intrinsic precursor (a glycolytic intermediate for methoxymalonate) or externally supplemented compounds (benzoate or cinnamate for benzoyl-CoA). Although the yield of soraphen A in *S. lividans* (less than 0.3 mg l^{-1}) was six to ten times lower than the original titers reported for So ce26, fermentation optimization, strain improvement and metabolic engineering could be expected to increase soraphen production to levels where the flexibility of the heterologous expression system could truly be exploited to produce novel soraphen derivatives.

Acknowledgements

The work described here was supported by and carried out at the facilities of the Ciba Geigy – Novartis Agribusiness – Syngenta succession line of companies. The authors acknowledge the contributions of the other former members of the Natural Product Genetics group: D. Steven Hill, Philip E. Hammer, James Beck, Jennifer Zawodny, Stephanie Money, Amber Gaudreau; and the contributions of Thomas Schupp (Novartis Pharma A.G.), and Makoto Ono, An Hu, Racella McNair, James Pomes and Belle Abrera (Syngenta Biotechnology, Inc.).

References

1. Gerth, K.; Bedorf, N.; Irschik, H.; Hofle, G.; Reichenbach, H. *J. Antibiot. (Tokyo)* **1994**, *47*, 23-31.
2. Vahlensieck, H. F.; Pridzun, L.; Reichenbach, H.; Hinnen, A. *Curr. Genet.* **1994,** *25*, 95-100.
3. Baltz, R. H. *SIM News* **2005**, *55*, 5-16.

4. Abu-Elheiga, L.; Matzuk, M. M.; Abo-Hashema, K. A.; Wakil, S. J. *Science* **2001**, *291*, 2613-2616.
5. Abu-Elheiga, L.; Oh, W.; Kordari, P.; Wakil, S. J. *Proc. Natl. Acad. Sci. USA* **2003**, *100*, 10207-10212.
6. Shen, Y.; Volrath, S. L.; Weatherly, S. C.; Elich, T. D.; Tong, L. *Mol. Cell* **2004**, *16*, 881-891.
7. Zirkle, R.; Ligon, J. M.; Molnár, I. *Microbiology (Reading, Engl.)* **2004**, *150*, 2761-2774.
8. Zirkle, R.; Ligon, J. M.; Molnár, I. *J. Biosci. Bioeng.* **2004**, *97*, 267-274.
9. Ligon, J.; Hill, S.; Beck, J.; Zirkle, R.; Molnár, I.; Zawodny, J.; Money, S.; Schupp, T. *Gene* **2002**, *285*, 257-267.
10. Schupp, T.; Toupet, C.; Cluzel, B.; Neff, S.; Hill, S.; Beck, J. J.; Ligon, J. M. *J. Bacteriol.* **1995**, *177*, 3673-3679.
11. Walsh, C. T. *ChemBioChem* **2002**, *3*, 124-134.
12. Leadlay, P. F. *Chem. Biol.* **1997**, *1*, 162-168.
13. Haydock, S. F.; Aparicio, J. F.; Molnar, I.; Schwecke, T.; Khaw, L. E.; Koenig, A.; Marsden, A. F. A.; Galloway, I. S.; Staunton, J.; et al. *FEBS Lett.* **1995**, *374*, 246-248.
14. August, P. R.; Tang, L.; Yoon, Y. J.; Ning, S.; Müller, R.; Yu, T.-W.; Taylor, M.; Hoffmann, D.; Kim, C.-G.; Zhang, X.; Hutchinson, C. R.; Floss, H. G. *Chem. Biol.* **1998**, *5*, 69-79.
15. Hill, A. M.; Harris, J. P.; Siskos, A. P. *Chem. Commun. (Cambridge, United Kingdom)* **1998**, 2361-2362.
16. Yu, T. W.; Bai, L.; Clade, D.; Hoffmann, D.; Toelzer, S.; Trinh, K. Q.; Xu, J.; Moss, S. J.; Leistner, E.; Floss, H. G. *Proc. Natl. Acad. Sci. USA* **2002**, *99*, 7968-7973.
17. Wu, K.; Chung, L.; Revill, W. P.; Katz, L.; Reeves, C. D. *Gene* **2000**, *251*, 81-90.
18. Hill, A. M.; Thompson, B. L.; Harris, J. P.; Segret, R. *Chem. Commun. (Cambridge, United Kingdom)* **2003**, 1358-1359.
19. Wilkinson, C. J.; Frost, E. J.; Staunton, J.; Leadlay, P. F. *Chem. Biol.* **2001**, *8*, 1197-1208.
20. Reichenbach, H.; Höfle, G. In *Drug discovery from nature*; Grabley, S., Thiericke, R., Eds.; Springer-Verlag: Berlin Heidelberg, 1999, pp 149-179.
21. Jaoua, S.; Neff, S.; Schupp, T. *Plasmid* **1992**, *28*, 157-165.
22. Pradella, S.; Hans, A.; Sproer, C.; Reichenbach, H.; Gerth, K.; Beyer, S. *Arch. Microbiol.* **2002**, *178*, 484-492.
23. Reichenbach, H. *Environ. Microbiol.* **1999**, *1*, 15-21.
24. Julien, B.; Fehd, R. *Appl. Environ. Microbiol.* **2003**, *69*, 6299-6301.
25. Stephens, K.; Hartzell, P.; Kaiser, D. *J. Bacteriol.* **1989**, *171*, 819-830.
26. Hartzell, P.; Kaiser, D. *J. Bacteriol.* **1991**, *173*, 7615-7624.
27. Spormann, A. M. *Microbiol. Mol. Biol. Rev.* **1999**, *63*, 621-641.

28. Pfeifer, B. A.; Khosla, C. *Microbiol. Mol. Biol. Rev.* **2001,** *65,* 106-118.
29. Pfeifer, B.; Hu, Z.; Licari, P.; Khosla, C. *Appl. Environ. Microbiol.* **2002,** *68,* 3287-3292.
30. Murli, S.; Kennedy, J.; Dayem, L. C.; Carney, J. R.; Kealey, J. T. *J. Ind. Microbiol. Biotechnol.* **2003,** *30,* 500-509.
31. Egland, P. G.; Gibson, J.; Harwood, C. S. *J. Bacteriol.* **1995,** *177,* 6545-6551.
32. Murakami, T.; Holt, T. G.; Thompson, C. J. *J. Bacteriol.* **1989,** *171,* 1459-1466.
33. Kuhstoss, S.; Rao, R. N. *J. Mol. Biol.* **1991,** *222,* 897-908.
34. Henderson, D. J.; Brolle, D. F.; Kieser, T.; Melton, R. E.; Hopwood, D. A. *Mol. Gen. Genet.* **1990,** *224,* 65-71.
35. Grund, E.; Kutzner, H. J. *J. Basic Microbiol.* **1998,** *38,* 241-255.
36. Gerth, K.; Pradella, S.; Perlova, O.; Beyer, S.; Muller, R. *J. Biotechnol.* **2003,** *106,* 233-253.
37. Bentley, S. D.; Chater, K. F.; Cerdeno-Tarraga, A. M.; Challis, G. L.; Thomson, N. R.; James, K. D.; Harris, D. E.; Quail, M. A.; Kieser, H.; Harper, D.; Bateman, A.; Brown, S.; Chandra, G.; Chen, C. W.; Collins, M.; Cronin, A.; Fraser, A.; Goble, A.; Hidalgo, J.; Hornsby, T.; Howarth, S.; Huang, C. H.; Kieser, T.; Larke, L.; Murphy, L.; Oliver, K.; O'Neil, S.; Rabbinowitsch, E.; Rajandream, M. A.; Rutherford, K.; Rutter, S.; Seeger, K.; Saunders, D.; Sharp, S.; Squares, R.; Squares, S.; Taylor, K.; Warren, T.; Wietzorrek, A.; Woodward, J.; Barrell, B. G.; Parkhill, J.; Hopwood, D. A. *Nature (London)* **2002,** *417,* 141-147.
38. Kaneko, M.; Ohnishi, Y.; Horinouchi, S. *J. Bacteriol.* **2003,** *185,* 20-27.
39. Schneider, S.; Mohamed, M. E.; Fuchs, G. *Arch. Microbiol.* **1997,** *168,* 310-320.

Chapter 16

Engineering Starter Units in Aromatic Polyketides

Wenjun Zhang and Yi Tang*

Department of Chemical and Biomolecular Engineering, University of California, Los Angeles, CA 90095
***Corresponding author: yitang@ucla.edu**

The starter unit represents an attractive site in the aromatic polyketide scaffold for introducing alternative chemical functionalities. Most aromatic polyketide synthases (PKSs) are primed by acetate through the decarboxylative condensation of malonyl-ACP. Nonacetate-primed PKSs have been sequenced and characterized. Precursor-directed biosynthesis and heterologous recombination of initiation and elongation modules have led to the engineered biosynthesis of completely new aromatic polyketide scaffolds containing novel starter units, as well as the regioselective modification of known aromatic polyketides.

Polyketides are a large family of complex and structurally diverse natural products, most of which are produced as second metabolites by bacteria and fungi (*1*). Polyketides have a broad range of biological activities, including antibiotics, antitumor agents, immunosuppressants, antiparasitic agents, etc. As in fatty acid biosynthesis, polyketides are constructed through repetitive Claisen condensations of acyl-CoAs. In contrast to fatty acid, however, polyketides exhibit greater variety in structure with respect to the choice of acyl-CoA building blocks, carbon chain lengths, degrees of reduction, backbone cyclization, and other post-condensation enzymatic modifications (*2*). On the basis of biosynthetic mechanisms, PKSs can be classified into three groups. Type I PKSs consist of large, multifunctional megasynthases with catalytic domains in a linear modular fashion. Type III PKSs consist of homodimeric ketosynthases and do not utilize acyl carrier proteins (ACPs). Type II PKSs are composed of dissociated, monofunctional proteins. Type II PKSs synthesize aromatic, multicyclic polyketides via iterative Claisen condensations of multiple malonyl-CoA derived extender units, followed by regiospecific cyclodehydrations and modifications (*3-5*). Genetic analyses of type II PKSs have revealed that each gene PKS cluster consists of a minimal PKS, as well as a collection of auxiliary enzymes (*6,7*). Minimal PKSs include the ketosynthase (KS, or KS_{α}), the chain-length factor (CLF, alternatively referred to as KS_{β}), the ACP and the malonyl-CoA:ACP acyltransferase (MAT, shared between fatty acids synthase (FAS) and PKSs).

The type II minimal PKS exhibits exceptionally stringent specificity towards malonyl-ACP as the extender unit. No alternative extender units have been observed in the backbones of aromatic polyketides. As a result, the primer unit of an aromatic polyketide is an attractive position for introducing novel building blocks (*8*). A number of labs have focused on broadening the natural diversity of aromatic polyketides by targeted manipulations of starter units (*9*). However, most aromatic PKSs, including actinorhodin (*act*) (*10*), tetracenomycin (*tcm*) (*11*), and pradimicin (*pms*) (*12*) PKSs, are primered with an acetyl unit, which arises through decarboxylation of the corresponding malonyl-ACP (*13,14*). Decarboxylation of malonyl-ACP affords an acetyl-ACP that can be transacylated onto the active site cysteine of KS to initiate chain elongation. Structural analysis, accompanied by feeding studies, have revealed a few aromatic PKSs that utilize starter units other than acetate (Figure 1). For example, daunorubicin (*dnr*) is primed with a propionate unit (*15*); frenolicin (*fren*) is primed with a butyrate unit (*16*); and, R1128 (*zhu*) is primed with alkyl units of varying lengths (*17*). In these cases an additional ACP (for *fren* and R1128 PKSs), a ketosynthase (KSIII) and an acyl transferase (AT) homologue are commonly present, composing a standalone *initiation module*. Nonacetate initiated chain elongation is also observed during enterocin (*enc*) (*18*) and oxytetracycline (*oxy, otc*) (*19*) biosynthesis, which are primed with benzoate and malonamate, respectively.

Advances in genetic engineering have enabled the rational alteration of non-natural starter units with several powerful strategies, such as precursor-directed biosynthesis and combinatorial manipulation of initiation and elongation modules. In these cases PKSs have been engineered to accept non-natural starter units, leading to the biosynthesis of novel polyketides (*9*). In the following sections, we will discuss several different strategies that led to the engineering biosynthesis of aromatic polyketides containing nonacetate starter units.

Aryl-Priming

A number of important natural products, including the aromatic polyketides enterocin, wailupemycins (from marine bacterium *Streptomyces maritimus* (*18*)), and thermorubin (*20*) (from bacterium *thermoactinomyces vulgari)*, have aryl functionalities as the starter units (benzoate for enterocin and wailupemycin; salicylate for thermorubin). The polyketide backbones of enterocin and wailupemycins A-C are synthesized by iterative type II PKS from an uncommon benzoyl-CoA primer unit and seven malonate extender units (Figure 2). The gene cluster (~21 kb) of enterocin PKS has been cloned and contains 20 open reading frames (ORF) (*21*). Moore and coworkers discovered that the centrally located minimal PKS genes *enc*ABC are flanked by a number of genes involved in the benzoate starter unit biosynthesis. The starter unit benzoyl-CoA is biosynthesized from phenylalanine in multiple steps. The key enzyme phenylalanine ammonia lyase (EncP) catalyzes the first step by deaminating phenylalanine into *trans*-cinnamic acid. After the benzoyl-CoA starter unit is formed, it is then loaded onto the *enc* PKS to initiate elongation. An alternative route of starter unit biosynthesis is through the direct feeding of benzoic acid, which is converted into benzoyl-CoA by a monofunctional CoA-ligase, EncN.

The one-step, phenylalanine-independent route of benzoyl-CoA biosynthesis presents an opportunity for precursor-directed biosynthesis, which enables incorporation of unnatural starter units by feeding biosynthetic precursor analogues to the fermentation broth. Feeding experiments with fluorinated benzoic acids resulted in modest yields of mono- and difluorinated enterocins in the bacterium *S. hygroscopicus* No. A-5294 (*22*), suggesting that the *enc* PKS can accept unnatural aryl starter units. To bypass the phenylalanine-dependent route of benzoyl-CoA biosynthesis and to suppress the wild type benzoate priming of the *enc* minimal PKS, Moore and coworkers inactivated the phenylalanine ammonia lyase EncP in the *enc* PKS in *Streptomyces maritimus* by a double crossover homologous recombination (*23*). The resulting strain did not produce the benzoate-primed polyketides enterocin and wailupemycins D-G in the absence of exogenously supplied benzoic acid. A series of novel aryl acids, such as *p*-fluorobenzoic acid, 2-thiophenecarbotclic acid, 3-thiophenecarbotclic acid and cyclohex-1-enecarbotclic acid were administered to the mutant strain. The aryl acids were converted into the corresponding aryl-CoAs by CoA ligase EncN and were utilized as starter units in place of benzoyl-CoA. Moore and coworkers were able to recover enterocin and wailupemycin derivatives containing the unnatural aryl starter units (Figure 2).

Alkyl-Priming

A common priming mechanism of nonacetate chain initiation is utilized by the PKSs of daunorubicin, frenolicin and R1128, where the starter molecules are short-chain alkyl groups in place of the common acetate.

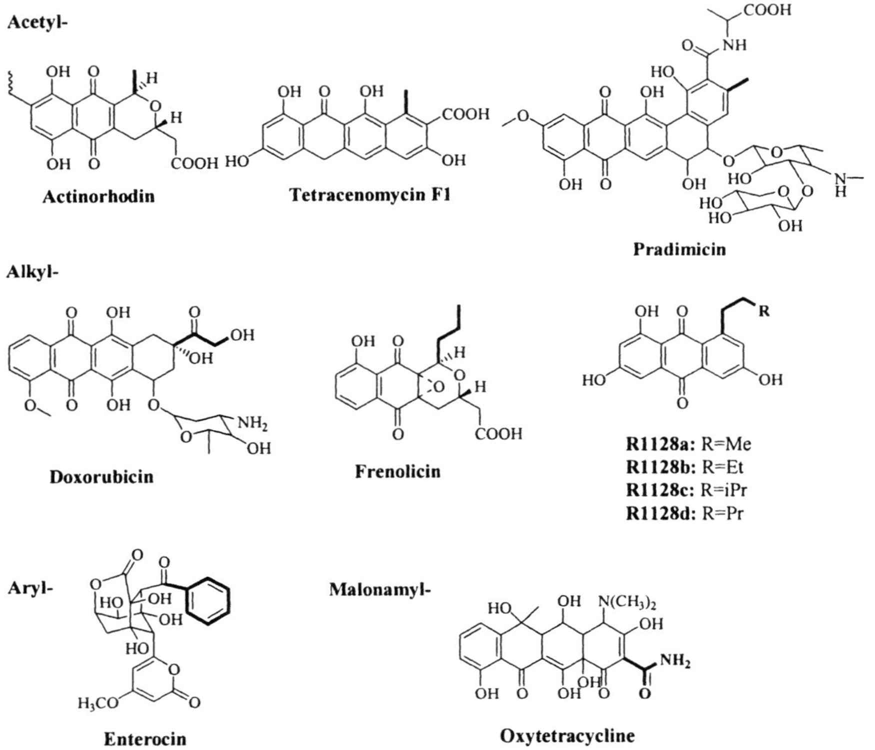

Acetyl-

Actinorhodin

Tetracenomycin F1

Pradimicin

Alkyl-

Doxorubicin

Frenolicin

R1128a: R=Me
R1128b: R=Et
R1128c: R=iPr
R1128d: R=Pr

Aryl-

Enterocin

Malonamyl-

Oxytetracycline

Figure 1. Examples of natural aromatic polyketides. The starter units are indicated in bold.

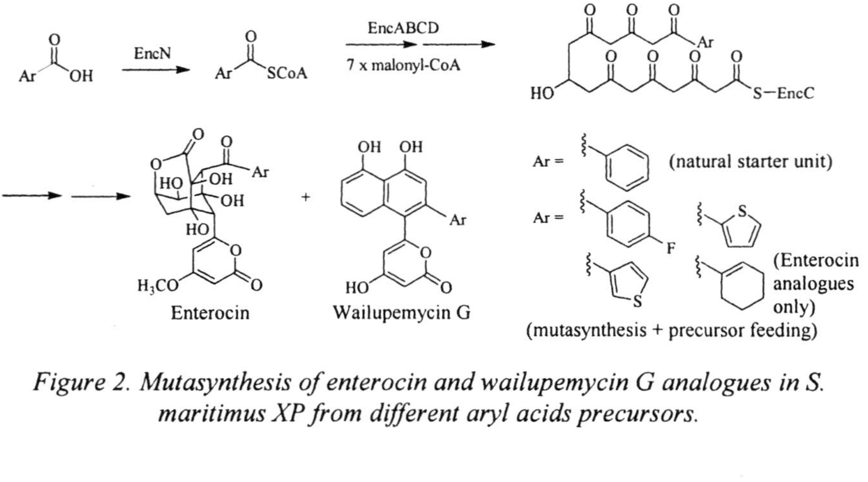

EncN

EncABCD

7 x malonyl-CoA

Enterocin + Wailupemycin G

Ar = (natural starter unit)

Ar = (Enterocin analogues only)

(mutasynthesis + precursor feeding)

Figure 2. Mutasynthesis of enterocin and wailupemycin G analogues in S. maritimus XP from different aryl acids precursors.

Doxorubicin (*dox*) and daunorubicin (*dnr*) are important antitumor anthracyclines produced by *Streptomyces peucitus*. They are each assembled from a propionyl-CoA starter unit and nine malonyl-CoA extender units. Propionyl-CoA is derived from multiple sources, such as amino acid catabolism and degradation of odd-numbered fatty acids (*24*). Sequence analyses of the *dnr* and *dox* gene clusters by the Hutchinson (*15*) and Strohl (*25*) groups revealed that DpsC (a KSIII homologue that lacks an active-site cysteine) and DpsD (a proposed AT homologue) were involved in chain initiation. Subsequently, it was found that DpsC works as the "fidelity factor" for the propionyl starter unit and contributes to the selection of propionyl-CoA as the starter unit, while DpsD was not essential for propionate priming (*26*).

The anthraquinones R1128 A-D are non-steroidal estrogen receptor antagonists produced by *Streptomyces* sp. R1128 (*27*). This series of compounds are primed with various alkyl side chains (including butyrate, pentanoate, hexanoate and iso-hexanoate). Each of the nonacetate primer units is derived from a short chain acyl-CoA. Nucleotide sequence analysis of the biosynthetic gene cluster by the Khosla group (*17*) revealed among the 14 ORFs found in the R1128 PKS cluster, a priming KSIII (ZhuH), an ACP_p (ZhuG) and an AT (ZhuC) are involved in starter unit biosynthesis and incorporation. ZhuH was found to catalyze the condensation of a short chain alkylacyl-CoA with malonyl-ZhuG to yield β-ketoacyl-ZhuG. The condensed product can then be subjected to β-ketoreduction, dehydration and enoylreduction to yield a saturated, alkylacyl-ZhuG. Since the genes encoding the reductive cycle are absent from the R1128 gene cluster, the R1128 PKS must recruit ketoreductase (KR), dehydratase (DH) and enoylreductase (ER) from the endogenous FAS. Therefore, the actions of ZhuH, KR, DH and ER essentially 1) elongate the short acyl-CoA precursor by two methylene units, and 2) replace the acyl carrier from Coenzyme A to ZhuG. For example, propionyl-CoA and isobutyryl-CoA are converted into pentanoyl-ZhuG and isohexanoyl-ZhuG, respectively.

In vitro titration experiments showed that ketosynthases involved in primer unit biosynthesis (i.e. KSIII) and polyketide chain elongation (i.e. KS-CLF) have orthogonal ACP specificities. KS-CLFs are able to employ ACPs from different minimal PKSs to support polyketide turnover with comparable catalytic efficiencies, while KSIIIs are specific toward priming ACP_ps only (*28,29*). Minimal ACPs cannot support the condensation reaction catalyzed by KSIII, while ACP_p cannot support chain elongation catalyzed by KS-CLFs. Therefore, the additional ACP_p (ZhuG) present in R1128 PKS is indispensable in primer unit biosynthesis.

The AT homologue ZhuC in the R1128 PKS was found to be an acetyl-ACP thioesterase that is required to suppress acetate priming. *In vitro* PKS reconstitution experiments with ZhuC showed that its addition to the *act* and *tcm* minimal PKSs significantly attenuated the synthesis of acetate-primed polyketides (*30*). ZhuC was shown to catalyze the rapid hydrolysis of acetyl-ACP to yield acetate and holo-ACP (k_{cat} >150 min^{-1}). In contrast, ZhuC does not hydrolyze medium-chain alkylacyl-ACPs. The catalytic efficiency (k_{cat}/K_m) of ZhuC towards acetyl- and propionyl-ACP is >100 times higher than for hexanoyl- and octanoyl-ACPs. Thus, the role of ZhuC is clear: it is a proofreading enzyme that selectively hydrolyzes acetyl-ACP species that can compete with the alkylacyl-ACP species for priming the minimal PKS. In the absence of ZhuC, the more abundant acetyl-ACP can prevent the loading of alkylacyl-ACP and yield acetate primed polyketides. ZhuC effectively cleanses the acetyl-ACPs to allow exclusive priming of the KS-CLF by alkylacyl-ACP.

We can therefore conclude that the R1128 initiation module consists of three unique proteins: ZhuG, ZhuH and ZhuC. ZhuG is a dedicated ACP_p that carries alkylacyl starter units of various lengths to initiate polyketide assembly. Alkylacyl-ZhuG can directly prime the KS-CLF, without the involvement of additional acyltransferases. ZhuH is the gatekeeping KSIII enzyme that selects short-chain CoAs. ZhuC serves as an acetyl-ACP thioesterase that attenuates the acetyl priming pathway in favor of non-acetate priming. Transferring the alkylacyl moiety to ZhuAB (KS-CLF) completes polyketide priming and commits the PKS to the biosynthesis of alkylacyl-primed polyketides. The priming and editing mechanisms for R1128 PKS are shown in Figure 3.

Frenolicin is an anti-malarial agent produced by *Streptomyces roseofulvus*. It is derived from a butyryl starter unit and seven malonyl-CoA extender units. Analysis of the frenolicin biosynthetic gene cluster reveals a similar priming mechanism as in R1128 biosynthesis (*16*). The dissociated initiation module of *fren* PKS consists of FrenI (KSIII), FrenJ (the ACP_p in addition to the minimal PKS ACP FrenN) and FrenK (homologous to ZhuC), each of which is an essential component in the biosynthesis and incorporation of the butyryl starter unit by the *fren* PKS.

Using *Streptomyces coelicolor* strain CH999 as a heterologous host, the R1128 initiation module was coexpressed with numerous minimal PKS modules. We hypothesized that efficient interactions between the heterologus modules can afford novel polyketides bearing nonacetate starter units. The first test was to see if the exclusively primed *act* minimal PKS can interact with alkylacyl-ZhuG *in vivo*. The *act* minimal PKS is a C16 synthase and synthesizes an acetate-primed octaketide. When coexpressed with ZhuG, ZhuC and ZhuH, a new polyketide was produced as the major product. (Figure 4A) (*31*). The backbone of the compound was determined to be a novel hexaketide bearing aliphatic starter units. Both pentanoate and isohexanoate starter units were incorporated into the backbone, in good agreement with the substrate specificity of ZhuH. In the presence of *act*III KR, the C9 position of the pentanoate-primed hexaketide is regioselectively reduced and the backbone spontaneous cyclized to form YT46. YT46 exhibits a bicyclic scaffold previously not observed among aromatic polyketides. Substrate feeding studies using ^{13}C-propionic acid confirmed the origin of the pentanoate primer unit.

We next assayed whether the *tcm* minimal PKS can similarly interact with the R1128 initiation module (Figure 4C). The *tcm* minimal PKS is known as an

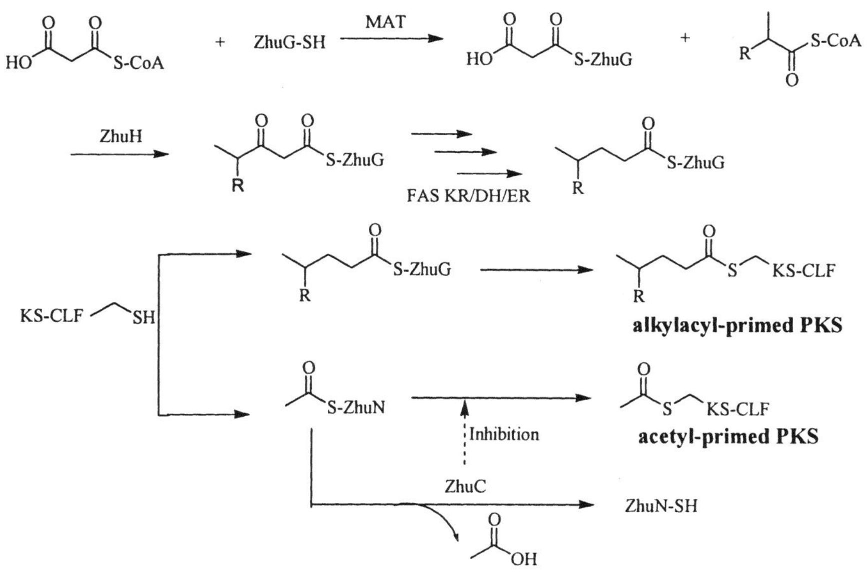

Figure 3. Priming and editing mechanisms for the R1128 initiation module. The acyl group carried by ZhuG is shuttled to the KS-CLF of the minimal PKS. The acyl-primed KS-CLF is then able to elongate the starter unit into a full length polyketide. ZhuC serves as an acetyl-ACP thioesterase.

acetate-primed decaketide synthase. Coexpression of the heterologous initiation and minimal PKS modules, along with the *actIII* KR, afforded a reduced, alkylacyl-primed octaketide backbone. Spontaneous cyclization of the nascent polyketide yielded YT85, the alkyl-primed version of mutactin. Tailoring enzymes such as cyclases were able to process the unnatural polyketides efficiently. For example, equipping the *tcm* minimal PKS and R1128 initiation module with cyclases and aromatases (CYC/ARO) specific for reduced or unreduced octaketides resulted in the transformation of the nascent polyketides into the expected anthraquinones YT127 and YT128, respectively. Both of the alkylacyl-primed anthraquinones are more potent as cytotoxic agents against the breast cancer cell line MCF-7 than their acetate primed counterparts. In addition, installing the alkylacyl side chains in the anthraqunone scaffolds brought forth inhibitory activities against glucose-6-phosphate translocase, an attractive target for the treatment of Type II diabetes (*31*).

Our findings suggest 1) the R1128 initiation module can operate in a heterologous host to afford nonacetate starter units in the forms of alkylacyl-

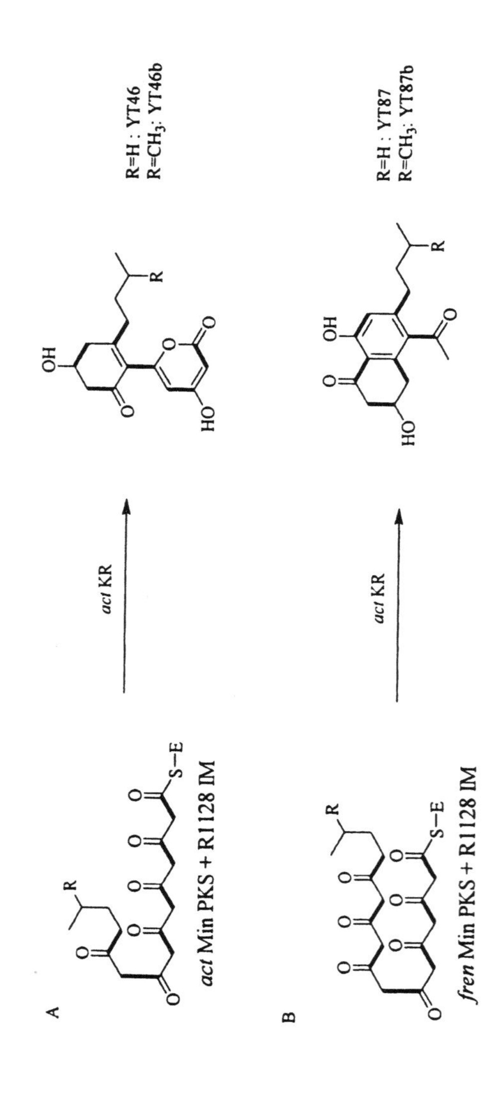
A
act Min PKS + R1128 IM
act KR
R=H : YT46
R=CH$_3$: YT46b
B
fren Min PKS + R1128 IM
act KR
R=H : YT87
R=CH$_3$: YT87b

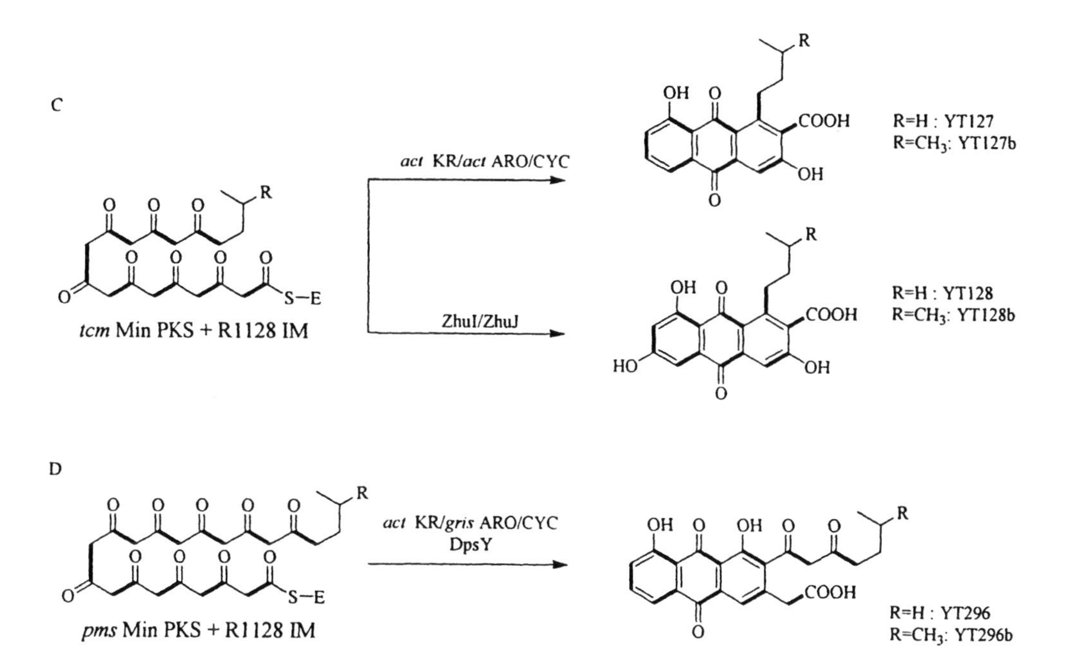

Figure 4. Novel aromatic polyketides produced by act (A), fren (B), tcm (C) and pms (D) minimal PKSs in the presence of the R1128 initiation module.

ZhuG; and 2) different minimal PKSs can interact with the alkylacyl starter unit and elongate it to afford new polyketides. The polyketides can bear completely novel cyclization patterns (YT46), as well as regioselectively modified analogs of known polyketides (YT85). Therefore, combining bimodular aromatic PKSs is a general mechanism for priming aromatic polyketide backbones with nonacetate precursors. The biosynthesis of YT46, YT85, YT127 and YT128 also reveals important biochemical properties of the PKS enzymes, including the KS-CLF, ketoreductase and cyclases. It is evident that the minimal PKS controls polyketide chain length by counting the number of atoms incorporated into the backbone, rather than the number of elongation cycles. For example, when the decaketide-specific *tcm* PKS was primed with a C_6 alkyl starter unit, the minimal PKS is effectively converted into an octaketide synthase: it performed seven additional iteration cycles, compared to nine iterations performed with acetate priming. The final number of carbons in the backbone remains stable at C_{20}. Incorporation of bulky, nonacetate primer units is compensated by a corresponding reduction in the number of condensation cycles. In contrast, auxiliary PKS enzymes recognize specific functional groups in the backbone, not overall chain length nor the identities of the starter unit. For example, the *act* KR only selectively reduced the C-9 keto group, despite the differences at the starter unit position. The *act* ARO and CYC were both able to process the alkylacyl-primed substrates regardless of starter unit identity, as long as the number of ketide units remained constant.

In addition to octaketide synthases (*act*) and decaketide synthases (*tcm*), a nonaketide synthase, such as that from the *fren* PKS (Figure 4B) (*32*), can also be primed with an alkylacyl starter unit to become an alkylacyl-primed heptaketide synthase. A novel compound YT87 was synthesized by an *S. coelicolor* strain coexpressing the R1128 initiation module and the *fren* minimal PKS. YT87 is an alkylacyl-primed heptaketide reduced at C-7. A dodecaketide synthase, such as that from the *pms* PKS (Figure 4D) (*33*), can likewise become a decaketide synthase when primed with an alkylacyl starter unit. Alkylacyl-primed analogues of SEK15, RM20b (acetyl-primed decaketides) were detected in the *pms* minimal PKS and R1128 initiation bimodular constructs. Coexpression of additional decaketide-specific tailing enzymes, including the *gris* ARO/CYC and DpsY cyclase, produced alkylacyl aklanonic acid YT296, an analogue of a key biosynthetic intermediate in the formation of several anthracycline pharmaceuticals. These results further confirm the versatile features of bimolecular aromatic PKSs identified above. It is noteworthy that elongation modules from dodecaketide-specific spore pigment PKSs were unable to interact with the R1128 initiation module, which indicates a possible fundamental incompatibility between antibiotic and spore pigment biosynthesis in the actinomycetes bacteria.

Short chain acyl-CoA substrates utilized by the R1128 initiation module are derived mainly through amino acid catabolism. The catabolic pathways involve transamination (catalyzed by branched-chain amino acid transaminases) to convert the amino acid into its corresponding α-ketoacid, followed by decarboxylation (catalyzed by acyl-CoA dehydrogenase) to yield the corresponding acyl-CoA. Metabolites derived from amino acid (both natural and unnatural) feedings were efficiently utilized by the R1128 initiation module to yield YT127 analogues (*32*), demonstrating the considerable potential for crosstalk between amino acid catabolism and aromatic polyketide biosynthesis.

The decaketide-specific *tcm* minimal PKS was reconstituted *in vitro* and several metabolically inaccessible alkylacyl-CoAs were examined as sources of primer units (*30*). In the presence of the R1128 proofreading enzyme ZhuC, the *tcm* minimal PKS was readily primed by propionyl-CoA to yield propionyl-primed decaketides; butyryl-CoA to yield the butyryl-primed nonaketides; hexanoyl-CoA to yield hexanoyl-primed octaketides; octanoyl-CoA to yield octanoyl-primed heptaketides; and decanoyl-CoA to yield decanoyl-primed hexaketides. These findings suggested that the minimal PKS can tolerate and extend unnatural starter units of varying chain lengths.

The alkylacyl starter units incorporated into aromatic polyketides are chemically inert and cannot be easily utilized as orthogonal reactive handles for semisynthetic modifications. To further expand the repertoire of primer units that can be introduced into aromatic polyketides using bimodular PKSs, initiation modules that can synthesize alternative starter units are required. They can be achieved through protein engineering efforts. KSIII homologues found in initiation modules serve as gatekeepers in primer unit selection and are therefore attractive targets for protein engineering. The X-ray crystal structure of ZhuH has already been solved and has led to the identification of a binding pocket for the acyl-CoA moiety (*34*). Altering the size and polarity of the gatekeeping residues on enzymes using rational mutagenesis and directed evolution may enlarge the repertoire of acyl-CoA moieties recognized by KSIII enzymes. In addition, the starter unit library could potentially be expanded through the use of other initiation modules. Recently, a novel aromatic PKS priming mechanism was found in hedamycin (*35*). The sequenced gene cluster suggested that the biosynthesis of pluramycins involves an iterative type I PKS system for the generation of a novel starter unit. Hedamycin therefore serves as an ideal model for studying starter unit communications between type I and type II PKSs.

Malonamyl-Priming

An unusual priming mechanism is observed in the biosynthetic pathway of tetracycline, one of the most important drugs in treating infectious diseases (*36*). Tetracyclines are aromatic polyketides synthesized by soil-borne actimonyces using type II PKS. Chlorotetracycline and oxytetracycline are biosynthesized by *Streptomces aurefaciens* and *Streptomyces rimosus,* respectively. The tetracycline carbon skeleton is assembled from ten malonate-derived building blocks through iterative condensations catalyzed by the minimal PKS, with an amide unit at one terminus of the polyketide backbone. This amide moiety is universal among all tetracycline derivatives and is important for the metal-chelating properties of tetracyclines. Substrate feeding and mutation studies (*37-39*) suggested that the amide unit stems from an intact malonamyl unit, which is likely derived from a yet unidentified malonyl precursor. Biosynthesis of the polar, amide starter unit is especially interesting from an biocombinatorial perspective, since most of other aromatic polyketides are primed by chemically inert, aliphatic and aromatic starter units. The polar amide unit can serve as a useful reaction handle for orthogonal semisynthetic modifications of polyketides. Cloning, sequencing and heterologous expression of tetracycline biosynthetic genes are expected to shed some light on the nature of the novel malonamate starter unit.

The *oxy* gene cluster from *S. rimosus* has been sequenced with a combination of shotgun and cosmid walking techniques (*40*). The gene cluster was previously mapped to be between the two resistant genes *otr*A and *otr*B (*41,42*), within which a total of 21 ORFs were identified. The functions of these genes are assigned based on sequence homology to known enzymes in type II PKSs. Interestingly, no homologues of KSIII or ACP_p are present in the *oxy* cluster. Their absence indicates the involvement of a novel chain initiation mechanism. Two ORFs (*oxy*P and *oxy*D) encoding enzymes may be involved in the biosynthesis of the malonamyl starter unit. OxyP shows high sequence homology to acetyl-ACP thioesterases and likely ensures malonamyl priming of the *oxy* PKS through elimination of the competing acetyl-ACP species. The heterologous expression experiments by Hopwood, Khosla and coworkers (*19*) showed that the minimal *oxy* PKS can be reaily primed by an acetate starter unit and can synthesize an acetate-derived polyketide backbone. OxyD, which is encoded by *oxy*D immediately downstream of the minimal PKS genes, shows high sequence identity to the type II asparagine synthases. Asparagine synthase catalyzes the transamination between glutamine and aspartic acid, using ATP as a cofactor. It is hypothesized that OxyD directly amidates malonyl substrate to yield malonamyl starter unit in an ATP-dependent, two-step reaction.

OxyD, together with the minimal *oxy* PKS, were reconstituted in the engineered host CH999 (Figure 5). Coexpression of immediate tailoring enzymes OxyJ (KR) afforded amidated polyketide WJ35 with a novel isoquinolone scaffold. This result indicates that OxyD is the only enzyme required to biosynthesize and to insert an amide starter unit into the polyketide backbone in the heterologous host. The predominant yield of amidated polyketides indicates that the *oxy* minimal PKS prefers the incorporation of malonamyl-OxyC to acetyl-OxyC, even in the absence of OxyP. It is noteworthy that *oxy* initiation module did not interact with heterologous KS-CLF. The *act*

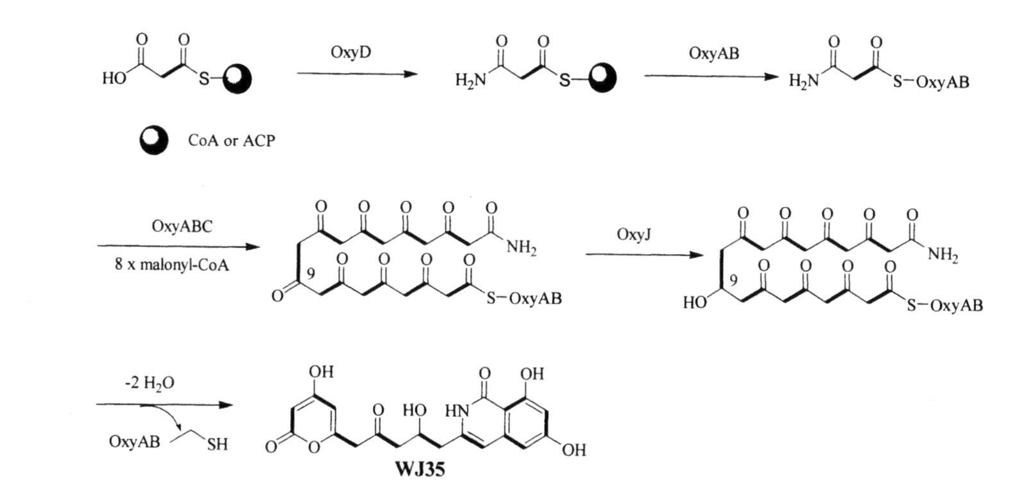

Figure 5. In vivo biosynthesis of WJ35 by the extended oxy minimal PKS with OxyD and OxyJ (KR).

and the *tcm* KS-CLF were each coexpressed in the presence of *act* KR, OxyD and OxyC in CH999. No amidated polyketides were recovered from these strains. Their lack of interaction are in sharp contrast to the broad compatibility between the R1128 initiation module and heterologous minimal PKSs. Therefore, in order to efficiently introduce the novel manolamate starter unit into other important aromatic polyketides, additional mechanistic investigations of OxyD and *oxy* minimal PKS are required.

References

1. Katz, L.; Donadio, S. *Annu Rev Microbiol* **1993**, *47*, 875-912.
2. Hopwood, D. A.; Sherman, D. H. *Annu. Rev. Genet.* **1990**, *24*, 37-66.
3. Rawlings, B. J. *Nat Prod Rep* **1999**, *16*, 425-484.
4. Rawlings, B. J. *Nat Prod Rep* **1997**, *14*, 523-556.
5. Staunton, J.; Weissman, K. J. *Nat Prod Rep* **2001**, *18*, 380-416.
6. McDaniel, R.; Ebert-Khosla, S.; Hopwood, D. A.; Khosla, C. *Science* **1993**, *262*, 1546-1550.
7. McDaniel, R.; Ebert-Khosla, S.; Hopwood, D. A.; Khosla, C. *Nature* **1995**, *375*, 549-554.
8. Koppisch, A. T.; Khosla, C. *Biochemistry* **2003**, *42*, 11057-11064.
9. Moore, B. S.; Hertweck, C. *Nat. Prod. Rep.* **2002**, *19*, 70-99.
10. Dreier, J.; Shah, A. N.; Khosla, C. *J. Biol. Chem.* **1999**, *274*, 25108-25112.
11. Bao, W.; Wendt-Pienkowski, E.; Hutchinson, C. R. *Biochemistry* **1998**, *37*, 8132-8138.
12. Dairi, T.; Hamano, Y.; Igarashi, Y.; Furumai, T.; Oki, T. *Biosci Biotechnol Biochem* **1997**, *61*, 1445-1453.
13. Dreier, J.; Khosla, C. *Biochemistry* **2000**, *39*, 2088-2095.
14. Bisang, C.; Long, P. F.; Cortes, J.; Westcott, J.; Crosby, J.; Matharu, A. L.; Cox, R. J.; Simpson, T. J.; Staunton, J.; Leadlay, P. F. *Nature* **1999**, *401*, 502-505.
15. Hutchinson, C. R. *Chem. Rev.* **1997**, *97*, 2525-2536.
16. Bibb, M. J.; Sherman, D. H.; Omura, S.; Hopwood, D. A. *Gene* **1994**, *142*, 31-39.
17. Marti, T.; Hu, Z. H.; Pohl, N. L.; Shah, A. N.; Khosla, C. *J. Biol. Chem.* **2000**, *275*, 33443-33448.
18. Moore, B. S.; Hertweck, C.; Hopke, J. N.; Izumikawa, M.; Kalaitzis, J. A.; Nilsen, G.; O'Hare, T.; Piel, J.; Shipley, P. R.; Xiang, L.; Austin, M. B.; Noel, J. P. *J. Nat. Prod.* **2002**, *65*, 1956-1962.
19. Fu, H.; Ebert-Khosla, S.; Hopwood, D.; Khosla, C. *J. Am. Chem. Soc.* **1994**, *116*, 6443-6444.
20. Craveri, R.; Coronelli, C.; Pagani, H.; Sensi, P. *Clin Med (Northfield Il)* **1964**, *71*, 511-521.

21. Piel, J.; Hertweck, C.; Shipley, P. R.; Hunt, D. M.; Newman, M. S.; Moore, B. S. *Chem Biol* **2000**, *7*, 943-955.
22. Kawashima, A.; Seto, H.; Kato, M.; Uchida, K.; Otake, N. *J Antibiot (Tokyo)* **1985**, *38*, 1499-1505.
23. Kalaitzis, J. A.; Izumikawa, M.; Xiang, L.; Hertweck, C.; Moore, B. S. *J. Am. Chem. Soc.* **2003**, *125*, 9290-9291.
24. Weissman, K. J.; Bycroft, M.; Staunton, J.; Leadlay, P. F. *Biochemistry* **1998**, *37*, 11012-11017.
25. Ye, J.; Dickens, M. L.; Plater, R.; Li, Y.; Lawrence, J.; Strohl, W. R. *J Bacteriol* **1994**, *176*, 6270-6280.
26. Rajgarhia, V. B.; Priestley, N. D.; Strohl, W. R. *Metab. Eng.* **2001**, *3*, 49-63.
27. Hori, Y.; Abe, Y.; Ezaki, M.; Goto, T.; Okuhara, M.; Kohsaka, M. *J. Antibiot.* **1993**, *46*, 1055-1062.
28. Meadows, E. S.; Khosla, C. *Biochemistry* **2001**, *40*, 14855-14861.
29. Tang, Y.; Lee, T. S.; Kobayashi, S.; Khosla, C. *Biochemistry* **2003**, *42*, 6588-6595.
30. Tang, Y.; Koppisch, A. T.; Khosla, C. *Biochemistry* **2004**, *43*, 9546-9555.
31. Tang, Y.; Lee, T. S.; Khosla, C. *PLoS Biol.* **2004**, *2*, 227-238.
32. Tang, Y.; Lee, T. S.; Lee, H. Y.; Khosla, C. *Tetrahedron* **2004**, *60*, 7659-7671.
33. Lee, T. S.; Khosla, C.; Tang, Y. *J Am Chem Soc* **2005**, *127*, 12254-12262.
34. Pan, H.; Tsai, S.; Meadows, E. S.; Miercke, L. J.; Keatinge-Clay, A. T.; O'Connell, J.; Khosla, C.; Stroud, R. M. *Structure* **2002**, *10*, 1559-1568.
35. Bililign, T.; Hyun, C. G.; Williams, J. S.; Czisny, A. M.; Thorson, J. S. *Chem Biol* **2004**, *11*, 959-969.
36. Chopra, I.; Hawkey, P. M.; Hinton, M. *J Antimicrob Chemother* **1992**, *29*, 245-277.
37. Thomas, R.; Williams, D. J. *J. Chem. Soc. Chem. Commun.* **1983**, 128-130.
38. Thomas, R.; Williams, D. J. *J. Chem. Soc. Chem. Commun.* **1983**, 677-679.
39. Petkovic, H.; Thamchaipenet, A.; Zhou, L. H.; Hranueli, D.; Raspor, P.; Waterman, P. G.; Hunter, I. S. *J. Biol. Chem.* **1999**, *274*, 32829-32834.
40. Zhang, W. J.; Wojcicki, W. K.; Tang, Y. *J. Am. Chem. Soc.* **2005**, *in preparation.*
41. Butler, M. J.; Friend, E. J.; Hunter, I. S.; Kaczmarek, F. S.; Sugden, D. A.; Warren, M. *Mol. Gen. Genet.* **1989**, *215*, 231-238.
42. Doyle, D.; McDowall, K. J.; Butler, M. J.; Hunter, I. S. *Mol Microbiol* **1991**, *5*, 2923-2933.

Chapter 17

Aspergillus flavus Genomics for Discovering Genes Involved in Aflatoxin Biosynthesis

Jiujiang Yu and Thomas E. Cleveland

Southern Regional Research Center, Agricultural Research Service, U.S. Department of Agriculture, 1100 Robert E. Lee Boulevard, New Orleans, LA 70124

Aflatoxins are polyketide-derivatives initially synthesized by condensation of carbon units into decaketide norsolorinic acid. Aflatoxins are toxic and the most carcinogenic natural compounds. In order to better understand the molecular mechanisms that control aflatoxin production, identification of genes involved in aflatoxin biosynthesis employing a genomics strategy in *Aspergillus flavus* was carried out. Sequencing and annotation of *A. flavus* expressed sequence tags (EST) identified 7,218 unique EST sequences. Genes that are involved in or potentially involved in aflatoxin formation were identified from these ESTs. Gene profiling using microarray has thus far identified hundreds of genes that are highly expressed under aflatoxin-producing conditions. Primary annotation of the *A. flavus* whole genome sequence data showed that there are over two dozens of unique polyketide synthase genes. Further investigations on the functional involvement of these genes in aflatoxin biosynthesis are underway. The results are expected to provide information for developing novel strategies to control aflatoxin contamination.

Aflatoxins are toxic, mutagenic, and the most carcinogenic secondary metabolites to animals and humans (1, 2, 3). They are produced primarily by *Aspergillus flavus* and *A. parasiticus*. Sterigmatocystin (ST) and dihydrosterigmatocystin (DHST) are the penultimate precursors of aflatoxins produced by various *Aspergilli* including *A. nidulans*. Since ST and DHST are aflatoxin precursors, any discussion on aflatoxin biochemical pathway, homologous genes and regulatory mechanism, applies to ST. The technological breakthroughs in genomics, such as high throughput sequencing, makes it possible to study the genetics and regulatory mechanism of aflatoxin biosynthesis at a genome scale. It will certainly promote revolution in our understanding on these filamentous fungi for controling aflatoxin contamination in food and feed. In this chapter, we report the current status of the *Aspergillus flavus* genomics programs; EST, microarrays, and whole genome sequencing, in discovering genetic components in aflatoxin biosynthesis.

Biosynthesis of Aflatoxins

The aflatoxins, B_1, B_2, G_1 and G_2 (AFB_1, AFB_2, AFG_1 and AFG_2) are the major four toxins among at least 16 structurally related toxins known to date. Aflatoxin B_1 is the most potent toxin and carcinogen known. Aflatoxins are polyketide-derived compounds. Their structures are composed of bis-furan-containing dihydrofuranofuran and tetrahydrofuran moieties (rings) fused with a substituted coumarin as shown in Figure 1. The research on aflatoxin originated from the incidence of "Turkey-X" disease in 1960 when 100,000 turkeys died due to aflatoxin contamination in peanut-meal feed (4). Now, we have a farely good understanding on the genetics and biochemistry of aflatoxin biosynthesis (1, 2, 3, 6-46). As many as 15 structurally-defined aflatoxin intermediates have been identified in the aflatoxin biosynthetic pathway starting with the acetate, the polyketide precursors. At least 23 enzymatic steps involved in the aflatoxin biosynthesis have been characterized or proposed and 29 clustered aflatoxin pathway genes and transcripts and 4 sugar cluster genes have been cloned or characterized (Figure 1, 42, 44, 45).

The early stage of aflatoxin biosynthetic pathway

The aflatoxin biosynthesis can be described in two major stages: the early and later stages. The early stage aflatoxin intermediates are colored pigments (brick red, yellow, or orange in color), which covers from acetate to VERA.

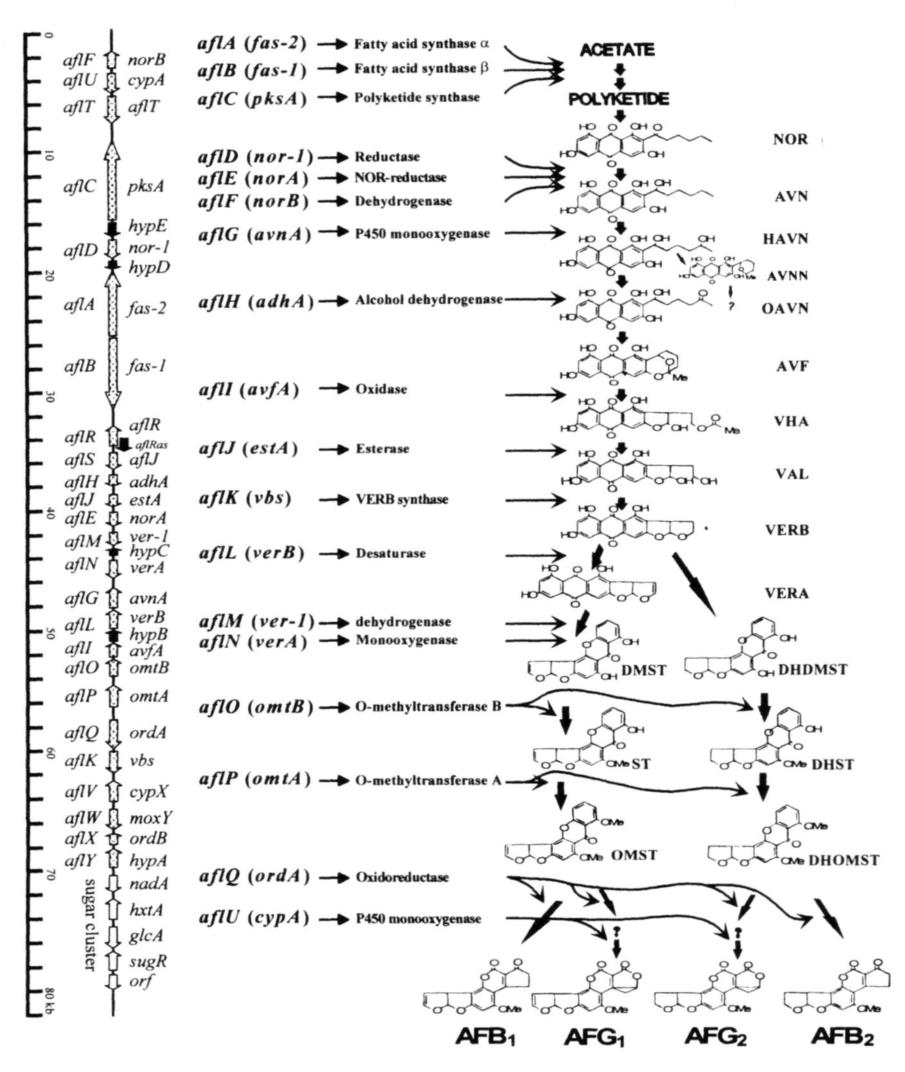

Figure 1. Caption on page 249. →

Figure 1. Clustered genes (A) and the aflatoxin biosynthetic pathway (B). The generally accepted pathway for aflatoxin biosynthesis is presented in panel B. The clustered genes with their new (on the left of the vertical line) and old (on the right) names are shown in panel A. The vertical line represents the 82 kb aflatoxin biosynthetic pathway gene cluster and sugar utilization gene cluster in A. parasiticus and A. flavus. Arrows along the vertical line indicate the aflatoxin pathway genes and the direction of gene transcription. The solid filled arrows represent transcripts identified recently. The aflRas represent aflR gene antisense transcript. The orf represents an Open Reading Frame with unknown function. The ruler on the far left indicates the relative sizes of these genes in kilobase pairs. Arrows in panel B indicate the connections from the genes to the enzymes they encode, from the enzymes to the bioconversion steps they are involved in, and from the intermediates to the products in the aflatoxin bioconversion steps. Abbreviations: NOR, norsolorinic acid; AVN, averantin; HAVN, 5'-hydroxy-averantin; OAVN, oxoaverantin; AVNN, averufanin; AVF, averufin; VHA, versiconal hemiacetal acetate; VAL, versiconal; VERB, versicolorin B; VERA, versicolorin A; DMST, demethylsterigmatocystin; DHDMST, dihydrodemethylsterigmatocystin; ST, sterigmatocystin; DHST, dihydrosterigmatocystin; OMST, O-methylsterigmatocystin; DHOMST, dihydro-O-methylsterigmatocystin; AFB_1, aflatoxin B_1; AFB_2, aflatoxin B_2; AFG_1, aflatoxin G_1; and AFG_2, aflatoxin G_2.

The later stage intermediates are toxins which are colorless under normal light and fluorenscent under UV light, which covers from DMST and DHDMST to the four aflatoxins. The synthesis of aflatoxins starts from acetate generated from the TCA cycle. Acetate is used as the basic building block by condensation to form polyketides. It is believed that two fatty acid synthases (FAS-1 and FAS-2) and a polyketide synthase (PKS) are involved in the synthesis of polyketide from acetate units (5, 47, 48, 49). Sequence analyses have shown that there are two large genes, *aflA* (*fas-2*) and *aflB* (*fas-1*) in the aflatoxin pathway gene cluster encoding for fatty acid synthase-1 (FASα) and fatty acid synthase-2 (FASβ), respectively (24, 44). The cloning of the *aflC* (*pksA*) gene encoding a PKS for the synthesis of polyketide in *A. parasiticus* (11, 50) and in *A. nidulans* (*wA* and pks^{ST}) (51, 52) confirmed that a polyketide synthase is required for aflatoxin and ST biosynthesis. The gene that encodes a PKS was cloned The predicted amino acid sequences of these PKSs contain typical four conserved domains common to other known PKS proteins: β-ketoacyl synthase (KS), acyltransferase (AT), acyl carrier protein (ACP), and thioesterase (TE). The gene for PKS from *A. parasiticus* (*pksA* or *pksL1*) was systematically renamed *aflC* (37, 44) in the aflatoxin pathway gene cluster and its homolog in *A. nidulans* was *stcA* (53).

Norsolorinic acid (NOR) is the first stable aflatoxin intermediate in the pathway (8, 9). The predicted conversion product from PKS in the aflatoxin pathway is believed to be a noranthrone. The conversion for noranthrone to norsolorinic acid (NOR) is poorly defined, but it has been proposed to occur by a noranthrone oxidase (54), a monooxygenase (5), or spontaneously (55). After the cloning and characterization of the *aflD* (*nor-1*) gene encoding a ketoreductase for the conversion of NOR to AVN (56, 57), it was identified that there are possibly three genes (*aflD=nor-1*, *aflE=norA* and *aflF=norB*) responsible for the conversion from NOR to AVN since the NOR-accumulating mutants were always leaky without completely blocking aflatoxin biosynthesis (8, 44, 58).

Averantin (AVN) is converted to 5' hydroxyaverantin (HAVN) (30, 38, 59) by a cytochrome P450 monooxygenase *aflG* (*avnA*). The cloning and characterization of a gene, *aflH* (*adhA*), encoding an alcohol dehydrogenase in *A. parasiticus* (15) demonstrated that AVN might be converted directly to AVF or indirectly to AVF through an intermediate substrate averufanin (AVNN). Yabe et al. (30, 31) proposed that AVNN was considered a shunt metabolite and not an aflatoxin intermediate. Recent studies demonstrated that oxoaverantin (OAVN) was a new aflatoxin intermediate between HAVN and averufin (AVF) (35, 60, 61, 62).

The conversion from averufin (AVF) to versiconal hemiacetal acetate (VHA) is catalyzed by an averufin oxidase encoded by *aflI* (*avfA*) (41). VHA was converted to versiconal (VAL) by an esterase (29, 63, 64, 65). VAL is converted to Versicolorin B (VERB) by a VERB synthase (31). The gene, *aflK* (*vbs*), was cloned, characterized and expressed (66, 67, 68). It was demonstrated that the versicolorin B synthase catalyzes the side chain cyclodehydration of racemic VHA to VERB. This is a key step in the aflatoxin formation since it closed the bisfuran ring of aflatoxin for binding to DNA. VERB is converted to versicolorin A (VERA) by a desaturase encoded by *aflL* (*verB*) (31).

The later stage of aflatoxin biosynthetic pathway

In the later stage of aflatoxin biosynthetic pathway, the VERB is a critical branch point leading either to AFB_1 and AFG_1 or to AFB_2 and AFG_2 formation. VERB is converted to VERA and demethyldihydrosterigmatocystin (DMDHST). VERA is converted to demethylsterigmatocystin (DMST). *aflM* (*ver-1*) is a key gene for a ketoreductase in aflatoxin biosynthesis (69) for the conversion of VERA to DMST. The gene *aflN* (*verA*), cloned in *A. parasiticus*, may also be involved in the same step from VERA to DMST (44, 45). DMST is converted to sterigmatocystin (ST) and DMDHST to dihydrosterigmatocystin (DHST) by a methyltransferase encoded by the *aflO* (*omtB*) gene (28, 33, 41, 70, 71). ST is converted to *O*-methylsterigmatocystin (OMST) and DHST to dihydro-*O*-

methylsterigmatocystin (DHOMST) by an *O*-methyltransferase encoded by *aflP* (*omtA*) gene (28, 36, 72).

OMST is converted to aflatoxin B_1 (AFB_1) and aflatoxin G_1 (AFG_1) and DHOMST to aflatoxin B_2 (AFB_2) and aflatoxin G_2 (AFG_2) by a P-450 monooxygenase encoded by *aflQ* (*ordA*) (28, 34, 39, 73) in *A. parasiticus* and *A. flavus*. Expression and substrate feeding in yeast system demonstrated that an additional enzyme is required for the G-group aflatoxin (AFG_1 and AFG_2) formation (39). This enzyme is also a cytochrome P450 monooxygenase encoded by a gene, *aflU* (*cypA*) for the conversion from OMST to AFG_1 and DHOMST to AFG_2 (74). It was proposed that formation of G-group aflatoxins requires the CypA-catalyzed epoxidation of the double bond in this intermediate (74).

The aflatoxin biosynthetic pathway gene cluster

The clustering organization of aflatoxin pathway genes was first evidenced by the linkage of the *aflD* (*nor-1*) and *aflM* (*ver-1*) genes in a cosmid clone with the regulatory gene *aflR* and an aflatoxin pathway gene, *uvm8* (now named *aflB* =*fas-1*) in the middle (25). By mapping overlapping cosmid clones in *A. parasiticus* and *A. flavus*, a consensus cluster map consisting of at least nine aflatoxin pathway genes or open reading frames (ORF) was established (37).

Exploitation of the genes within the gene cluster identified at least 29 genes or transcripts including the regulatory genes, *aflR* and *aflS* (*aflJ*) and the four sugar cluster genes (44, 45). These genes are packed within approximately 82 kb DNA regions in the *A. parasiticus* and *A. flavus* genomes (Figure 1, panel A). The sugar cluster marks one end and the *aflF* (*norB*) might possibly mark the other end of this cluster (44, 45; Figure 1, panel A). The *nadA* gene was classified into the sugar cluster previously. Based on recent studies this gene may be involved in aflatoxin synthesis since it is expressed in phase with the aflatoxin pathway genes (75, 76). The importance of the sugar utilization genes next to the aflatoxin pathway genes may partially explain the induction of aflatoxin production by glucose rich media. This indicates a possible link between primary and secondary metabolism.

In *A. parasiticus*, partial duplication of the aflatoxin gene cluster, consisting of seven duplicated genes, *aflR2, aflJ2, adhA2, estA2, norA2, ver1B, omtB2*, was reported (17, 77). Four (*aflR2, aflJ2, adhA2, estA2*) of these genes are found to be intact. However, the genes within this partial duplicated cluster, due possibly to the chromosome location (78, Yu et al. unpublished), were found to be nonfunctional under normal conditions.

The presence of an aflatoxin and ST gene cluster suggests that there may be some evolutionary advantages to the organism. The cluster organization assures that all of these genes are available for transcription and translation at the same time. It is an efficient mechanism in utilizing energy resources available. The importance of gene cluster in fungal adaptation and secondary metabolite production is to be further investigated by characterizing more secondary metabolism pathways and their gene clusters.

There is a positive regulatory gene, *aflR*, for activating pathway gene transcription, which is located in the middle of the aflatoxin gene cluster (10, 23). The *aflR* gene has been shown to be required for transcriptional activation of most, if not all, of the structural genes (12, 13, 14, 20, 21, 26, 79, 80, 81) by binding to the palindromic sequence 5'-TCGN5CGA-3' in the promoter region of the structural genes in *A parasiticus, A. flavus* (20, 21) and in *A. nidulans* (81). *aflS* (*aflJ*) and an *aflR* antisense sequence (*aflRas*) were also found to be involved in the regulation of transcription (18, 82). The exact regulatory roles by *aflS* (*aflJ*) and *aflRas* are to be studied.

Gene profiling by *Aspergillus flavus* genomics strategy

Studies on aflatoxin biosynthesis in *A. flavus* and *A. parasiticus* led to the cloning of 29 clustered genes within a 82kb DNA region on the chromosome (44). In order to understand the mechanism of aflatoxin biosynthesis at the genome level, the *A. flavus* EST, microarrays and whole genome sequencing programs have been conducted. The goal of *A. flavus* genomics is to understand the relationships between primary and secondary metabolism, between aflatoxin production and fungal survival, genetic regulation of aflatoxin production by *aflR* and the regulator proteins that regulate *aflR* and the evolution of *Aspergillus* section *Flavi* for devising strategies of reducing and eliminating aflatoxin contamination in food and feed (83).

Aspergillus flavus EST

A. flavus wild type strain NRRL 3357 (ATCC# 20026) was selected for making the EST library. RNAs were purified from fungal mycelia grown under eight aflatoxin-supportive and non-supportive media conditions. The 5′ ends of 26,110 cDNA clones from a normalized cDNA expression library were sequenced at The Institute for Genomic Research (TIGR). After comparison and assembly of overlapping sequences, 7,218 unique sequences were identified, which represent an estimated 60% of the predicted 12,000 functional genes in the *A. flavus* genome (84). These EST sequences have been released to the public at the NCBI GenBank Database (http://www.ncbi.nlm.nih.gov/). The Gene Ontology data were compiled to construct the *A. flavus* gene index which can be accessed at the TIGR web site (http://www.tigr.org/tdb/tgi/).

From the EST database, almost all of the aflatoxin pathway genes known in the aflatoxin pathway gene cluster in *A. parasiticus* and *A. flavus* have been identified. In addition, four new transcripts expressed in the EST library (Figure 1, labeled as *hypB*, *hypC*, *hypD*, and *hypE*) were identified in the cluster. Other categories of genes identified could be potentially involved directly or indirectly in aflatoxin production such as in global regulation, signal transduction, pathogenicity, virulence, and fungal development (Table I, 84). Global regulation of aflatoxin formation is of great interest but is also the least known aspect in aflatoxin biosynthesis. The genes for mitogen-activated protein kinase (MAPK), MAPK kinase (MAPKK) and MAPKK kinase (MAPKKK) in stress responses could be good candidates in this category (Table I). The homolog of a regulatory gene, *laeA* for loss of *aflR* expression in *A. nidulans* (85), was also found (NAGEM53TV) to be expressed in *A. flavus* cDNA library. Many genes encoding hydrolytic enzymes in the ESTs could be highly expressed virulence or pathogenicity factors during fungal invasion of *A. flavus* into crop plants. The oxidative stress may be another factor that triggered aflatoxin biosynthesis in *A. parasiticus* (86, 87). Secondary metabolism is often correlates with fungal developmental processes such as sporulation and sclerotia formation (16, 88, 89). Several genes involved in fungal development and conidiation were identified in the *A. flavus* EST library (84).

Aspergillus flavus microarrays

Gene profiling using microarrays is a powerful tool to detect a whole set of genes transcribed under specific conditions and to study the biological functions of these interested genes, gene expression and regulation, and to identify factors involved in plant-microbe (crop-fungus) interaction (75, 76, 90). A genomic DNA amplicon microarray was constructed at the Institute for Genomic Research (TIGR). A total of 9,445 pairs of sequence specific primers to the known unique ESTs were made (2 pairs of primers were made for each of the long ESTs). PCR amplification was performed using *A. flavus* genomic DNA as a template. The successful amplicons plus 31 aflatoxin pathway genes, representing 5002 unique ESTs, were arrayed in triplicate onto Telechem Superamine aminosilane coated slides. Profiling of genes that are potentially involved in aflatoxin formation under aflatoxin-producing vs. non-producing conditions is being performed. In addition to the genes directly involved in aflatoxin formation, hundreds of genes have been found to be significantly up or down regulated. These genes are either hypothetical proteins with unknown function or potential regulatory factors. Using TIGR MeV software program, the candidate genes, that expressed across several experiments, were further screened out. Gene profiling using microarray can effectively narrow down the target list of potential genes. Since the putative functions of these genes are derived from homology to existing database, additional validation and functional studies are needed to confirm the functional identities of these candidate genes.

Table I. Genes Potentially Involved in Aflatoxin Formation Identified

EST ID	*hit accession #*	*putative function*
Genes putatively involved in regulation and signal transduction		
TC7659	SP|O94321	Multistep phosphorelay regulator 1
TC10096	SP|Q03172	Zinc finger protein 40
TC5059	SP|O13724	Zinc-finger protein zpr1
TC11661	SP|P32586	Protein-tyrosine phosphatase 2
TC8756	PIR|H87338	sensor histidine kinase
TC9598	GB|BAB47691.1	transcriptional regulator
TC11834	GB|AAD24428.1	MAP protein kinase (MAPK)
TC9026	GB|AAM82166.1	MAP kinase kinase (MAPKK)
TC11700	GB|AAL77223.1	Bck1-like MAP kinase kinase kinase
NAGEM53TV	GI|37622141	methyltransferase (*laeA*)
Genes putatively involved in virulence and pathogenicity		
NAGBB06TV	GB|AAF40140.1	β (1-3) glucanosyltransferase Gel3p
NAGBM43TV	GB|BAA34996.1	oligo-1,4 - 1,4-glucantransferase
NAGCR76TV	GB|AAM77702.1	endoglucanase
TC10675	GB|AAC49904.1	mixed-linked glucanase precursor
TC11738	GB|CAD24293.1	β-galactosidase
TC11835	GB|BAC07256.1	cellobiohydrolase D
TC8959	GB|AAK58059.1	glucan 1,3 β-glucosidase-like protein
Genes putatively involved in stress response and antioxidation		
NAGAG45TV	GB|AAK54753.1	thiol-specific antioxidant
NAGCF11TV	GB|CAA60962.1	oxidative stress resistance
TC8386	GB|BAC56176.1	Cu,Zn superoxide dismutase
TC10087	GB|AAK17008.1	Mn-superoxide dismutase
TC10342	SP|P29429	Thioredoxin
TC9135	PIR|T48748	probable glutaredoxin 8D4.220
TC10000	GB|AAQ84041.1	peroxisomal-like protein

***Aspergillus flavus* Whole Genome Sequencing**

The *Aspergillus flavus* whole genome sequencing by a shotgun approach was carried out at TIGR led by Dr. William C. Nierman with the funding from USDA, National Research Initiative awarded to Professor Gary Payne, North Carolina State University, Raleigh, North Carolina, USA. The Food and Feed Safety Research Unit of Southern Regional Research Center, USDA/ARS, provided funding for fine finishing and gene calling. Currently the sequencing has been completed to its target of 5X coverage. Physical gaps are being closed by a newly developed autoclosure technique to achieve the same quality equivalent to 10X coverage of shot-gun sequencing alone (Nierman and Feldblyum, Personal communication). Primary assembly indicated that the *A. flavus* genome consists of 8 chromosomes and the genome size is about 36.3 Mbp. Preliminary annotation of the *A. flavus* genome sequence data with the help of *A. flavus* EST data indicated that the estimated functional genes in the *A. flavus* genome are 13,071 (http://www.aspergillusflavus.org/genomics). This is consistent with karyotyping studies and previous estimates. The A. flavus genome sequence data have been released to the NCBI GenBank database (http://www.ncbi.nlm.nih.gov) and is also available through the Aspergillus flavus web page (http://www.aspergillusflavus.org/genomics) for blast search analysis and for download.

A. flavus genomics provided a rapid and effective method for identification of genes potentially involved in aflatoxin formation and infection of crops by *A. flavus*. The availability of the *A. oryzae* whole genome sequence (91), a close relative of *A. flavus*, which is used in industrial fermentation for enzyme production that produce no aflatoxins, is very helpful for identifying genes specifically for aflatoxin formation and for pathogenesis through comparative genomics. Genes responsible for the biosynthesis of secondary metabolites such as aflatoxins are those encoding PKSs, non-ribosomal peptide synthethases (NRPS), cytochrome P450 monooxygenases, fatty acid synthases (FAS), carboxylases, dehydrogenases, reductases, oxidases, oxidoreductases, epoxide hydrolases, oxygenases, and methyltransferases (44, 45, 92). While PKSs, NRPSs, and cytochrome P450 enzymes are believed to be the key enzymes, the exact biological role of NRPS in the biosynthesis of secondary metabolites is unclear. Primary annotation revealed that there exist over two dozens of PKSs, many NRPS, and numerous cytochrome P450 enzymes in the *A. flavus* genome. Within the aflatoxin biosynthetic pathway gene cluster there is a single gene encoding the PKS and at least 5 genes encoding cytochrome P450 monooxygenases (Figure 1). The function of the PKS in the aflatoxin gene cluster is specific for aflatoxin biosynthesis and no other PKS is known to be involved in aflatoxin biosynthesis. These PKSs in the *A. flavus* genome could play very important roles in the formation of other mycotoxins (aflatrem or cyclopiazonic acid) or pigments. Further elucidation of their functions by targeted mutagenesis could assist in devising novel strategies for reducing and eliminating mycotoxin contamination of food and feed.

References

1. Bennett, J. W. *Final Report Cooperative Agreement*, **1987**, 58-7B30-3-556.
2. Bennett, J. W.; Klich, M. *Clin. Microbiol. Rev.* **2003**, *16,* 497-516.
3. Bennett, J. W.; Kale, S.; Yu, J. In *Foodborne Diseases.* Shabbir, S. Ed.; Humana Press: Totowa, NJ, 2006; (In press).
4. Lancaster, M. D.; Jenkins, F. P.; Phillip, J. M. *Nature* **1961**, *192,* 1095-1096.
5. Bhatnagar, D.; Ehrlich, K. C.; Cleveland, T. E. In *Handbook of Applied Mycology: Mycotoxins in Ecological Systems.* Bhatnagar, D.; Lillehoj, E. B.; Arora, D. K. Eds.; Marcel Dekker: New York, NY, 1992; pp 255-286.
6. Bhatnagar, D.; Yu, J.; Ehrlich, K. C. In *Fungal Allergy and Pathogenicity. Chem. Immunol.* Breitenbach, M.; Crameri, R.; Lehrer, S. B. Eds.; Basel: Karger, 2002; Vol. 81, pp 167-206.
7. Bhatnagar, D.; Ehrlich, K. C.; Cleveland, T. E. *Appl. Microbiol. Biotechnol.* **2003**, *61,* 83-93.
8. Bennett, J. W. *J. Gen. Microbiol.* **1981**, *124,* 429-432.
9. Bennett, J. W.; Chang, P. K.; Bhatnagar, D. *Adv. Appl. Microbiol.* **1997**, *45,* 1-15.
10. Chang, P.-K.; Cary, J. W.; Bhatnagar, D.; Cleveland, T. E.; Bennett, J. W.; Linz, J. E.; Woloshuk, C. P.; Payne, G. A. *Appl. Environ. Microbiol.* **1993**, *9,* 3273-3279.
11. Chang, P.-K.; Cary, J. W.; Yu, J.; Bhatnagar, D.; Cleveland, T. E. *Mol. Gen. Genet.* **1995**, *248,* 270-277.
12. Chang, P.-K.; Ehrlich, K. C.; Yu, J.; Bhatnagar, D.; Cleveland, T. E. *Appl. Environ. Microbiol.* **1995**, *61,* 2372-2377.
13. Chang, P.-K.; Yu, J.; Bhatnagar, D.; Cleveland, T. E. *Mycopathologia* **1999**, *147,* 105-112.
14. Chang, P.-K.; Yu, J.; Bhatnagar, D.; Cleveland, T. E. Appl. Environ. Microbiol. **1999,** *65,* 2058-2512.
15. Chang, P.-K.; Yu, J.; Ehrlich, K. C.; Boue, S. M.; Montalbano, B. G.; Bhatnagar, D.; Cleveland, T. E. *Appl. Environ. Microbiol.* **2000**, *66,* 4715-4719.
16. Chang, P.-K.; Bennett, J. W.; Cotty, P. J. *Mycopathologia* **2001**, *153,* 41-48.
17. Chang, P.-K.; Yu, J. *Appl. Microbiol. Biotechnol.* **2002,** *58,* 632-636.
18. Chang, P.-K. *Mol. Genet. Genomics* **2003**, *268,* 711-719.

19. Cleveland, T. E.; Cary, J. W.; Brown, R. L.; Bhatnagar, D.; Yu, J.; Chang, P.-K. *Bull. Inst. Compr. Agric. Sci.* Kinki Univ. Japan. **1997**, *5,* 75-90.
20. Ehrlich, K. C.; Cary, J. W.; Montalbano, B. G. *Biochim. Biophys. Acta* **1999**, *1444,* 412-417.
21. Ehrlich, K. C.; Montalbano, B. G.; Cary, J. W. *Gene* **1999**, *230,* 249-257.
22. Minto, R. E.; Townsend, C. A. *Chem. Rev.* **1997**, *97,* 2537-2555.
23. Payne, G. A.; Nystrom, G. J.; Bhatnagar, D.; Cleveland, T. E.; Woloshuk, C. P. *Appl. Environ. Microbiol.* **1993**, *59,* 156-162.
24. Payne, G. A.; Brown, M. P. *Annu. Rev. Phytopathol.* **1998**, *36,* 329-362.
25. Trail, F.; Mahanti, N.; Rarick, M.; Mehigh, R.; Liang, S. H.; Zhou, R.; Linz, J. E. *Appl. Environ. Microbiol.* **1995**, *61,* 2665-2673.
26. Woloshuk, C. P.; Foutz, K. R.; Brewer, J. F.; Bhatnagar, D.; Cleveland, T. E.; Payne, G. A. *Appl. Environ. Microbiol.* **1994**, *60,* 240814.
27. Yabe, K.; Ando, Y.; Hamasaki, T. *Appl. Environ. Microbiol.* **1988**, *54,* 2101-2106.
28. Yabe, K.; Ando, Y.; Hashimoto, J.; Hamasaki, T.; *Appl. Environ. Microbiol.* **1989**, *55,* 2172-2177.
29. Yabe, K.; Ando, Y; Hamasaki, T. *J. Gen. Microbiol.* **1991**, *137,* 2469-2475.
30. Yabe, K.; Nakamura, Y.; Nakajima, H.; Ando, Y.; Hamasaki, T. *Appl. Environ. Microbiol.* **1991**, *57,* 1340-1345.
31. Yabe, K.; Hamasaki, T. *Appl. Environ. Microbiol.* **1993**, *59,* 2493-2500.
32. Yabe, K.; Matsuyama, Y.; Ando, Y.; Nakajima, H.; Hamasaki, T. *Appl. Environ. Microbiol.* **1993**, *59,* 2486-2492.
33. Yabe, K.; Matsushima, K.; Koyama, T.; Hamasaki, T. *Appl. Environ. Microbiol.* **1998**, *64,* 166-171.
34. Yabe, K.; Nakamura, M.; Hamasaki, T. *Appl. Environ. Microbiol.* **1999**, *5,* 3867-3872.
35. Yabe, K.; Nakajima, H. *Appl. Microbiol. Biotechnol.* **2004**, *64,* 745-55.
36. Yu, J.; Cary, J. W.; Bhatnagar, D.; Cleveland, T. E.; Keller, N. P.; Chu, F. S. *Appl. Environ. Microbiol.* **1993**, 59, 3564-3571.
37. Yu, J.; Chang, P.-K.; Cary, J. W.; Wright, M; Bhatnagar, D.; Cleveland, T. E.; Payne, G. A.; Linz, J. E. *Appl. Environ. Microbiol.* **1995**, *61,* 2365-2371.
38. Yu, J.; Chang, P.-K.; Cary, J. W.; Bhatnagar, D.; Cleveland, T. E. *Appl. Environ. Microbiol.* **1997**, *63,* 1349-1356.
39. Yu, J.; Chang, P.-K.; Cary, J. W.; Ehrlich, K. C.; Montalbano, B.; Dyer, J. M.; Bhatnagar, D.; Cleveland, T. E. *Appl. Environ. Microbiol.* **1998**, *64,* 4834-4841.
40. Yu, J.; Chang, P.-K.; Bhatnagar, D.; Cleveland, T. E. *Appl. Microbiol. Biotechnol.* **2000**, *53,* 583-590.

41. Yu, J.; Woloshuk, C. P.; Bhatnagar, D.; Cleveland, T. E. *Gene* **2000**, *248,* 157-167.
42. Yu, J.; Chang, P.-K.; Bhatnagar, D.; Cleveland, T. E. *Biochim. Biophys. Acta* **2000**, *1493,* 211-214.
43. Yu, J.; Chang, P.-K.; Bhatnagar, D.; Cleveland, T. E. *Mycopathologia.* **2002**, *156,* 227-234.
44. Yu, J.; Chang, P.-K.; Ehrlich, K. C.; Cary, J. W.; Bhatnagar, D.; Cleveland, T. E.; Payne, G. A.; Linz, J. E.; Woloshuk, C. P.; Bennett, J. W. *Appl. Environ. Microbiol.* **2004**, *70,* 1253-1262.
45. Yu, J.; Bhatnagar, D.; Cleveland, T. E. *FEBS Lett.* **2004**, *564,* 126-130.
46. Yu, J. In *Fungal Biotechnology in Agricultural, Food, and Environmental Applications.* Arora, D. K., Ed.; Marcel Dekker: New York, NY, 2004; Vol. 21, pp 343-361.
47. Townsend, C. A.; Christensen, S. B.; Trautwein, K. *J. Am. Chem. Soc.* **1984**, *106,* 3868-3869.
48. Watanabe, C. M. H.; Wilson, D.; Linz, J. E.; Townsend, C. A. *Chem. Biol.* **1996**, *3,* 463-469.
49. Brown, D. W.; Adams, T. H.; Keller, N. P. *Proc. Natl. Acad. Sci.* USA, **1996**, *93,* 14873-14877.
50. Feng, G. H.; Leonard, T. J. *J. Bact.* **1995**, *177,* 6246-6254.
51. Mayorga, M. E.; Timberlake, W. E. *Mol. Gen. Genet.* **1992**, *235,* 205-212.
52. Yu, J.-H.; Leonard, T. J. *J. Bact.* **1995**, *177,* 4792-4800.
53. Brown, D. W.; Yu, J.-H.; Kelkar, H. S.; Fernandes, M.; Nesbitt, T. C.; Keller, N. P.; Adams, T. H.; Leonard, T. J. *Proc. Natl. Acad. Sci.* USA **1996**, *93,* 1418-1422.
54. Vederas, J. C.; Nakashima, T. T. *J. Chem. Soc. Chem. Commun.* **1980**, *4,* 183-185.
55. Dutton, M. F. *Microbiol. Rev.* **1988**, *52,* 274-295.
56. Chang, P.-K.; Skory, C. D.; Linz, J. E. *Curr. Genet.* **1992**, *21,* 231-233.
57. Trail, F.; Chang, P.-K.; Cary, J.; Linz, J. E. *Appl. Environ. Microbiol.* **1994**, *60,* 4078-4085.
58. Cary, J. W.; Wright, M.; Bhatnagar, D.; Lee, R.; Chu, F. S. *Appl. Environ. Microbiol.* **1996**, *62,* 360-366.
59. Bennett, J. W.; Lee, L. S.; Shoss, S. M.; Boudreaux, G. H. *Appl. Environ. Microbiol.* **1980**, *39,* 835-839.
60. Sakuno, E.; Yabe, K.; Nakajima, H. *Appl. Environ. Microbiol.* **2003**, *69,* 6418-6426.
61. Sakuno, E.; Wen, Y.; Hatabayashi, H.; Arai, H.; Aoki, C.; Yabe, K.; Nakajima, H. *Appl. Environ. Microbiol.* **2005**, *71,* 2999-3006.
62. Wen, Y.; Hatabayashi, H.; Arai, H.; Kitamoto. H. K.; Yabe, K. *Appl. Environ. Microbiol.* **2005**, 71, 3192-8.
63. Yao, R. C.; Hsieh, D. P. H. *Appl. Microbiol.* **1974**, *28,* 52-57.

64. Bennett, J. W.; Lee, L. S.; Cucullu, A. F.; *Bot Gaz*, **1976**, *137*, 318-324.
65. Kusumoto, K.; Hsieh, D. P. *Can. J. Microbiol.* **1996**, *8*, 804-810.
66. Silva, J. C.; Minto, R. E.; Barry, C. E.; Holland, K. A.; Townsend, C. A. *J. Biol. Chem.* **1996**, *271*, 13600-13608.
67. Silva, J. C.; Townsend, C. A. *J. Biol. Chem.* **1996**, *272*, 804-813.
68. McGuire, S. M.; Silva, J. C.; Casillas, E. G.; Townsend, C. A. *Biochem.* **1996**, *35*, 11470-11486.
69. Skory, C.D.; Chang, P.-K.; Cary, J.; Linz, J. E. *Appl. Environ. Microbiol.* **1992**, *58*, 3527-3537.
70. Kelkar, H. S.; Keller, N. P.; Adams, T. H. *Appl. Environ. Microbiol.* **1996**, *62*, 4296-4298.
71. Motomura, M; Chihaya, N; Shinozawa, T; Hamasaki, T; Yabe, K. *Appl. Environ. Microbiol.* **1999**, *65*, 4987-4994.
72. Keller, N. P.; Dischinger, J. H. C.; Bhatnagar, D.; Cleveland, T. E.; Ullah, A. H. J. *Appl. Environ. Microbiol.* **1993**, *59*, 479-484.
73. Prieto, R.; Woloshuk, C. P. *Appl. Environ. Microbiol.* **1997**, *63*, 1661-1666.
74. Ehrlich, K. C.; Chang, P.-K.; Yu, J.; Cotty, P. J. *Appl. Environ. Microbiol.* **2004**, *70*, 6518-6524.
75. Price, M. S.; Conners, S. B.; Tachdjian, S.; Kelly, R. M.; Payne, G. A. *Fungal Genet. Biol.* **2005**, *42*, 506-518.
76. Price, M. S.; Yu, J.; Bhatnagar, D.; Cleveland, T. E.; Nierman, W. C.; Payne, G. A. Ph.D. Thesis, North Carolina State University, Raleigh, NC 27695-7567. **2005**.
77. Liang, S.-H.; Skory, C. D.; Linz, J. E. *Appl. Environ. Microbiol.* **1996**, *62*, 4568-4575.
78. Chiou, C. H.; Miller, M.; Wilson, D. L.; Trail, F.; Linz, J. E. *Appl. Environ. Microbiol.* **2002**, *68*, 306-315.
79. Ehrlich, K. C.; Montalbano, B. G.; Bhatnagar, D.; Cleveland, T. E. *Fungal Genet. Biol.* **1998**, *23*, 279-287.
80. Flaherty, J. E.; Payne, G. A. *Appl. Environ. Microbiol.* **1997**, *63*, 3995-4000.
81. Yu, J.-H.; Butchko, R. A.; Fernandes, M.; Keller, N. P.; Leonard, T. J.; Adams, T. H. *Curr. Genet.* **1996**, *29*, 549-555.
82. Meyers, D. M.; O'Brian, G.; Du, W. L.; Bhatnagar, D.; Payne, G. A. *Appl. Environ. Microbiol.* **1998**, *64*, 3713-3717.
83. Yu, J.; Proctor, R. H.; Brown, D. W.; Abe, K.; Gomi, K.; Machida, M.; Hasegawa, F.; Nierman, W. C.; Bhatnagar, D.; Cleveland, T. E. In *Appl. Mycol. Biotechnol.* Arora, K. D.; Khachatourians, G. G. Eds., Elsevier Science, 2004; Vol. 4, pp. 249-283.
84. Yu, J.; Whitelaw, C. A.; Nierman, W. C.; Bhatnagar, D.; Cleveland, T. E. *FEMS Lett.* **2004**, *237*, 333-340.
85. Bok, J. W.; Keller, N. P. *Eukaryot. Cell.* **2004**, *3*, 527-535.

86. Mahoney, N.; Molyneux, R. J. *J. Agric. Food Chem.* **2004**, *52,* 1882-1889.
87. Kim, J. H.; Campbell, B. C.; Yu, J.; Mahoney, N.; Chan, K. L.; Molyneux, R. J.; Bhatnagar, D.; Cleveland, T. E. *Appl. Microbiol. Biotechnol.* **2005**, *7,* 807-815.
88. Bennett, J. W.; Leong, P. M.; Kruger, S.; Keyes, D. *Experientia*, **1986**, *42,* 841-851.
89. Calvo, A. M.; Wilson, R. A.; Bok, J. W.; Keller, N. P. *Microbiol. Mol. Biol. Rev.* **2002**, *66,* 447-459.
90. O'Brian, G. R.; Fakhoury, A. M.; Payne, G. A. *Fungal Genet. Biol.* **2003**, *39,* 118-127.
91. Machida, M.; Denning, D. W.; Nierman, W. C.; Yu, J.; Bennett, J. W.; Bhatnagar, D.; Cleveland, T. E.; et al. *Nature*, **2005**, *438*, 1157-1161.
92. Keller, N. P.; Watanabe, C. M. H.; Kelkar, H. S.; Adams, T. H.; Townsend, C. A. *Appl. Environ. Microbiol.* **2000**, *66,* 359-362.

Indexes

Author Index

Subject Index

C

D

E

F

G

H

I

K

L

M

N

O

R

S

T

S

T

U

V

W

X

Y

Z